Human Factors in Simulation and Training

Human Factors in Simulation and Training: Theory and Methods covers theoretical concepts on human factors principles as they apply to the fields of simulation and training in the real world.

This book discusses traditional and nontraditional aspects of simulation and training. Topics covered include simulation fidelity, transfer of training, limits of simulation and training, virtual reality in the training environment, simulation-based situation awareness training, automated performance measures, performance assessment in simulation, adaptive simulation-based training, and scoring simulations with artificial intelligence.

This book will be a valuable resource for professionals and graduate students in the fields of ergonomics, human factors, computer engineering, aerospace engineering, and occupational health and safety.

Human Factors in Simulation and Training

Theory and Methods

Second Edition

Edited by
Dennis Vincenzi, Mustapha Mouloua,
P. A. Hancock, James A. Pharmer,
and James C. Ferraro

CRC Press is an imprint of the
Taylor & Francis Group, an **informa** business

Front cover image: Sergey Ryzhov/Shutterstock

Second edition published 2024
by CRC Press
2385 NW Executive Center Drive, Suite 320 Boca Raton, FL 33431

and by CRC Press
4 Park Square, Milton Park, Abingdon, Oxon, OX14 4RN

CRC Press is an imprint of Taylor & Francis Group, LLC

First edition published by CRC Press 2019

Library of Congress Cataloging-in-Publication Data

Names: Vincenzi, Dennis A., editor.
Title: Human factors in simulation and training : theory and methods / edited by Dennis A. Vincenzi, Mustapha Moloua, Peter A. Hancock, James Pharmer, and James C. Ferraro.
Description: Second edition. | Boca Raton : CRC Press, [2023] | Includes bibliographical references and index. |
Identifiers: LCCN 2023001393 (print) | LCCN 2023001394 (ebook) | ISBN 9781032512525 (hbk) | ISBN 9781032512532 (pbk) | ISBN 9781003401360 (ebk)
Subjects: LCSH: Simulation methods. | Occupational training. | Human engineering.
Classification: LCC T57.62 .H845 2023 (print) | LCC T57.62 (ebook) |
DDC 620.8/2--dc23/eng/20230113
LC record available at https://lccn.loc.gov/2023001393
LC ebook record available at https://lccn.loc.gov/2023001394

ISBN: 978-1-032-51252-5 (hbk)
ISBN: 978-1-032-51253-2 (pbk)
ISBN: 978-1-003-40136-0 (ebk)

DOI: 10.1201/9781003401360

Typeset in Nemilov
by Deanta Global Publishing Services, Chennai, India

Contents

Preface

As we look toward the future, we find that it is neither totally random nor totally predictable. This is a truth that persists despite the years that have passed since the publication of the previous edition of this book. We maintain that, if the future were completely predictable, there would be no point looking forward because we would already know what was to come. If it were completely random, we would not bother because we could not know anything systematic about forthcoming events. That life lies between these two polar extremes gives us both the motivation to try to understand the future and the belief that we can do so, at least to a useful degree. Indeed, the triumphs of science encourage us to believe that we are making "progress" in so far as our predictions of the future are concerned. At least in relation to many physical processes, these are growing more accurate as the years progress. And, of course, the more we can know about the future, the more we can generate rational courses of action based on this understanding. This respective confluence of ideas encourages us to develop theories, models, methodologies, and other such instruments to continue to improve our predictive capabilities.

However, although certain forms of prediction work well for some of the simpler physical processes, there are many forms of complex interaction in which our predictive capacities are at present rudimentary at best. Unfortunately, many of these complex processes—global warming, for example—may prove so dangerous to our species that we cannot afford to assert predictions that are radically incorrect. Flawed prediction here can spell our end. As a consequence, we are in ever greater need of technologies that allow us to generate and refine predictions as well as exploring alternative potentialities found to be countertheses and antitheses to these various propositions. One such technology is simulation.

As a tool, simulation is an aid to the imagination. It allows us to create, populate, and activate possible futures and explore the ramifications of these developed scenarios. However, in common with all tools, it performs its task only to the degree that it is open to facile interaction with the user. One can imagine that on many occasions a poor simulation with its impoverished or wildly inaccurate outcomes might be of even more harm than good. Thus, as with all tools and technologies, we certainly need the application of the branch of science that turns user–machine antagonism into user–machine synergy. That branch of science is human factors. Hence, the focus of this present work is on human factors issues as they pertain to simulation in support of training and predicting humans to do certain tasks. In general, these issues revolve around two central themes, represented uniquely across two books.

The theme of this book concerns the methods by which simulation technology is utilized, in addition to theories surrounding its potential shortcomings and how best to get the most out of simulated training and assessment. Indeed, there are many techniques and insights from the behavioral sciences that can help us to construct better ways to create and visualize possible futures. In this work, we have solicited chapters that deal with a wide variety of topics, beginning with theory and methods, in areas ranging from traditional training to augmented reality to virtual reality. Areas of coverage include fields related to healthcare and aviation. This theory-based book will focus on human factors aspects of simulation and training ranging from the history of simulators and training devices to future trends in simulation from both a civilian and military perspective. The chapters in the book discuss methods of utilizing simulation technology to guide decision-making and evaluate performance. Chapters will be comprised of in-depth discussions of specific issues, including fidelity, interfaces and control devices, transfer of training, simulator sickness, effects of motion in simulated systems, and virtual reality.

Dennis A. Vincenzi
Mustapha Mouloua
P. A. Hancock
James A. Pharmer
James C. Ferraro

Editors

Dennis A. Vincenzi earned his doctoral degree in 1998 from the University of Central Florida in Human Factors and Applied Experimental Psychology and has over 25 years of experience as a Human Factors researcher. He has been employed by Embry-Riddle Aeronautical University from 1999 to 2004, where he held the position of Assistant Professor in the Department of Human Factors and Systems in Daytona Beach. In 2004, Dr. Vincenzi left Embry-Riddle to work for the United States Navy as a Senior Human Factors Engineer at the Naval Air Warfare Center Training Systems Division (NAWCTSD) in Orlando, FL. His duties included performing Human Factors research involving simulation and training system development for a variety of Navy sea and air platforms, including the F/A 18 Hornet and Super Hornet, F-35 JSF, Los Angeles, Ohio, and Virginia class submarines, and Littoral Combat Ship (LCS). He was also heavily involved in research involving pilot selection, human performance, and ground control station design for a number of Navy, Marine Corps, and Special Operations Command Unmanned Aerial Systems (UAS). Since returning to Embry-Riddle Aeronautical University in 2012, Dr. Vincenzi has been involved in research related to UAS and regulatory requirements within the NAS and has been heavily involved in the development of an experimental gesture-based interface used for investigating user preference, usability, and functionality issues related to interface design in virtual environments. Dr. Vincenzi is currently the Program Chair for the Master of Science in Human Factors program at Embry-Riddle Aeronautical University.

Mustapha Mouloua is Professor of Psychology and Director of the Transportation Research Group at the University of Central Florida (UCF), Orlando, FL. He earned his Ph.D. (1992) and M.A. (1986) degrees in Applied/Experimental Psychology from the Catholic University of America, Washington, DC. Before joining the faculty at UCF in 1994, he was Postdoctoral Fellow at the Cognitive Sciences Laboratory of the Catholic University of America from 1992 to1994, where he studied and researched several aspects of humanautomation interaction topics sponsored by NASA, and the Office of Naval Research (ONR). He has over over 30 years of experience in the teaching and research related to complex humanmachine systems. His research interests include vigilance and sustained attention, cognitive aging, human performance assessment, humanautomation interaction, pilotalerting systems interaction, automation and workload in aviation systems, simulation, and training in transportation systems. Dr. Mouloua made over 300 conference presentations with his undergraduate and graduate students, as well as his professional colleagues. He also has about 200 research publications and scientific reports published in journals and proceedings, such as *Experimental Aging Research*, *Human Factors*, *Ergonomics*, *Perception and Psychophysics*, *Journal of Experimental Psychology: Human Perception and Performance*, *International Journal of Aviation Psychology*, *Journal of Cognitive Engineering and Decision Making*, *Proceedings of the Human*

Factors and Ergonomics Society, *Applied Ergonomics*, *Transportation Research Part F: Traffic Psychology and Behaviour*, *Ergonomics in Design*, *Transportation Research Record*, and *International Journal of Occupational Safety and Ergonomics*. Together with his colleagues Raja Parasuraman and Robert Molloy, he was the winner of the Jerome Hirsch Ely Award of the Human Factors and Ergonomics Society in 1997. He was previously Director of the Applied/Experimental and Human Factors Psychology doctoral program (20082017). At UCF, Dr. Mouloua earned eight prestigious Teaching and Research Awards and was inducted into the UCF College of Sciences Millionaire Club for procuring over $1 million in research funds. He was awarded a UCF "Twenty Years' Service" award in 2014, was awarded the UCF International Golden Key and Honorary member status in 2011, and his research was selected to be among the top 30 best published research articles in the last 50 years by the Human Factors and Ergonomics Society in 2008.

P. A. Hancock, D.Sc., Ph.D., is Provost Distinguished Research Professor in the Department of Psychology and the Institute for Simulation and Training, as well as at the Department of Civil and Environmental Engineering and the Department of Industrial Engineering and Management Systems at the University of Central Florida (UCF). At UCF in 2009, he was created the 16th ever University Pegasus Professor (the Institution's highest honor) and in 2012 was named 6th ever University Trustee Chair. He directs the MIT^2 Research Laboratories. He is the author of over 1,100 refereed scientific articles, chapters, and reports as well as writing and editing more than 25 books. He has been continuously funded by extramural sources for every one of the forty years of his professional career. This includes support from NASA, NSF, NIH, NIA, FAA, FHWA, NRC, NHTSA, DARPA, NIMH, and all of the branches of the US Armed Forces. He has presented or been an author on over 1,200 scientific presentations. In association with his colleagues Raja Parasuraman and Anthony Masalonis, he was the winner of the Jerome Hirsch Ely Award of the Human Factors and Ergonomics Society for 2001, the same year in which he was elected a Fellow of the International Ergonomics Association. In 2006, he won the Norbert Wiener Award of the Systems, Man and Cybernetics Society of the Institute of Electrical and Electronics Engineers (IEEE), being the highest award that Society gives for scientific attainment. He is a Fellow and past President of the Human Factors and Ergonomics Society and a Fellow and twice past President of the Society of Engineering Psychologists as well as being a former Chair of the Board of the Society for Human Performance in Extreme Environments. Most recently he has been elected a Fellow of the Royal Aeronautical Society (RAeS) and in 2016 was named the 30th Honorary Member of the Institute of Industrial and Systems Engineers (IISE). He currently serves as a member of the United States Air Force, Scientific Advisory Board (SAB), and has also served on the US Army Science Board (ASB). He is also a Fellow of AAAS and IEEE.

James Pharmer is the Chief Scientist for the Research, Development, Test, and Evaluation (RDT&E) Department and the Head of the Experimental and Applied Human Performance Research and Development (R&D) Division at the Naval Air

Warfare Center Training Systems Division (NAWCTSD) in Orlando, Florida.. He is a Naval Air Warfare Center Aviation Division (NAWCAD) Fellow and has over 20 years of experience in training and human performance R&D for advanced military systems across a variety of warfare domains. His work includes conducting R&D and direct participation on systems acquisition teams to support human systems integration (HSI) implementation for Navy ships, aircraft, and systems. He chairs multiple working groups to develop HSI policy, processes, and education. He holds a doctoral degree in Applied Experimental Human Factors Psychology from the University of Central Florida and a master's degree in Engineering Psychology from the Florida Institute of Technology.

James C. Ferraro is a human factors research scientist specializing in simulation and game-based assessment of human performance in complex systems. He earned his Ph.D. in Human Factors and Cognitive Psychology from the University of Central Florida (UCF) in 2022 and his M.A. in Applied Experimental and Human Factors Psychology from UCF in 2019. Dr. Ferraro has led and contributed to a number of research efforts in support of government-sponsored (NAVAIR, USAF) projects to improve training and selection of personnel in various occupations. Areas include air traffic control, tactical urban warfare, explosive ordnance disposal, special forces rotary wing operations, and unmanned aircraft operations. His research on topics such as pilot/operator attentional strategies, trust in automated systems, and predictors of individual performance has been presented at local, regional, and international conferences (Human Factors and Ergonomics Society Annual Meeting, International Symposium on Aviation Psychology, Conference on Applied Human Factors and Ergonomics) and published in multiple academic journals (Ergonomics, Applied Ergonomics). He is the technical editor of the two-volume book set *Human Performance in Automated and Autonomous Systems* (2019) and the co-author of published book chapters pertaining to human monitoring of automated systems and the role of trust in unmanned vehicle operations. Dr. Ferraro is currently a Senior Research Scientist with Adaptive Immersion Technologies, based in Tampa, FL.

Contributors

Joshua Andrews
Modern Hire
Raleigh, NC

Peyton Bailey
Design Interactive
Orlando, FL

Elizabeth Blickensderfer
Embry-Riddle Aeronautical University
Daytona Beach, FL

Michael Brannick
University of South Florida
Tampa, FL

Meredith Carroll
Florida Institute of Technology
Melbourne, FL

Deborah Sater Carstens
Florida Institute of Technology
Melbourne, FL

Nicklas Dahlstrom
Lund University School of Aviation
Lund, Sweden

Suvranu De
FAMU-FSU College of Engineering
Tallahassee, FL

John Deaton
Florida Institute of Technology
Melbourne, FL

Cali Fidopiastis
Design Interactive
Orlando, FL

T. Chris Foster
Naval Air Warfare Center Aircraft Division (NAWCAD)
Patuxent River, MD

Jared Freeman
Aptima, Inc.
Arlington, VA

Trysha Galloway
The Learning Chameleon, Inc.
Culver City, CA

Michael Geden
Modern Hire
Raleigh, NC

Carter Gibson
Modern Hire
Birmingham, AL

Steve Hall
Winter Springs, FL

P. A. Hancock
University of Central Florida
Orlando, FL

Claire L. Hughes
Design Interactive
Orlando, FL

Cullen D. Jackson
Beth Israel Deaconess Medical Center
Boston, MA

Anna Johansson
Beth Israel Deaconess Medical Center, Department of Medicine
Boston, MA

Joan H. Johnston
United States Army Combat Capabilities Development Command Soldier Center
Natick, MA

Joseph R. Keebler
Embry-Riddle Aeronautical University
Daytona Beach, FL

John L. Kleber
Embry-Riddle Aeronautical University
Daytona Beach, FL

Nick Koenig
Modern Hire
Rogers, AR

Elizabeth H. Lazzara
Embry-Riddle Aeronautical University
Daytona Beach, FL

Michael G. Lilienthal
EWA Government Systems Inc.
Herndon, VA

Dahai Liu
Embry-Riddle Aeronautical University
Daytona Beach, FL

Nikolas D. Macchiarella
Embry-Riddle Aeronautical University
Daytona Beach, FL

Shem Malmquist
Florida Institute of Technology
Melbourne, FL

Phillip M. Mangos
Adaptive Immersion Technologies
Tampa, FL

Jacqueline McSorley
Embry-Riddle Aeronautical University
Daytona Beach, FL

William F. Moroney
University of Dayton
Dayton, OH

Maria Chaparro Osman
Aptima Inc.
Orlando, FL

Henry L. Phillips IV
Soar Technology, Inc.
Pensacola, FL

Di Qi
Chapman University
Orange, CA

Summer Rebensky
Aptima, Inc.
Dayton, OH

Ernesto Ruiz
Florida Department of Health
Orlando, FL

William J. Salter
Strategic Solutions Consulting
Harvard, MA

Richard J. Simonson
Embry-Riddle Aeronautical University
Daytona Beach, FL

Kay M. Stanney
Design Interactive
Orlando, FL

Emily M. Stelzer
MITRE Corporation
McLean, VA

Ronald Stevens*
UCLA School of Medicine, Brain Research Institute
Los Angeles, CA,
and
The Learning Chameleon, Inc.
Culver City, CA,

Dennis A. Vincenzi
Embry-Riddle Aeronautical University
Daytona Beach, FL

Emily E. Wiese
Blueprint Test Preparation
Manhattan Beach, CA

Ann Willemsen-Dunlap
JUMP Simulation and Education Center
Peoria, IL

Kimberly N. Williams
Embry-Riddle Aeronautical University
Daytona Beach, FL

Jiahao Yu
Embry-Riddle Aeronautical University
Daytona Beach, FL

* The editors would like to pay their respects to Dr. Ron Stevens, who sadly passed away prior to publication of this book project. We are very grateful for his contributions and dedication to the field of human factors in simulation and training.

1 Human Factors in Simulation and Training

An Overview

T. Chris Foster, William F. Moroney, Henry L. Phillips IV, and Michael G. Lilienthal

CONTENTS

DOI: 10.1201/9781003401360-1

INTRODUCTION

Simulation is pervasive. Indeed, we probably received a simulator during the first days, if not hours, of our lives. Interestingly, it was called a "pacifier." Although the name might have been provided by parents seeking a respite, the pacifier is both a simulator and a stimulator (see Appendix B for definitions). It simulated the nipple of a mother's breast or a feeding bottle and stimulated a sucking/rooting reflex in the infant, which improved the child's muscle tone and provided a sense of comfort.

Young children simulate as they play with their toys, athletes envision their success, and would-be fighter pilots do battle in the safety of their homes, whereas actual and would-be politicians rule simulated cities. For engineers, educators, and trainers, simulation is a standard tool of the trade. Business relies heavily on simulation as part of its planning process. Simulation is *not* a relatively modern development attributable to technological and mathematical advances. It has a much longer history.

We use or have used simulation in many ways in our daily life. Consider the following examples (modified from Raser, 1969):

- Make-believe: a child playing with blocks and model toys of various scales. It is interesting to note that as individuals age, the level of fidelity they expect increases. Thus, for most young children, scale is irrelevant, and they see no inconsistency in playing with toy vehicles of various sizes. For an adult, scale is critical, when creating a diorama.
- Artificial: as in an "artificial Christmas tree."
- Substitution: margarine (as a butter substitute) or clothing made of natural or synthetic material in lieu of animal fur.
- Imitation: faux leather purses or wallets made of a synthetic material.
- Deception: simulated storefronts as used in movies and on theatrical stages, or inflated military vehicles positioned where they can be observed by aerial reconnaissance or satellites.
- Mimicry: animal calls used by hunters to lure other animals; vibration device built into an infant's bouncy seat to serve as a surrogate for parental rocking motion.
- Metaphor: Shakespeare's play *Henry V* is a morality play applicable to modern warfare.
- Analogy: simulated sunlight used to reduce seasonal adaptive disorders.
- Representation.
  - Mathematical: a graph representing the dynamics of stock markets.
  - Logical: a "You-are-here" map.
  - Physical: ranges from toys (scaled representations) to tinker-toy-like representations of chemical bonds or DNA and holographic images.
  - Mental: our perception or belief of how a piece of software works.
  - Visual: computer-driven monitors/displays such as Oculus Rift and HoloLens.

SIMULATION: THE PERFECT STORM

The second decade of the 21st century could be described as a perfect storm in which multiple factors converged to increase the use of simulators for training, including the following.

Technological advances:

- Major technological innovations in the capabilities of visual, auditory, and tactile technologies facilitate immersion in the learning process. Today we have multiple untethered head-mounted displays (HMDs) such as Microsoft's HoloLens (https://www.youtube.com/watch?v=uIHPPtPBgHk) and Apple's Apple Glass (https://www.youtube.com/watch?v=JLNvSYr4eeI). Wearable computing has arrived.
- Virtual Reality (VR) "finally beginning to come of age, having survived the troublesome stages of the famous 'hype cycle'—the Peak of Inflated Expectation, even the so-called Trough of Disillusionment" (Ruben & Gray, 2020). Despite claims in the popular press of superior training capacities, VR and traditional training are essentially equivalent. However, Extended Reality (XR) offers the additional advantage of safety, affordability (Kaplan et al., 2019), increased presence and immersion as well as the psychological dimension of flow (Kaplan et al., 2020), and portability.
- The development of artificial neural networks, AI and intelligent tutors, allows for virtual adversaries and trainers. In a 2020 simulation, an intelligent adversary defeated a highly qualified F-16 pilot in five rounds of simulated Air Combat Maneuvering (ACM) (https://www.thedrive.com/the-war-zone/35888/ai-claims-flawless-victory-going-undefeated-in-digital-dogfight-with-human-fighter-pilot). With respect to trainers, VIPER (Virtual Instructor Pilot Referee) is an AI tutor designed "to guide and provide feedback to students in practice sessions using virtual reality" (Natali et al., Aug 2020). Live, virtual, and constructive entities can now be combined to fully exercise a system's capabilities (https://www.lockheedmartin.com/en-us/news/features/2016/the-future-of-training-blends-real-and-virtual-worlds.html).
- Commercial off-the-shelf (COTS) computer displays and networking technologies were pivotal to accessing cutting-edge training simulations (Wilson, 2018). Schultz (2014) describes how Boeing capitalized on an operator's gaming experience by using a common Xbox game console controller to control the US Army's high-energy laser mobile demonstrator.
- Access to broadband, while still not universal, has increased.

Increased computation power:

- The fusion of inexpensive computation power, affordable broadband, and wireless networks in a persistent cloud-based environment that is available 24/7.
- The exponential decrease in computation and storage costs over the past decades: 1 MB of memory cost $1 million in 1967, and by 2017 it cost $.02 (Mearian, 2017), which enabled the creation of realistic visualizations and dynamic human–computer interactions.
- Reduced size, weight, and power requirements are embedded in these decreased costs.

New generation of leaders and learners:

- Millennials (born between 1981 and 1996) and Generation Z (born 1997–2012) have imprinted on technology-based platforms for education, accessing information, gaming, and entertainment. They are more willing to accept technological change than their predecessors. Indeed, they expect these changes since they have grown up with the Internet, electronic gaming, online research, instantly accessing data and videos. Lectures and linear learning are things of the past to these digital natives.
- Today's upper management and middle management decision-makers were the first generation of "tech-savvy" personnel during the 1990s and early 2000s. They recognized the cost-benefits of simulation in selection and training. They are now positioned to facilitate/mandate administrative changes favoring simulation. Consider agencies such as such as the FAA, which now accepts time spent on qualified training devices as contributing to time spent flying the actual aircraft (https://www.faa.gov/news/updates/?newsId=85426). The US Navy is now certifying submarine bridge crews in a Submarine Bridge Trainer (SBT) (https://www.navaltoday.com/2017/06/06/us-navy-opens-new-submarine-bridge-trainer/, https://www.dvidshub.net/news/236453/realistic-submarine-bridge-trainer-opens-trident-training-facility-bangor). AI-based Learning Management Systems (LMS) and eLearning allow on-site and remote employees to learn online. Properly designed simulations increase employee engagement and performance (https://elearningindustry.com/ai-based-humanlike-trainers-change-corporate-learning).

Innovations in education and training:

- The changing nature of education:
  - Increased use of electronic platforms and software as educational tools.
  - Acceptance of distance learning, which was, in part, imposed by the COVID-19 pandemic.
  - Focus on performance and demonstration of learning as opposed to class hours.
  - Recognition of individual differences in learning rates.
  - Gradual introduction of avatar tutors into the learning process.
  - Immersion and deliberate practice recognized as key contributors in the development of expertise.
  - Today's teachers are being trained in how to effectively utilize emerging simulation technologies and incorporate them into hybrid teaching, which involves some online and some in-person students.
- "Training anytime, anyplace" has become a mantra. Many simulations are now portable across platforms (including handheld devices), albeit with some loss of capability and resolution, as they connect wirelessly to cyberspace. A good example would be the US Navy's RRL (Ready, Relevant Leaning) program (https://www.netc.navy.mil/RRL), which is designed to

deliver "the right training in the right way." The teaching tools used range from: "YouTube-like videos to more complex, immersive simulators and virtual trainers." BFTT (Battle Force Tactical Training) provides stimulation/simulation of shipboard tactical equipment to facilitate team training. This allows crews to practice on the actual equipment they would be using in real combat. Multiple vessels at a variety of locations can participate in coordinated training exercises (AN/USQ-T46. January 15, 2020).

- Simulation has blended training, mission planning, rehearsal, and execution into one continuum.
- Educators in high-risk environments such as the medical, chemical, nuclear, and aerospace industries now have tools that better enable them to deliver realistic training in safe environments, which increases training transfer while reducing overall risk. These tools are cost-effective since they require fewer resources to serve multiple learners. They also allow for experimentation and testing, which could not ethically be performed on humans or animals.

Global acceptance of simulation/gaming:

- Higher-level simulations are no longer the sole domain of technologists; with increased accessibility to more capable and affordable personal computers, simulations are now used daily by less technical people. Businesses use simulations for selection, training, and evaluation (Langley et al., 2016). In the workplace, simulations such as weather and climate forecasts, expected aircraft delays, stock market projections, mortgage cost estimates, and retirement projections are common tools. Models predicting COVID-19 hotspots and optimizing vaccine distribution have been accepted by the public.
- Gaming has become more common. Gaming simulations range from individual "first person" games such as Halo and Grand Theft Auto to massive multiplayer online (MMO)/MMO Role-Playing Games (MMORPG) games such as PlayerUnknown's Battle Grounds, Fortnite, Minecraft, World of Warcraft, etc. Students in grades 3–12 use Minecraft to learn about environmental challenges, food security, and global issues affecting agriculture (https://www.nasef.org/learning/farmcraft/?et_rid=40892865&et_cid=3725425).
- Self-contained gaming systems which integrate controls, displays, and a workstation into an area with a small footprint are emerging. Consider the concept gaming chair introduced at CES 2021. This chair provides self-contained, panoramic visuals in a 60 "rollout display, tactile feedback, and a transformable table for PC and console gaming" (https://www.razer.com/ca-en/concepts/razer-project-brooklyn). It goes beyond the current motion-simulating devices (inflatable seat pan cushions and backs, rumble motors, "butt kickers" and harness tighteners) described by Alexis (2020).

HUMAN FACTORS IN SIMULATION AND TRAINING: A BRIEF HISTORY

Where would society be if humans did not have the ability to simulate? Where would the world be if individuals such as Lincoln, Roosevelt, Churchill, Gandhi, Martin Luther King, Walt Disney, Bill Gates, and Steven Spielberg had not had dreams and visions (simulations) of a different world? Indeed, one could speculate that at some level the ability or drive to simulate is an essential part of "proactive" evolution. This ability is essential to human creativity, as Hofstadter (1979, p. 643) opined: "How immeasurably poorer our mental lives would be if we did not have this creative capacity for slipping out of the midst of reality into soft 'what ifs'!"

Schrage (2000, p. 13) stated, "We shape our models, and then our models shape us." To a large extent our internal models determine our perceptions, which in turn can influence and perhaps determine our behaviors in the real world. As individuals, we routinely utilize simulation when we engage in what is colloquially described as "wishful thinking" and in the psychological literature as "counterfactual thinking" (Roese & Olson, 1995). Counterfactual thinking is often characterized by conditional statements with an antecedent such as, "if only" and a consequence statement such as, "then" (Roese & Olson, 1995). Sternberg and Gastel (1989a, 1989b) describe counterfactual thinking as an important component of intelligence. Indeed, Tetlock and Belkin (1996) described counterfactual thinking in world politics, including the rise of Hitler, Western perceptions of Soviet politics, and the Cuban missile crisis.

Both today's computer-based simulation and the ubiquitous Internet evolved from Department of Defense (DoD) efforts, as have many other innovations, including emergency medical services, blood plasma, and lasers. During World War II, simulation supported part task flight training provided as a "blue box" (Figure 1.3), which used instruments within an enclosed cockpit, and had minimal physical displacement. During the 1970s, 1980s, and 1990s, the word simulation invoked images of moving-based simulators tossing cockpits about atop platforms supported by giraffe-like hydraulic legs. These full six degrees-of-freedom (DOF) hexapod motion simulators were housed in special air-conditioned buildings and supported by a highly skilled technical staff.

But research at the end of the 20th century and in the early part of the 21st century led to the decline of full six DOF hexapod motion simulators. While airlines and the DoD had invested heavily in motion-based simulators, the belief that "since the airplane has motion, flight simulators must also have motion" was severely challenged as researchers re-examined assumptions regarding requirements for motion.

Boothe (1994) argued that in simulators and flight training devices (FTDs), the emphasis is not just on accomplishing the required task; but for maximum "transfer of behavior," the task must be performed exactly as it would be in the aircraft. Thus, the same control strategies and control inputs must be provided in both the aircraft and the simulator. He believed that the emphasis should be on appropriate cues, as identified by pilots, who are the subject-matter experts. To achieve this end, Boothe emphasized replication of form and function, flight and operational performance, and perceived flying (handling) qualities. He noted that government and industry working groups utilize realism as their reference and safety as their justification.

However, Roscoe (1991) argued that "qualification of ground-based training devices for training needs to be based on their effectiveness for that purpose and not solely on their verisimilitude to an airplane" (p. 870). Roscoe concluded that pilot certification should be based on demonstrated competence, not hours of flight experience. Lintern et al. (1990, p. 870) argued further that for "effective and economical training, absolute fidelity is not needed nor always desirable, and some unreal-worldly training features can produce higher transfer than literal fidelity can." Caro (1988, p. 239) added, "The cue information available in a particular simulator, rather than stimulus realism per se, should be the criterion for deciding what skills are to be taught in that simulator." This position was reinforced in a 2020 study on the use of lethal force by Blacker, et al. Participants with varying levels of experience participated in both marksmanship and shoot/ don't shoot scenarios were trained on both a videogame and military-grade shooting simulator.

> Results supported the notion that shooting accuracy and decision making are independent components of performance. Individuals with firearms expertise outperformed novices on the military-grade simulator, but only with respect to shooting accuracy, not unintended casualties. Individuals with video game experience outperformed novices in the video game simulator, but again only on shooting accuracy.

The authors demonstrated that the capabilities of the less-expensive computer game simulation were adequate when evaluating the lethal-force-decision-making process. For an additional illustration of the importance of selecting the appropriate technology for training, see the parachute/water entry trainer in the mixed reality (MR) section of this chapter.

Thus, there are important differences of opinion regarding the specification of a simulator's capabilities. Scientists began to systematically re-examine assumptions regarding the requirements for motion. Burki-Cohen et al. (2007) at the Volpe National Transportation Systems Center, in a proof-of-concept study, evaluated the training value of a fixed-base flight simulator with a dynamic seat. They concluded: "This research did not find operationally relevant differences in performance or behavior of pilots tested in the FFS with motion after having been trained in the same FFS [Full Flight Simulator] with the motion system turned on or off—despite selection of maneuvers that require motion cues, at least theoretically." It made no difference whether the FFS represented a small turboprop "power house" or a sluggish four-engine jumbo jet, or whether the training in question was initial or recurrent training. Sparko et al. (2010), working with pilots with fewer than 500 hours of experience, concluded that there were no differences in the training success of pilots trained with either simulator type. In 2011. Bürki-Cohen et al. reported on a series of studies which addressed the following questions:

> 1) Are there maneuvers in airline-pilot training where platform motion cues, in addition to the visual cues from a wide-FOV OTW (Field-of- View/Out-the-Window) view and instruments, result in an operationally relevant improvement of transfer between the simulator and the airplane? 2) Do airline pilots need to be trained to avail themselves of motion cues? 3) Are motion cues from a hexapod platform representative of

> those experienced in the airplane? 4) Can alternative systems provide onset cues and perception of realism?

They concluded that the answer to the first three questions was No and the answer to the fourth question was a resounding Yes, based not only on their findings but also on the training provided on fixed base platforms by multiple smaller airlines. In addition to the science just discussed, the lower cost of training in simulators is also a driving factor.

According to Alexis (2020):

> A turnkey purchase price of a brand new and advanced fixed base simulator is in the region of £200,000 ($250,000, €220,000), whereas an Airbus A350 or Boeing 787 Level D simulator is approximately £15m ($18.5m, €16.5m). 1 hour in a fixed base simulator is approximately £120 ($150, €135) compared to 1 hour in a full motion simulator at £450 ($560, €500).

For an interesting comparison of commercial fixed base flight simulators visit https://simulatorreview.com/ In 2018, Losey described how the USAF is replacing legacy simulators, costing $4.5 million each, with VR simulators costing $15,000 each. For details on use of MR simulators, see Section Extended Reality: Augmented Reality, Virtual Reality, and Mixed Reality.

Within the past 20 years, the criticality of human factors in training and simulation has become more apparent in system development: from the conceptual phase to test and evaluation and deployment phase. Human factors impact both effectiveness and efficiency. As discussed above, military aviation was an early focus and lessons learned in that domain have transferred to domains such as civil aviation, energy-related industries, automotive systems, and healthcare. A quick perusal of the chapter titles in this book testifies to that fact. For a broad overview of the role of simulation in training, see Gawron, 2019, who describes: (1) the role of simulation in a training curriculum, (2) how to measure transfer of training from the simulator to the real world, and (3) the types of simulations and simulators used for training. Her report provides information on applications in environments ranging from medical facilities, driving, and aircraft to weightlessness. Let us now discuss the why of simulation, clarify the distinction between simulation and modeling, and describe the modeling and simulation (M&S) process.

WHY SIMULATE?

Before examining simulation in more detail, it is appropriate to ask, "Why did simulation evolve?" Simulation may have evolved for a variety of reasons, but its origins may be best attributed to the organism's strategy of accomplishing tasks with the least amount of effort thus conserving energy, avoiding overload, and maintaining homeostasis. This strategy is apparent even on a physiological basis, when our perceptual system filters data as we transform it into information. Thus, we may simulate because it is both effective and efficient.

The reasons we simulate may be reflected in an examination of our uses of simulations and simulators. Currently, simulators are frequently used for training of operators (aircrew, physicians, bus and truck drivers, ship navigators, nuclear power plant operators, etc.) and maintainers (troubleshooting by technicians). They are also used to maintain and evaluate individuals' proficiency level and allow individuals to qualify for a particular position or rating such as the pilot of a single-seat aircraft, surgeon, or nuclear power plant operator. Manufacturers routinely use digital prototypes for product research and development (e.g., automobile and aircraft cockpit prototypes). Systems engineers routinely use M&S as a prime means for managing the cost of developing, building, and testing increasingly complex systems. Test and evaluation experts rely on M&S to help design test scenarios/experiments and to support T&E execution and post-test analysis and reporting (see https://www.dau.edu/guidebooks/Shared%20Documents/Test_and_Evaluation_Mgmt_Guidebook.pdf and https://acqnotes.com/acqnote/careerfields/modeling-and-simulation-support). According to the Lindenberger Group (2017), businesses use simulations because they provide experiential learning with immediate feedback, increase knowledge retention, facilitate cooperation and encourage competition; provide a risk-free environment in which to fail, provide quantifiable training, and are less costly in terms of time and money. Educators use simulation to "demonstrate abstract concepts, allow interaction between users and simulated equipment, and provide users with feedback that allow users to improve their knowledge and skills. They are also cost-effective over the long-term" (Electronics Technician Training, Nov 16, 2018).

SIMULATION VERSUS MODELING

What is simulation? What is a model? Is there a difference between simulation and modeling? In common usage, the terms are interchangeable. Schrage (2000, p. 7) commented:

> Just what are the differences between models, simulations and prototypes? Once upon a time there was a fighting chance to answer this question simply and clearly. Today, technologies have conspired to turn any answer into a confusing jumble of semantics that obscure understanding.

Indeed, the terms are linked so closely that DoD Directive 5000.59-M (1998) provided a joint definition for M&S as "the use of models, including emulators, prototypes, simulators, and stimulators, either strategically or over time, to develop data as a basis for making managerial or technical decisions." Raser (1969, p. 6) describes a simulation as "a special kind of model, and a model is a special way of expressing theory." A "theory is a set of statements about some aspect of reality such as the past reality, present reality, or predicted reality. A theory attempts to describe the components of that reality and to specify the nature of the relationships among those components" (p. 6). Thus, for Raser, a model is a specific form of a theory, whereas a simulation is "an operating model that displays processes over time and that thus may develop dynamically" (p. 10).

It may be helpful to compare the usage of the terms modeling and simulation as nouns and verbs. According to DoD Instruction 5000.61 (Department of Defense, 2009), simulation (the noun) is defined as an executable implementation of a model, or execution of an implemented model, or a body of techniques for training, analysis, and experimentation using models, whereas model (the noun) is defined as a "physical, mathematical, logical, or other representation of a system, entity, phenomenon, or process."

This distinction is also reflected in differentiating between the verbs modeling and simulating. According to DoD Directive 5000.59 August 8, 2007, Incorporating Change 1, October 15, 2018, simulating is defined as "a method for implementing a model over time." Modeling is the "application of the standard, rigorous, structured methodology to create and validate a physical, mathematical, or otherwise logical representation of a system, entity, phenomenon, or process." According to the above definitions, the distinguishing feature is that a model is used to produce a simulation, whereas simulations implement models. Within the training community, which focuses on the outcome of modeling, the terms are often seen as interchangeable. But individuals who perceive modeling as a descriptive process see the model as the end product rather than the simulation of that model as the end product. However, as the terms are more commonly used interchangeably, the authors have also used the terms interchangeably throughout the remainder of this chapter.

The DoD, a leading developer and procurer of models and simulations, has developed fairly concise pragmatic definitions of simulation and modeling terms. Other professional organizations, such as IEEE and Military Operations Research Society (MORS), have also made efforts to standardize the terminology. Appendices A (Modeling) and B (Simulation) are provided to help clarify ambiguity associated with these terms as well as facilitate the reading and understanding of the material contained in the following chapters and related readings. An updated online modeling and simulation glossary is maintained at https://www.msco.mil/MSReferences/Glossary/MSGlossary.aspx

The Modeling and Simulation Process: Verification, Validation, and Accreditation

A computer model can exist and be valid only in a virtual world (e.g., an antigravity vehicle created for a video game) or can mimic some portion of the real world (e.g., motorcycle being designed for manufacture following the rules of physics). The model development process and the software development process have a similar structure: define the end state and the objectives (criteria for acceptance of the model), and follow a consistent system engineering process. In 2010, Johns Hopkins University Applied Physics Laboratory (JHU/APL) published a guide defining a systems engineering framework focused on the best practices for the development of credible stand-alone models and simulations (Morse et al., 2010).

Just as computers have evolved into faster computational, storage, and visualization tools, so too have software development processes modernized and become faster. Terms such as Agile, Lean, DevSecOps, computer factories, continuous delivery, and human-centered design appear as part of commercial best practices to rapidly develop, field, and continually upgrade software products, including simulations for training, experimentation, and the like (DoD Enterprise DevSecOps Reference Design Version 1.0 12 August 2019).

As the pace of developing, building, and deploying training and simulation systems continues to accelerate, the credibility of these models and simulations must be addressed. This is important for investors who use models and simulations to guide their purchase of stocks and bonds to realize profits and to aircrew who depend on transferring what they learn in the simulated flight environment of a six degree-of-freedom flight simulator to the real-world flight environment of a F-35 fighter aircraft.

To develop credible products, the process of developing a model must be scrutinized in parallel through the verification, validation, and accreditation (VV&A) process. Each of these terms is discussed in the following paragraphs.

Verification is the process for determining if the model does what the customer intended. That is, it is the process that examines the model implementation and its associated data to determine how accurately it represents the customer's conceptual description and specification for the model. It answers the question, "Does the model actually represent the design intent and accurately reproduce the model specifications?" This is typically done by examining the code and logic of the software to verify that it implements the conceptual model and that the model produces the correct results when needed.

Validation is the process for determining how accurately the model and its associated data match the real world within the context of the intended use of the model. It answers the question, "Does the model adequately depict the 'real' world that it was designed to represent?" Subject matter experts routinely "run the model through its paces" to test whether it acts and interacts in a simulated world as expected. Thus, a car test driver may be the ideal expert to determine whether the simulated handling characteristics of a new car model that only exists on a computer is an acceptable representation.

Accreditation is an official certification that there is sufficient evidence to show that the model is credible and suitable for a particular purpose. It answers the question, "Does the model provide an acceptable answer to my particular questions, or can it be used for my specific purpose?" Accreditation confirms that the simulation can meet the specific requirements developed in response to the objectives established at the beginning of the M&S development process.

VV&A are three intertwined processes that increase the credibility of the model, reduce risk, and increase the user's confidence in the M&S tool. These processes verify the model was made right, validate the right model was made, and accredit the model is right for the intended application (Balci, 1998). In essence, VV&A provides risk reduction, by ascertaining that the simulation supports the users' objectives. It calibrates the credibility of the model or simulation for an intended use.

There is no one cookie-cutter VV&A process that can be applied to the development of all models and simulations. VV&A processes must be tailored to the nature of the problem. How much and what parts of VV&A are employed (i.e., resources expended) are highly dependent on the importance of the decision that will be made based on the simulation. The VV&A process can make the model "transparent" enough for people to understand its assumptions and its limitations. As the demand for M&S continues to grow, the VV&A process continues to evolve and must become as agile as the software development processes are. As these techniques evolve, so does the VV&A lexicon, portions of which are provided for the interested reader in Appendix C. The Department of Defense Modeling and Simulation Enterprise VV&A site (https://wwa.msco.mil) has standardized documentation templates for V&V plans, the accreditation plans, and the reports for both in accordance with MIL-STD 3022. It also has a recommended best practice guide to assist those planning to do VV&A on their model or simulation. The site supports the DoD modeling and simulation VV&A instruction DoDI 5000.61.

For a general overview on VV&A processes, see Youngblood et al. (2000). There are many V&V methods available that range from informal reviews and walk-throughs to formal methods of Lambda Calculus and Proof of correctness (Balci, 1998; Balci et al., 2002; Petty, 2010).

ADVANTAGES AND DISADVANTAGES OF SIMULATION

Since M&S is so essential to many human endeavors, its advantages and disadvantages are described in the following sections.

ADVANTAGES

Cost-Effectiveness

Simulation and simulators are used primarily because they are cost-effective. They save both time and money while achieving some desired end. One objective of effective training systems is to provide the required training at the lowest possible cost. Simulation is a means for achieving that objective. Baudhuin (1987, p. 217) stated, "The degree of transfer from the simulator to the system often equates to dollars saved in the operation of the real system and in material and lives saved." Within the aviation community, the effectiveness of simulators is accepted as an article of faith. Indeed, the aviation industry could not function without simulators and FTDs, whose existence is mandated by FAA regulations (1991). As early as 1949, Williams and Flexman documented that even crude training devices (e.g., early simulators) produced payoffs with fewer trials, fewer flight hours, and fewer errors to qualify in the aircraft. The simulators were both cost- and training-effective devices (Flexman et al., 1972). In a very detailed analysis of cost-effectiveness, Orlansky and String (1977) reported that flight simulators for military training can be operated at between 5% and 20% of the cost of operating the aircraft being simulated; median saving is approximately 12%. They also reported that commercial airlines can amortize the cost of a simulator in less than nine months and the cost of an entire training facility in less than two years. Commercial airlines that use simulators accrue additional savings since they do not incur the loss of revenue associated with using aircraft for in-flight training.

In 1994, Beringer noted that since the cost of simulation has decreased as the capabilities of simulators have increased, today's question is more often phrased as, "If we can get more simulation for the same investment, what is the 'more' that we should ask for?" Thus, according to Beringer, cost is seen as facilitating, rather than prohibitive. The cost of computer-based simulation has decreased exponentially in the past 30 years, while processing speed and input/output capability have increased. In addition, the requirements for supporting infrastructure have decreased. The simulators described in the extended reality section of this chapter have a minimalist footprint and do not require special air-conditioning facilities or specially trained maintenance personnel.

Today's effectiveness questions are focused on how the required skills can be taught rapidly and inexpensively. In the healthcare area, Beal et al. (2017)'s meta-analysis of studies teaching medical students critical care medicine indicated that simulation was significantly more effective than other educational interventions or no intervention. While simulation improved skill acquisition, it was no better than other teaching methods in knowledge acquisition. They reported that their review was

> unable to address differences between types of simulation technology, the effect of duration or frequency of simulation teaching (the "dose" of simulation), the optimal timing by year of study, or retention of skills poststimulation. Further work is also needed to categorize the cost effectiveness of simulation-based teaching, because equipment and operational costs are high.
>
> **(p. 113)**

An earlier meta-analysis by Cook et al. (2012) reported that only five studies reported cost with comparison training. They concluded that simulation training was more costly (in money or faculty time) but more effective.

Availability

Many simulations are available 24/7, and do not require the physical presence of the objects being simulated or physical access to a simulator site. Many virtual reality/augmented reality simulations can be provided on self-contained systems such as Microsoft's HoloLens, Oculus Quest, and Varjo. The Extended Reality section of this chapter describes multiple training systems which could be available 24/7.

Simulators provide immediate access to a simulated location (described in the latitude and longitude) under specified environmental conditions (day, night, fog, sea state, etc.). Thus, the need for an actual physical presence at a location under specific environmental or operational conditions can be achieved. For example, simulators allow a student to complete an instrument landing system (ILS) approach and return immediately to the final approach fix (FAF) to commence the next ILS approach, without consuming time and fuel. Indeed, because simulators allow the instructor to "control reality," conflicting traffic in the landing approach can be eliminated to further increase the number of approaches flown per training session. In short, simulators provide more training opportunities than could be provided by an actual aircraft during the same time. Simulators also provide training in nonexistent aircraft or in aircraft where an individual's first performance in a new system is critical (consider the first flight in a single-seat aircraft and landings on asteroids or Mars).

Some surface vehicles and vessels are embedding simulations that allow training to take place in the actual vehicle (see https://www.netc.navy.mil/RRL https://www.tecom.marines.mil/Units/Divisions/Range-and-Training-Programs-Division/sandbox/DVTE/ and the capability described in the reduced environmental impact section in this chapter). In the case of remotely piloted air vehicles, a practice that is likely to become ever more prevalent is the use of the same operational terminal used for mission execution to perform training, as in the case of the US Navy's MQ-4 Triton Unmanned Aerial System (Lutz et al., 2017; Naval Technology, 2021).

Safety

Simulation provides a means for experiencing normal conditions in a safe and non-threatening environment. It also allows individuals to be exposed to controlled critical conditions that they hope they would never encounter, such as loss of control of a vehicle or egress during a fire. Simulation also provides an opportunity for initial qualification or requalification in a variety of workplaces such as control rooms of nuclear power plants and vehicles such as the space shuttle. For some tasks the ability to control the simulated environment is critical (e.g., hyperbaric chambers). Within the aviation domain, due to safety concerns, simulators may be the only way to teach some flight maneuvers or to expose aircrew to conditions that they are unlikely to experience under actual flight conditions (e.g., engine separation at takeoff, wind shear, loss of hydraulic systems, and engine fire). While UAV (unmanned air vehicle) controllers are usually located remotely from the vehicles under their control, they are responsible for safety of flight. Ribeiro et al. (2021) recommend creating virtual obstacles on Augmented Reality (AR) devices to reduce their fear of crashing in real environments.

Simulation is routinely used for crew resource management (CRM) research and training. CRM evolved from cockpit resource management (Wiener et al., 1993). The original objective of CRM was to reduce the number of aviation accidents and incidents by increasing the effectiveness of cockpit crew coordination and flight deck management. Since its introduction in 1978, CRM training has been expanded to include not just the cockpit flight crew but also cabin crew members, dispatchers, maintenance, and security personnel (Helmreich et al., 1990; Weigmann & Shappell, 1999; Salas et al., 2000; Weigmann & Shappell, 1999). CRM has also been incorporated into the training of medical teams, particularly in operating rooms (Helmreich & Schaefer, 1998). Gross et al. (2019) reviewed the use of CRM training in healthcare. The effectiveness of CRM was perhaps best demonstrated on January 14, 2009, when Capt. Chesley (Sully) Sullenberger landed a disabled aircraft in the Hudson River and the crew safely evacuated all passengers (Muhlenberg, 2011).

In addition, automation has increased the need for simulators, as Wiener and Nagel (1988, p. 453) commented, "It appears that automation tunes out small errors and creates the opportunities for larger ones." In automated glass (cathode-ray-tube-laden) cockpits, improvements in system reliability have reduced the probability and frequency of system problems, thus inducing a sense of complacency among the aircrew. However, when an unanticipated event occurs, the crew must be trained to respond rapidly and correctly. Simulators provide an opportunity for that type of training. The need for simulator training became apparent in the October 2018 and March 2019 losses of two 737Max aircraft carrying 436 people (https://transportation.house.gov/download/endsley-testimony, Kitroeff & Gelles (Jan. 8, 2020).

Surrogate Value

Simulation also reduces the usage of the actual system. Thus, with transportation systems, it reduces the exposure and number of hours on the actual vehicle, which in turn reduces mechanical wear and tear, as well as acquisition, sustainment, operations, and maintenance costs. To put that in perspective, in 2019, the estimated cost per flight hour of the F-35, USAF's most advanced fighter, was $35,000 (Insinna, 2019). Simulation also reduces infrastructure load (viz., highways or the National Airspace System). From an economic perspective, commercial airlines can use aircraft not required for training purposes on revenue-producing flights. The US Navy has a Blue and a Gold team which rotate on the same submarines. The use of submarine bridge trainers allows one crew to qualify bridge teams ashore while the other team is at sea (Gray, June 8, 2017).

Reduced Environmental Impact

Today, we have an emphasis on environmental concerns and the vehicles being simulated do not pollute, consume fuel, create noise, or leak hazardous substances (e.g., radiation). Neither do they damage people or property; indeed, simulated patients and bomb-damaged areas are repaired at the flick of a switch. Magnuson (2019) described a demonstration of how a live, virtual, constructive (LVC) training environment was used to "expand" the test range at Nellis Air Force Base. Advanced technologies allowed the pilot of a live airborne aircraft operating within the confines of the test range to participate in an exercise against virtual threats and targets that were presented on the aircraft's displays. These threats and targets behaved realistically and could be presented beyond the physical boundaries of the range. None of the simulated adversary aircraft consumed fuel and simulated ordinance did no physical damage, while the pilot was challenged in the simulated threat environment.

Improved Training Environment

Simulators incorporate instructional features that enhance student learning and facilitate instructor intervention. Gawron (2019) describes (1) the role of simulation in a training curriculum, (2) how to measure transfer of training from the simulator to the real world, and (3) how to know the types of simulations and simulators that are being used for training. Moroney and Moroney (2010, Table 19.1) provide a detailed listing of instructional features incorporated into modern simulators, including:

- Simulator instructor options such as preset/reset, crash/kill override, playback/replay, motion, and sound.

- Task conditions such as time of day, seasons, weather, and wind direction and velocity, as well as realism, aircraft stability, and instrument malfunction.
- Performance analysis/monitoring features such as automated performance measurement, debriefing aides, warnings, and advisories that a preset parameter is about to be or has been exceeded.

Standardized Training Environments

Simulators can provide identical flight dynamics and environmental conditions from training session to training session. Thus, the same task can be repeated until the required criteria are attained, and, indeed, until the task is overlearned (automated). Unlike the airborne instructor, the simulator instructor (SI) can focus on teaching the task without safety of flight responsibilities or concerns about violations of regulations. Thus, the instructor may deliberately allow a student to make mistakes such as illegally entering a terminal control area or flying below an assigned altitude.

Provide Data

Simulation provides opportunities for data collection, which are not available in the real world. Although the critical issue of collecting the right data is beyond the scope of this chapter, data collection permits the following:

- *Performance comparison*: As part of the diagnosis process, the student's performance can be compared with the performance criteria, as well as the performance of other students at the same stage of training.
- *Performance and learning diagnosis*: Having evaluated the student's performance, a teacher or instructor can gain some insight into the student's learning process and suggest new approaches in problem areas.
- *Performance evaluation*: Performance measurement can be used to evaluate the efficacy of different approaches to training a particular task.

Lack of Realism

Despite the emphasis on high fidelity and "realism," simulators are not realistic; rather they allow instructors to manipulate reality. In a sense, the lack of realism may contribute to their effectiveness. Lintern (1991, p. 251) stated that transfer could be enhanced by "carefully planned distortions of the criterion task." In addition, most instructional features found in simulators do not exist in the cockpit or the device being simulated. Indeed, if real cockpits had the same features as simulators, the "RESET" button would be used routinely, as it can undo and defy the laws of physics. The entertainment industry relies on this ability to suspend reality to move audiences to nonexistent worlds. Viewers readily suspended reality when the USS *Enterprise* on *Star Trek* goes to Warp Speed and when "repulsion technology" is used to power the "Land–Speeders" in *Star Wars*.

Disadvantages of Simulation

Does Not Necessarily Reflect Real-World Performance

Although performance in a simulation with reasonable fidelity is probably indicative of an individual's expected performance in the real world, we must recognize that performance in a simulator does not necessarily reflect how an individual will react in the real world. There are at least three reasons for this:

- Because there is no potential for an actual accident in a simulator, it would seem reasonable to expect that a trainee's stress level would be lower in a simulator. However, the stress level may be increased when an individual's performance is being evaluated or when they are competing for a position or a promotion.
- To the extent that team or individuals being evaluated or seeking qualification-in-type expect an emergency or unscheduled event to occur during their time in the simulator, their performance in a simulator may not reflect their in-flight performance. The trainee(s) would, in all probability, have reviewed operating procedures before the start of their period in the simulator. Nonetheless, it should be recognized that the student's review of procedures even in preparation for an evaluation is of value.

  Consider how difficult it would have been to induce the startle effect that contributed to the loss of Air France Flight 447 on a flight from Rio de Janeiro to Paris. Frozen airspeed sensors deprive crew and flight computers of airspeed data, and the crew induced an aerodynamic stall which resulted in the loss of 228 lives (https://humanfactors101.com/incidents/air-france-flight-447/, Geiselman et al., 2013).
- Performance (particularly vigilance and situational awareness) in a simulator, when being evaluated, rarely reflects the fatigue or boredom common to many workplaces. Therefore, performance in a simulator may be better than actual in-flight performance.

Surrogate Value

A possible downside to surrogate value within the military is that reduced utilization of the hardware will lead to fewer and less experienced maintenance personnel and reduced supply chain requirements. These apparent savings may also create personnel shortages and logistic problems when the operational tempo rises beyond the training level. To offset the reduced experience among maintainers, AR-enabled maintainer solutions have been developed to support refresher and deployed training needs (N193-D01, n.d.). Also see the AR for Aircraft Maintainers section later in this chapter.

User Acceptance

The acceptance and use of simulators is subject to the attitudes of simulator operators, instructors, trainees, and management. The increased use of computers, from preschool to college, has raised their acceptance in the education and training process.

However, trainees often voice the attitude that "they expected to be using the real system and not a simulation." User acceptance can be significantly influenced by a management policy that monitors the appropriate use of simulation. A critical element in this process is determining the appropriate performance metrics. Measuring effectiveness is a fairly complicated process that has performance measurement at its core. Lane's (1986) report is a "must read" for individuals interested in measuring performance in both simulators and the real world. Mixon and Moroney (1982) provided an annotated bibliography of objective pilot performance measures in both aircraft and simulators, whereas Gawron (2000) provided a compilation of measurement techniques. Readers interested in measuring transfer effectiveness are referred to Gawron (2019) and Boldovici (1987) for a discussion of sources of error and inappropriate analysis for estimating transfer effectiveness.

Overall, the advantages significantly outweigh any real or perceived disadvantages as evidenced by the general acceptance of simulators by management, trainees, and regulatory agencies.

A SAMPLING OF PROGRESS IN SIMULATION

Considering the many facets of simulation, we realized that any detailed discussion of all the areas in which simulation has been applied is beyond the scope of this chapter. Therefore, we examined two different human-related domains that have played a major role in the history of simulation—namely, wargaming and aviation. Wargaming, which originally focused on decision-making and strategy development, has evolved into distributed mission training at multiple locations on a distributed simulation network. Aviation has pushed the development of simulation technology from manually manipulated simulators to multiple live and virtual aircraft at different physical locations engaged in simulated air combat mission (ACM) training simulations. Manipulation of reality is perhaps the area in which the greatest growth has occurred in the past two decades. Therefore, we have devoted a section to the discussion of extended reality, which includes a section on the application of augmented reality in training recognition skills. We conclude this section with a discussion of unique simulators which will not be discussed in other chapters of this book.

WAR-GAMING

The ubiquitous Internet began as a Defense Advanced Research Project Agency (DARPA) effort to ascertain critical information that could successfully pass over multiple communication paths despite the destruction of some nodes. Many other innovations, including emergency medical services, blood plasma, and lasers, have warfare as their origin. Modern simulation is no exception, as it evolved from warfighting strategies.

Simulating battles through games can be traced back to the Hindu game Chaturanga and the Eastern game GO (Allen, 1989). Within the Western culture, chess is sometimes considered an early instantiation of simulation. Consider the elements involved in chess: each of the actors (knights, kings, queens, rooks, castles, etc.) has a role, predetermined start points, and specific rules of engagement with which they must comply. Chess involves both deterministic elements (rule compliance) and probabilistic elements (anticipation of the opponent's next move). The earliest known American chess book, *Chess Made Easy* by J Humphreys was published in 1802 and the first American Chess Congress in 1857 heralded a chess craze in the United States. No doubt, military officers also shared in that excitement for chess.

In 1882, Army Major William R. Livermore wrote "The American Kriegsspiel, A Game for Practicing the Art of War on a Topographical Map," which introduced war-gaming to the US military. Look-up tables were used to determine some outcomes (e.g., casualty rates were determined as a function of range, firepower, and duration of exposure), whereas a roll of the dice added the element of chance to the probability of the outcomes. The US Navy also adopted war-gaming. Starting almost 100 years ago between World War I and II, the Naval War College conducted 136 war-games (http://news.usni.org/2013/09/24/brief-history-naval-wargames, https://www.youtube.com/watch?v=PzwDWX8oQn8).

The Naval War College (Figure 1.1) developed wargaming without the benefit of computer hardware and software. The faculty had to rely on "human computers" who had a diversity of education and naval operational experience. The games were not mechanically nor electronically operated. Rather, they relied on paper to pass orders and messages and physical markers and ship models on floors to track progress through different theaters of battle. What it lacked in sophisticated modeling

FIGURE 1.1 War-gaming floor in Pringle Hall circa 1950 (from the U.S. Naval War College Archives) that looks like the games before World War II.

and simulation was made up for with brain power supplied by the faculty, umpires, and students. The immersion into the war was limited only by the imagination of the gamers who could visualize the battlespace. Although the process was basic, the insights gained from these simple models were useful. "Simulations" helped the War College students do the intellectual heavy lifting as they worked through the war games. The insights from the students and instructors were useful to the Navy leadership that had to plan for a potential war across the expanse of the Pacific.

This easily reconfigurable gaming laboratory provided some of the intellectual heavy lifting for the pre–World War II Navy's strategy and tactics and to understand emerging technological threats (e.g., aircraft and submarines). The insights gained were worked out in the more expensive fleet exercises where ideas were put to the reality test. The exercises allowed experimentation with and the refinement of the most promising tactics, techniques, and procedures using operational systems. The strategic and operational use of the new Navy aviation technology evolved on the floors of the Naval War College. Carrier aircraft evolved from spotters for battleship guns into an effective offensive weapon. A successful island-hopping strategy evolved from the original unworkable strategy of sailing an armada from the west coast to engage the Japanese fleet in a decisive battle at sea.

Naval officers had the freedom to think "out of the box" and conduct tests themselves as strategists, operators, and tacticians. Officers took real concerns they were faced with and experimented with different methods to overcome the challenges of the sheer size of the Pacific Ocean, the Japanese geographic advantages, and the Japanese technological advantages in aircraft and torpedoes. Even this noncomputerized war-gaming helped identify the shortfalls in how the fleet operated and how it would be vulnerable during resupply in the face of a modern peer adversary. The lessons learned were articulated as new requirements for the fast oilers, ammunition, and supply ships which had to keep up with the fleet of combatant ships. The war games showed the vulnerability of ports that could fall easy prey to the Japanese fleet and aircraft. Although the bombing at Pearl Harbor forced and aided the change, the war games were part of the evolution from a battleship gun club-centric fleet to an aircraft carrier offensive fleet. The Navy transformed itself from what worked in the past with battleship duels into the new dimension of offensive naval aviation.

The United States did not have a monopoly on war-gaming. In September 1941, the Japanese Naval War College played a war game for 11 days that ended with a surprise attack on Pearl Harbor. Allen (1989) reports that Vice Admiral Nagumo, who subsequently led the attack on Pearl Harbor, played himself during the game. Allen also reports that games played by Japanese war planners were used to convince Japanese politicians that Japan would win a war with the United States.

An intriguing discussion regarding the Battle of Midway and Japanese war games is provided by Perla (1990). On May 1–4, 1942, aboard the flagship Yamato, the invasion of Midway was simulated during a war game. As part of the simulation, US airplanes from Midway attacked the Japanese carriers and a die toss resulted in the sinking of two Japanese carriers. This loss was overruled by the presiding officer, Rear Admiral Ugaki, who determined that only one aircraft carrier was sunk and the other was only slightly damaged. After the conclusion of the war game, Ugaki inquired about a contingency plan in case a US Navy carrier task force might oppose the Japanese invasion. He received an ambiguous reply, which led him to caution that such a possibility should be considered (Perla, 1990). A month later, the real battle of Midway was fought, and the land-based planes did little damage to the Japanese fleet, although the unexpected US carrier task force effectively destroyed Japanese carrier-based airpower. With respect to the decision by Ugaki, Perla notes that the land-based aircraft did not score a single hit on Japanese carriers, thus validating Ugaki's inquiry. Regarding the Battle of Midway, Perla (1990) emphasizes that the outcome of the Japanese war game might have been different if the individual representing the commander of the American forces had behaved in a manner more characteristic of an American commander. The "fidelity" of play is an essential element so gamers will see it as a valid war game. How the gamers assess that fidelity of the game design and rules should not be overlooked.

The authors of the RAND study on the next-generation war game for the Marine Corps, Wong et al. (2019) have observed there are generational differences in the game design. Different experiences in civilian game technology shaped the players' expectations and their comfort level with gaming fidelity. A key concept in their study was that gamers developed different paradigms in adulthood largely based on the commercial games available in adolescence. This affects how they view game design, immersion, and gameplay.

The baby boomers had exposure to mostly social board games which gave them an understanding of the rules, patterns of hex games, and other manual tabletop war games with little need for further explanation. Little of their gaming experience was digital during their adolescence and early adulthood.

Generation X and millennials had more common experience with personal computer (PC), console, and social media games. They came of age on PC and console games. Generation X popular consoles were made by companies like Atari and Nintendo, which had games like Space Invaders (1983), Super Mario Brothers (1983), and The Legend of Zelda (1986). These adolescents saw the beginnings of handheld consoles like the Nintendo Gameboy. Most of their gaming experience was less social interacting than the board games of the Baby Boomers and the technology the Millennials would use.

Millennials had access to gaming consoles with increased video animation and computational power such as the Xbox and the PlayStation along with sophisticated PC games that use Internet connectivity. Millennials played in real time in massively multiplayer online games such as Everquest (1999) and the World of Warcraft (2004). The latest generation (Generation Z, iGen, or digital natives) are forming their own construct which will most likely reflect the characteristics of mobile, handheld but networked gaming systems. These are increasingly using augmented and virtual reality systems in games such as Pokemon Go (2016). The computation power of the Nintendo Switch provides a hybrid platform that is both a portable handheld and a networked game console.

As the Marine Corps and other military departments develop their next-generation war games, it is most likely they will move to an increased level of immersion moving away from the baby boomer dominant board game-style format. That strong immersive experience is one of the hallmarks of educational war games whose priorities are to train and educate the gamers. The military will continue to leverage emerging technologies such as AR, VR, and gaming networks to construct a highly immersive gaming environment. Scenario, decision, adjudication, and visualizations tools will accelerate the achievement of rapidly constructed immersive education and training environment all based on basic war-gaming design and analysis concepts and lessons learned from over a hundred years of gaming both in the military and in the commercial sectors.

Readers interested in more details on war games can consult the 32-page war games bibliography prepared by Aegis Research Corporation (2002), http://citeseerx.ist.psu.edu/viewdoc/download?doi=10.1.1.473.2503&rep=rep1&type=pdf and the RAND Corporation website on wargaming (https://www.rand.org/topics/wargaming.html) and the University of Toronto bibliography (https://www.utm.utoronto.ca/asc/game-enhanced-learning-bibliography). Caffrey's (2019) book, *On Wargaming: How Wargames Have Shaped History and How They May Shape the Future*, provides many insights; also see https://usnwc.libguides.com/wargaming.

Online Gaming

When the World Wide Web was made available to the public in 1993, it could only be accessed on a dial-up modem on a phone line with a 56-kbps data transfer rate. So slow that a single song could take 10 minutes to download. Cisco's 2020 Annual Broadband report estimated the average bandwidth in North America is greater than 70 Mbps (over 12,500 times faster); and that bandwidth is expected to have doubled by the end of 2023. Online gaming has kept up with the advances in the computational power of personal computers, gaming consoles, and increased bandwidth. Developers can now offer games and experiences for an individual or for a group of people, who can be collocated or separated geographically. The online experience can be quests or challenges where you can win or lose or the opportunity to explore and carry out a virtual life without the pressures of a combat scenario. Some platforms enable gamers (without programming skills) to create characters, structures, machines, and even their own environment. One example, *The Sims*, available since 2000, has ten million active monthly users. Gamers can interact with other persons in the game world through avatars they create. These avatars can start as infants, teenagers, or adults with different personalities, lifestyles, and physical traits that can gain skills and "mature" from their experiences in the game world. The *Sims* user experience (UX) development team goes well beyond writing code and addresses systems issues such as branding, design, usability, and function. They emphasize pleasure, aesthetics, and fun while making the in-game experience as similar as possible to real-life experiences. Players can try different or idealized versions of themselves in a no-pressure environment, where there are no repercussions for their actions (Mueller, 2021).

Educators have experimented using online virtual worlds as educational tools. One such tool Second Life (SL), has been used to create popular virtual worlds for education and for the study of virtual worlds in education research.

For instance, Sweigart and Hodson-Carlton (2013) used SL avatars to improve nursing student interview skills; Esteves et al. (2011) taught computer programming; and Nestler et al. (2013) taught physical security aspects of cybersecurity. Texas A&M Veterinary Medicine had a class role-play a Veterinary Emergency Team's hurricane disaster response using professors, students, and avatars working in the SL virtual learning environment (https://www.youtube.com/watch?v=I5VXf5Ky5ms). For education, digital games are being used extensively in kindergarten through high school to teach "critical 21st-century skills" (http://glasslabgames.org/). In a review of the effectiveness, Young et al. (2012) reported some positive effects on language learning, history, and physical education (exergames) but no positive effects on science and math learning. However, Weiss et al. (2006) reported that kindergarten students taught with multimedia in cooperative learning or with multimedia in individual learning did better in math than students without either type of training.

Using similar technology, hands-on virtual labs are becoming prevalent and providing unique learning opportunities. One example is the Computer Emergency Response Team (CERT) Simulation, Training, and Exercise Platform (STEPfwd) (https://stepfwd.cert.org/vte.lms.web). Tang et al. (2012) described using Sustain City, for undergraduate-level training in science and engineering. Hiskins et al. (2011) discussed virtual simulation to train high school students to design electric cars. Basu et al. (2013) described an enactment, or E-World, a multi-agent simulation, to teach science to high school students. MIT's Open Courseware site (http://ocw.mit.edu/index.htm) is an excellent source of multiple educational simulations. Moreno-Ger et al. (2009) described several low-cost platforms for educational game development. Ulrich et al. (2017) developed a gamified nuclear power plant simulation.

However, Picciano et al. (2012) expressed concerns with the quality of instruction, policies for funding such courses, and attendance requirements for online learning. Eshet-Alkalai et al. (2010) also discussed challenges in delivering different instructional technologies effectively. Educational researchers have noted that the explosive growth in online learning has not been followed by many studies on their effectiveness. Carnahan (2012) compared a traditional classroom setting where there is face-to-face contact between the student and the teacher, with learning in a virtual world where the student interacted with the teacher and other students via an avatar that represented him. The research showed no significant differences in academic achievement by the seventh-grade STEM students in either environment. Student satisfaction was higher with the virtual classroom. This could encourage greater engagement by the students in a game-like entertaining environment. D'Angelo et al. (2013) metanalyses of 59 articles showed overall that computer-based interactive simulations improved training more than instruction without the simulations; however, the findings were restricted to science, math, and engineering (STEM) students. More studies for different age groups from kindergarten to graduate education and for non-STEM students are needed. Reduction in equipment cost and increased Internet connectivity are expected to accelerate the adoption of virtual reality/augmented reality systems for online gaming and education.

Aviation

US Army Signal Corps' Specification No. 486 (1907) for its first "air flying machine" has one very straightforward user-centric requirement: it should be sufficiently simple in its construction and operation to permit an intelligent man to become proficient in its use within a reasonable time. Apparently, the "intelligent man" needed help or the flying machine was not simple enough; less than three years later, Haward (1910), as quoted in Rolfe and Staples (1986), described an early flight simulator as "a device which will enable the novice to obtain a clear conception of the workings of the control of an aeroplane, and of the conditions existent in the air, without any risk personally or otherwise" (p. 15). As early as 1910, the Wright Brothers used a "kiwi bird" flight simulator in their training school at Huffman Prairie (Bernstein, 2000). Students learned rudimentary flight control while seated in a defunct Wright Type B Flyer mounted atop a trestle. Motion was induced by means of a motor-driven cam. With respect to cost, Bernstein reports Bernard Whelan as saying

> It was the only thing that I know that cost more then than it does today. It cost sixty dollars an hour to take flight training at the Wright School. That's a dollar a minute. And they didn't sign you up for anything less than four hours.
>
> **(p. 123)**

A similar device is described by Adorian et al. (1979). The Antoinette trainer (Figure 1.2) required a student to maintain balanced flight while seated in a barrel (split the long way) equipped with short "wings." The barrel, with a universal joint at its base, was mounted on a platform slightly above shoulder height so that instructors could push or pull on these "wings" to simulate "disturbance" forces. The student's task was to counter the instructors' inputs and align a reference bar with the horizon by applying appropriate control inputs through a series of pulleys.

FIGURE 1.2 One of the earliest rudimentary trainers, an Antoinette Trainer (ca. 1910).

A more dynamic simulator was utilized by the French Foreign Legion. They realized that an airframe with minimal fabric on its wings would provide trainees with insight into the flight characteristics of the aircraft while limiting damage to the real aircraft and the student (Caro, 1988). In 1917, Winslow, as reported in Rolfe and Staples (1986), described this device as a "penguin" capable of hopping at about 40 mi (65 km) per hour. Although of limited use, it was a considerable improvement from the earlier flight training method of self-instruction in which trainees practiced solo until they learned basic flight maneuvers. Instructors would participate in the in-flight training after the trainees had, through trial and error, learned the relationship between input and system response (Caro, 1988). Apparently, the legionnaires understood the value of a skilled flight instructor.

The origins of modern flight simulators can be traced to 1929, when Edward A. Link received a patent for his generic three-degrees-of-freedom (yaw with limited pitch and roll), ground-based flight simulator (Figure 1.3). His initial trainer with stubby wings and rudders was designed to demonstrate simple control surface movements and make them apparent to an instructor. Later a hood would be added over the cockpit which, with the appropriate instruments in the cockpit, allowed the device to be used for instrument flight training. Link based his design on the belief that the trainer should be as analogous to the operational setting as possible. Through the use of compressed air that actuated bellows, the trainer could pitch, yaw, and roll, enabling student pilots to gain insight into the relationship between stick inputs and movement in three flight dimensions. A patent for an aviation-training machine

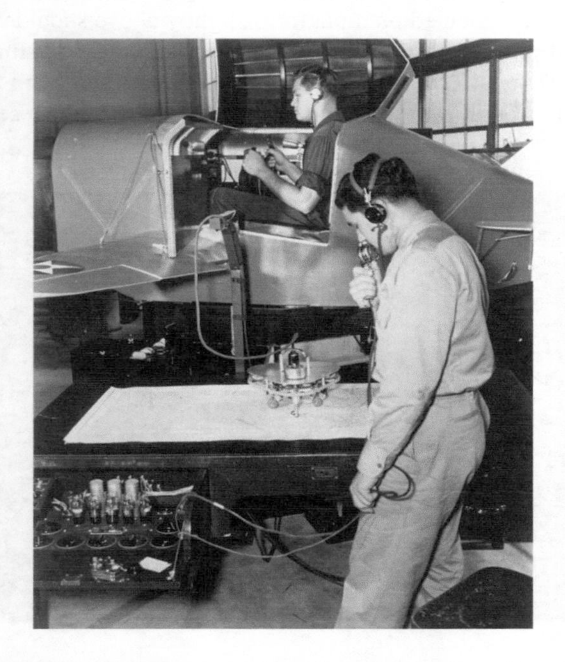

FIGURE 1.3 "Blue Box" link trainer and associated instrument table.

also using "an air-operated motor" was granted to Levitt L. Custer of Dayton, Ohio, in 1930. Link's invention was originally marketed as a coin-operated amusement device (Fischetti & Truxal, 1985); however, the value of Link's simulator was recognized when the US Navy and Army Air Corps began purchasing trainers in 1934. Flight instructors, watching from outside the "Blue Box," would monitor the movements of the ailerons, elevator, and rudder to assess the student's ability to make the control inputs necessary for various flight maneuvers. A plotting table (shown in Figure 1.3) was also developed that allowed instructors to observe the ground track and speed of the aircraft.

When the United States entered World War II, there were over 1,600 trainers in use throughout the world. The number of trainers increased as the Allied forces rushed to meet the demand for pilots. During the war years, approximately 10,000 link trainers were used by the US military (Caro, 1988; Stark, 1994). In 1943, an operational flight trainer for the US Navy's PBM-3 aircraft was produced by Bell Telephone Laboratories (Pohman & Fletcher, 1999). This trainer used analog circuitry to solve flight equations in real time and presented the results using the actual controls and instruments available in the aircraft. During World War II, considerable assets were devoted to the development of electronic digital computers. In 1944, with US Navy funding, the Massachusetts Institute of Technology (MIT) undertook the development of an electronic flight simulator (Waldrop, 2001). Until then, development efforts had focused on batch-processing-type computers, which solved an equation and then waited for the next equation. However, the MIT developers realized that unlike previous computers, the one named Whirlwind had to respond in real time to the constantly changing pilot's inputs and the dynamic response of the simulated aircraft. Thus, they built the first real-time digital computer, which was the basis of the modern PC. Its performance in 1951 was equivalent to the 1980 TRS-80 PC; however, its vacuum-tube electronics required an area approximately the size of a small house, with unique electrical and cooling requirements.

After the war, simulations developed for military use were adapted by commercial aviation. Loesch and Waddell (1979) reported that by 1949, the use of simulation had reduced airline transition flight training time by half. Readers interested in details on the intriguing history of flight simulation may consult the excellent three-volume history entitled *50 Years of Flight Simulation* (Royal Aeronautical Society, 1979). Jones et al. (1985) provide an excellent overview of the state of the art in simulation and training through the early 1980s. Moroney and Moroney (2010) update that literature.

Following World War II and throughout the 1950s, aircraft diversity and complexity created the need for aircraft-specific simulators, that is, simulators that represent a specific aircraft in instrument layout, performance characteristics, and flight handling qualities. Successful representation of instrument layout and performance characteristics was readily accomplished; however, the accurate reproduction of flight-handling qualities was a more challenging task (Loesch & Waddell, 1979). Exact replication of flight is based on the unsupported belief that higher fidelity simulation would result in greater transfer of training from the simulator to the actual aircraft. This belief has prevailed for many years and continues today. However, almost 70 years ago, researchers were questioning the need for duplicating every aspect of flight in the simulator (Miller, 1954; Stark, 1994).

Within the context of education, Spannaus (1978) lists three characteristics of simulations: (1) they are based on a model of reality, (2) the objectives must be at the level of application, and (3) the participants must deal with the consequences of their decisions. He believes that for an activity to be called a simulation, students cannot be just observers but must also be involved. Today's educators and trainers call this process "active learning and immersion."

Ricci et al. (1996) provide experimental support for the advantages of active learning, specifically computer-based gaming, in knowledge acquisition and retention.

Caro (1979, p. 84) emphasized that a flight-training simulator's purpose was "to permit required instructional activities to take place." However, from his 1979 examination of existing simulators, simulator design procedures, and the relevant literature, Caro concluded that "designers typically are given little information about the instructional activities intended to be used with the device they are to design and the functional purpose of those activities" (p. 84). Fortunately, some progress has been made in this area. Today, as part of the system development process, designers (knowledgeable about hardware and software), users and instructors (knowledgeable about the tasks to be learned), and trainers and psychologists (knowledgeable about skill acquisition and evaluation) interact as a team in the development of training systems (Stark, 1994). The objective of this development process is to maximize training effectiveness while minimizing the cost and time required to reach the training objective (Stark, 1994). The XR section of this chapter contains some good examples of this process.

Salas et al. (1998) propose that, to fully exploit the tremendous progress in simulation technology, we should reduce the emphasis on fidelity and realism, and focus on enhancing the learning of complex skills by bridging the gap between training research findings and the use of training technology. They challenge the following assumptions: (1) simulation is all you need, (2) more fidelity is better, and (3) if the aviators like it, it is good. They propose that the emphasis should shift from technology to learning, specifically, to a trainee-centered approach with a more holistic consideration of the training process.

Modern flight simulators are used for initial and advanced training as well as proficiency maintenance and evaluation. They have become an integral part of the research, development, test, and evaluation (RDT&E) cycle. With the development of the glass and digital cockpit with its onboard computers, and associated recording media (including the black box), simulators are now used in both reconstructing aircraft accidents and in some cases identifying appropriate recovery procedures.

During the 1990s, flight simulation was a worldwide industry with many competitors (Sparaco, 1994), and sales of $3 billion per year for commercial airlines and $2.15 billion per year for the DoD. Individual simulators ranged in price from $3000 for a basic PC-based flight simulation with joystick controls up to an average of $10–13 million for a motion-based simulator (down from $15–17 million in the early 1990s). Today, the number of competitors has been reduced as the aerospace industry has consolidated. This consolidation has been attributed to advances in technology and the emphasis on leaner and more efficient core-business-focused organizations (Wilson, 2000). The Flight Simulator Market is expected to reach $8.03 billion by 2026, according to Global News Reports and Data (https://www.globenewswire.com/news-release/2019/07/11/1881533/0/en/Flight-Simulator-Market-to-Reach-USD-8-03-Billion-By-2026-Reports-And-Data.html). The highest growth rate is expected to be in the area of UAVs (unmanned air vehicles). The reduction in hardware cost for display technologies and the reduced emphasis on simulating motion has changed the traditional focus from hardware to software. Major suppliers in this industry are now focusing on how to incorporate the Internet and XR into training systems. For details on the use of XR, see the following section.

EXTENDED REALITY: AUGMENTED REALITY, VIRTUAL REALITY, AND MIXED REALITY

Introduction to XR

This section provides a brief overview of XR technologies and how they are reshaping the way we both do our jobs and train to do our jobs. Extended Reality is a blanket term encompassing VR, AR, and MR. XR technology has wide applicability in sectors, including, but not limited to, medical, aeronautical, maritime, and automotive industries, and disparate applications ranging from engineering design, manufacturing, training, operational applications, maintenance, and sustainment. Its suitability will necessarily depend on the fidelity requirements of the use case (i.e., specific application) and the capability of the XR technology to support those requirements. While there is not complete agreement over the definitions of VR, AR, and MR technologies, Milgram and Kishino (1994) developed a Reality-Virtuality (RV) Continuum, which is helpful for conceptualizing their relations. A variation of their model, referred to here as the XR Reality-Virtuality Continuum, has been adapted to include XR and VR (Figure 1.4).

While the term Real Environment (RE) is clear intuitively, other terms in the XR Continuum require some definition to ensure a common understanding. First, Virtual Environment (VE) can be defined as "the representation of a computer model or database which can be interactively experienced and manipulated by the virtual

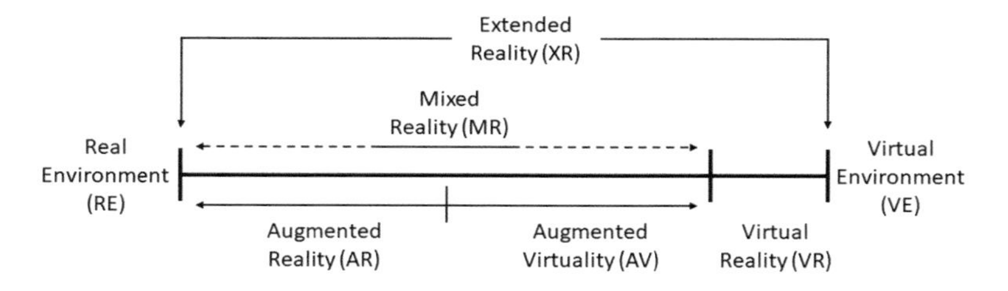

FIGURE 1.4 Extended Reality (XR) Reality–Virtuality (RV) continuum. (Adapted from Milgram and Kashino, 1994, and Doolani et al., 2020.)

environment participant(s)" (Barfield & Furness, 1995, p. 4). Mazuryk and Gervautz (1996, p. 5) explain that in a VE "system a computer generates sensory impressions that are delivered to the human senses." So, a virtual environment requires (1) a computer model, (2) a representation of that model which stimulates the user's senses (e.g., visual, auditory, haptic), (3) a user or users, and (4) a way the user(s) can interact with the computer model. A perfect VE would result in complete immersion of the user within the VE such as in the book *Ready Player One* and the movie by the same name. This is not technically feasible, at least not today. Thus, there is an XR Reality–Virtuality Continuum between the real and virtual worlds.

The terms VR and VE are often used interchangeably; however, as indicated above, current technology does not allow perfect immersion in a VE. Therefore, it is useful to distinguish between the two. VR has been defined in various ways as described in the Kaplan et al. (2020) meta-analysis. Boud et al. (1999, p. 32) define VR as a "three-dimensional computer-generated environment, updating in real time, and allowing human interaction through various input/output devices." This definition aligns very closely with that of a VE. Heim (1998, p. 221) defines VR as a "technology that convinces the participant that he or she is actually in another place by substituting the primary sensory input with data produced by a computer." The key distinction between the two definitions is Heim's (1998) emphasis on the technology that allows immersion in the VE. This is a useful distinction. A final component to bear in mind when considering VR technology is that the focus is on the VE and intrusion of the RE is a distraction to be overcome. So, stated succinctly, VR technology seeks to immerse the user(s) in a VE and detach the user(s) from the RE.

AR, Augmented Virtuality (AV), and MR technologies seek to blend the RE and the VE to achieve specific purposes. Milgram et al. (1995, p. 283) defined AR as "augmenting natural feedback to the operator with simulated cues." Drascic and Milgram (1996, p. 123) expanded on this definition stating that "AR describes that class of displays that consists primarily of a real environment, with graphic enhancements or augmentations." So, AR technologies enable the user to interact with the RE and overlay or otherwise add information from a VE to enhance the users experience with the RE.

AV can be conceptualized as the inverse of AR. That is, the user's experience is primarily that of the VE (i.e., computer-generated), which is augmented in some way by the real world (Drascic & Milgram, 1996; Milgram & Kishino, 1994). For example, including a representation of the user's hands in the image to aid in interactions with the VE. As a second example, a challenge in the adoption of VR technology has been users tripping over real-world objects while in the VE. Recent head-mounted displays (HMD) are working to overcome this by enabling users to map the RE in the VE (e.g., create borders in which the user can move without running into physical objects). In both instances the primary emphasis is on the VE, which has been enhanced with information from the RE (e.g., hand position/orientation, definition of the physical space and barriers).

Milgram et al. (1995, p. 283) define an MR environment as "one in which real-world and virtual-world objects are presented together within a single display." There is clearly significant overlap between this definition and those of AR and

AV environments in the preceding paragraphs. This is intentional as Milgram et al. (1995) conceptualized MR as a continuum in which RE and VE objects are displayed in a common view with greater or lesser emphasis on the RE or VE. Recently MR has begun to be used by some as distinct from AR and AV, while others continue to use it based on its earlier definition. This has created some confusion in terminology. For the present discussion, it is argued that while MR might include applications ranging from AR to AV, its value is in describing those instances in which the terms AR and AV do not adequately describe the use case of interest. That is, MR can be conceptualized as including a more balanced mix of the RE and the VE in which neither predominates and allowing purposeful interaction between the two. This view is consistent with the Reality-Virtuality Continuum proposed by Flavian et al. (2019). They distinguished MR from AR and AV referring to it as Pure MR and describing it as the complete merging of the VE and the RE. It is important to note that there is still significant overlap in the use of these terms and a review of the specific application is key to ensuring sufficient understanding of the technical approach being utilized.

Using this framework, the next section will briefly illustrate the history of VR in training. Then, we will review recent real-world applications of XR technologies in Training and Operational Support. Following this section, potential future use cases will be explored. Finally, technical challenges inhibiting widespread adoption will be highlighted.

Historical VR Devices

Space limitations preclude a detailed discussion of the evolution of XR devices; however, many of the advances in XR technology today were built on the innovations of decades of VR research. Before we turn to recent advances in XR technology, we are providing a comparison of three VR systems which illustrate the state-of-the-art in the 1960s, early 2000s, and 2020s. In the mid-1960s, Tom Furness of the Armstrong research laboratory at Wright-Patterson Air Force Base (Voices of VR Podcast, 2015) began the development of a large field of view HMD capable of providing partially overlapped fields to each eye. Known as the visually coupled airborne systems simulator (VCASS), it was the first virtual reality panoramic display ever constructed and tested. The innovative Farrand Optical Pancake Window™ optics provided an adjustable horizontal FOV of 100°–160° with a binocular overlap of 20°–60° and a constant vertical FOV of 60°. The display's instantaneous presentation was controlled by a state-of-the-art AC magnetic helmet-mounted tracker that could determine azimuth, elevation, and roll orientation, as well as x, y, and z helmet position information to a then unprecedented 14-bit resolution. The VCASS system was among the earliest virtual reality systems and could produce simulated air-to-air and air-to-ground mission scenarios, simulated cockpits, enhanced stereoscopic imagery effects, multiple same-scene views from different simulated altitudes/distances, and other special visual effects. Because of its physical appearance, VCASS (Figure 1.5) was also known as the "Bug That Ate Dayton." In 1986, Furness, who is sometimes called the "grandfather of VR," described its role in the Air Force's Super

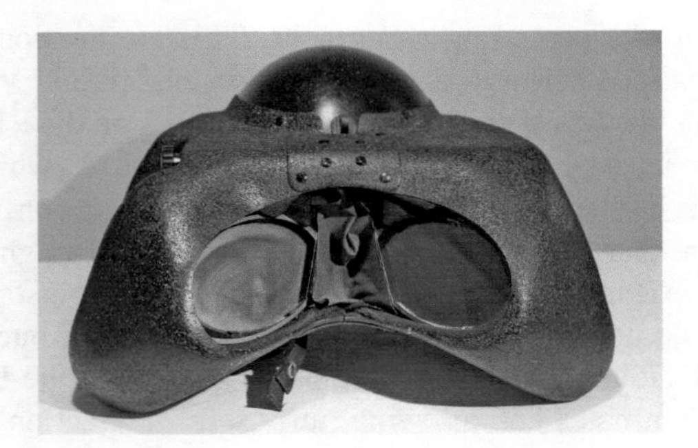

FIGURE 1.5 Frontal view of VCASS. Photo by Dean Kocian, approved for public release. (AFRL-2021-1547, 05-19-2021.)

Cockpit program and provided a list of human factor challenges inherent to such a device. Despite major technological progress, many of the challenges described by Furness still apply today.

In the late 1990s, researchers at the Naval Air Warfare Center Training Systems Division (NAWCTSD) developed a VR technology demonstration system for training ship handlers, called the Virtual Environment for Submarine Ship Handling Training (VESUB). This system was designed for integration with existing submarine piloting and navigation training simulators, and for evaluation of the effectiveness of VR as a training tool. Hayes et al. (1998) completed a training effectiveness evaluation on the tool and found it improved training in areas including checking range markers, issuing correct turning commands, contact management skills, reaction time for man overboard drills, and in using correct commands during emergency events. By 2001, the trainer (Figure 1.6) was deployed at the Naval Submarine School, and later expanded to all six Navy submarine training sites. It featured COTS hardware: the nVisor SX from NVIS, Inc, which provided a 60– diagonal field of view and was paired with an Intersense IS-900 Head Tracker. Images were managed by a system

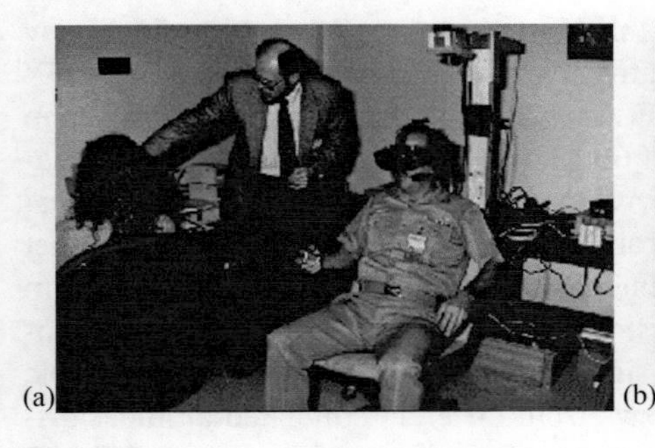

(a) (b)

FIGURE 1.6 Early VESUB system with headset; Bottom: VESUB system view from the virtual bridge. (From Wendland & Holland, 2006; Hayes, Vincenzi, Seamon, & Bradley, 1998 NAWCTSD PAO 1998071010, approved for public release.)

FIGURE 1.7 U.S. Air Force Tech. Sgt. Orson Lyttle participates in a virtual reality assignment with the Inter-European Air Forces Academy onboard Naval Air Station (NAS) Sigonella. (210506-N-GK686-1025 Naval Air Station (NAS) Sigonella, Italy. May 6, 2021.)

of four 500 MHz CPUs, which rendered up to 21,000 polygons at a refresh rate of 30 Hz, which is much lower than the common standard of 90 Hz, which is generally accepted as of this writing as a minimum for mitigating sim-sickness (VR Headset Authority, 2021), though no problems were reported by Hayes et al. (1998). VESUB was also equipped with a voice recognition system.

Today's commercial VR devices have become far more capable. VR can be used effectively on stand-alone systems or external computers that can power popular VR headsets such as Oculus Rift, HTC Vive, and Lenovo Mirage Solo. Current VR headset systems (Figure 1.7) are often paired with hand controllers or sensor-equipped gloves to improve how the user interacts with the virtual environment. Commercial headsets can also support wireless connections, dramatically increasing the range of possible use cases for such devices.

XR Applications in Training

Recent advances in XR technology have increased their viability in a range of potential applications. This section describes how this technology is already revolutionizing training.

VR in Medical Training

Medical domain training applications relying on XR technology vary widely. Tang et al. (2021) reported a VR blood-type and screen-training application designed to train blood-typing techniques that yielded evidence of enhanced readiness for trainees in comparison to a video-based training condition. Selvander and Asman (2012) performed a comparison of the effectiveness of a VR-based training module with a video-based training module for training ophthalmology students on the capsulorhexis procedure, one of the most difficult steps in cataract surgery involving the maneuver of instruments in and around the lens of the eye. They found that VR did not improve over video-based training in student maneuver of surgical instruments, but they also found that after over ten iterations in their experiment, students learning without the aid of VR were still lagging behind their peers trained in the VR training condition in avoidance of injury to the lens area of simulated patients. Anderson et al. (2018) evaluated the impact of the structure and distribution of VR training as a component of otorhinolaryngology training. These researchers compared performance of a medical student group that offered VR rehearsal as distributed practice prior to the course with a group that received the VR training during the course only and found that the additional distributed practice improved reaction times significantly and yielded higher overall scores on cadaveric dissection assessments. These results highlight the potential of VR training solutions to improve training outcomes.

AR in Medical Training

Nausheen and Bhupathy (2020) incorporated the Microsoft HoloLens AR device into Point of Care Ultrasound (POCUS) training, and found that it helped students track organ locations between the depictions on POCUS devices and the patient's anatomy. Training applications are not limited to surgical or procedural training; both Bork et al. (2020) and Kiourexidou et al. (2015) reported the use of AR to train different aspects of human anatomy. Bork and colleagues performed a comparison of an AR system with text-based training accompanied by three-dimensional fixed models, and found similar knowledge gains at course conclusion from both groups; these researchers found that collaborative AR systems yielded training advantages for mastery of topographic anatomy over single-user experiences, suggesting that system's ability to allow students to share visual perspectives with each other during instruction using the AR tools was likely to yield educational benefits relative to single-user systems.

AR in DoD Tactical Combat Casualty Care

The military is expanding the use of AR in medical training. For example, AR is beginning to play a key role in training for Tactical Combat Casualty Care (TCCC), which refers to care provided to warfighters at the point of injury by the personnel closest to them. These personnel will have far more limited equipment and options for treatment than are available at a traditional hospital. TCCC basics include the application of tourniquets to stop bleeding, nasal pharyngeal airway management, and treatment of tension pneumothorax, a condition in which air gets trapped in the chest cavity and collapses the lung, by needle decompression (Office of the Under Secretary of Defense for Personnel and Readiness, 2018). Two examples of recent Army-developed AR applications are provided in Figures 1.8 and 1.9.

The Army has developed TCCC training for battlefield responders using AR, which is used to simulate wounds, and a progressive physiology model representing changes in patient outcomes based on treatment provided. The use of AR in these devices makes it possible to model multiple types of wounds without modifying a

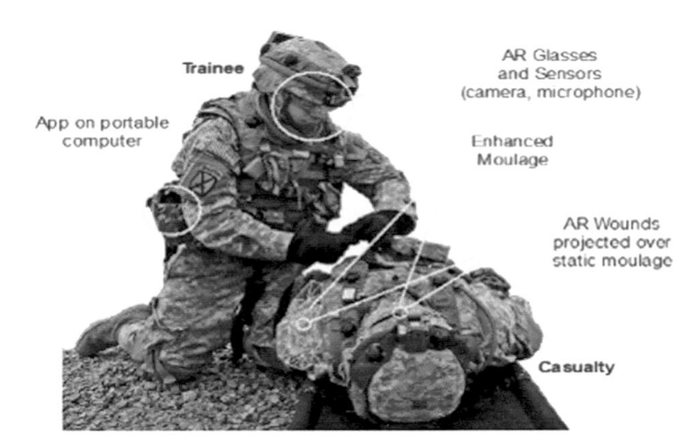

FIGURE 1.8 The Combat Casualty Care Augmented Reality Intelligent Training Systems (C3ARESYS) tool uses AR to improve learning outcomes in TCCC training. (Soar Technology, Inc. Used with permission.)

FIGURE 1.9 A visual overlay of a virtual watch/timer used in AUGMED training (Design Interactive, Inc.) https://www.dvidshub.net/image/6368782. The appearance of U.S. Department of Defense (DoD) visual information does not imply or constitute DoD endorsement.

mannequin, allows the trainee to see the results of their treatment, such as bleeding reductions based on the application of a tourniquet, and even more importantly, allows trainees to see how the simulated patient's condition is changing based on those treatments and their timing. Specific wounds can be changed between trainees where necessary without replacement of a specific mannequin. The use of AR in such a setting affords trainees the opportunity to interact with physical first aid kit tools or replicas, yielding critical experience at placing the devices with their own hands and applying appropriate pressure. The Combat Casualty Care Augmented Reality Intelligent Training Systems (C3ARESYS) tool (see Figure 1.8) is also capable of recognizing trainee interaction with physical devices like tourniquets as well as trainee-stated intentions recognized through natural language processing (A16-076, n.d.; Tanaka, Craighead, & Taylor, 2019; Tanaka, Craighead, Taylor, & Sottilare, 2019; Taylor et al., 2018).

The US Army also funded the development of AUGMED (see Figure 1.9), an AR-based TCCC trainer that accommodates both HoloLens and tablet-based interfaces, which incorporates explicit survival timelines as student prompts. This tool has been evaluated as part of the Combat Life Saver (CLS) course at Army Command Fort Indiantown Gap (FTIG) (Bolling, 2020). A third tool designed to enhance TCCC training by incorporating virtual overlays over a physical training manikin is the Virtual Patient Immersion Trainer (VPIT) described by Sushereba and Militello (2020). This device incorporates physiological injury progression and accompanying indicators, such as displacement of the trachea to one side and jugular vein distension due to trapped air in tension pneumothorax, allowing learners to track how these injury indications appear over time. The VPIT tool also uses eye-tracking to train gaze patterns, and to support performance evaluation to verify whether learners appear to have noticed a series of injury- and treatment-relevant cues before intervening. Empirical evidence of this trainer's effectiveness is not yet available.

As discussed above, tools like these are receiving increased attention, and the integration of AR in DoD medical training courses seems likely to increase. These applications are important as examples of AR being evaluated in a live training environment, though additional evidence of their training effectiveness is still needed (Kaplan et al., 2020).

VR Flight Training Devices

The DoD has used simulator-based training to support aviation training in one form or another since military flight began. Research has demonstrated that simulation-based training reduces the amount of flight time required to reach required performance levels (Kaplan et al., 2020). Today the DoD continues to demonstrate interest in the use of VR applications for aviation training. These devices offer portability, lower cost, easier maintenance, and afford an effective 360– field of view for the trainee, which is a feature of particular interest in jet training. A recent meta-analysis indicates that XR training has not yet demonstrated improved student performance following use of XR training devices; however, similarly it has not demonstrated a performance decrement (Kaplan et al., 2020). Given the other potential benefits of XR training devices, there is likely to be continued interest in efforts to develop and use XR devices in aviation training applications.

The US Navy recently completed an initial training evaluation of VR FTD for Primary Fixed-Wing training and Intermediate Strike training (McCoy-Fisher et al., 2019). The evaluation included the study of a T-6B VR FTD and two T-45C VR FTD variants (Figures 1.10 and 1.11). While specific components varied, in general each system was composed of a VR HMD, computer, throttle and control stick (referred to as Hands on Throttle and Stick (HOTAS), rudder pedals, monitor, and base with integrated seat. The device enabled student pilots to fly a range of maneuvers and missions leveraging VR technology in a VE. The results indicated that the devices

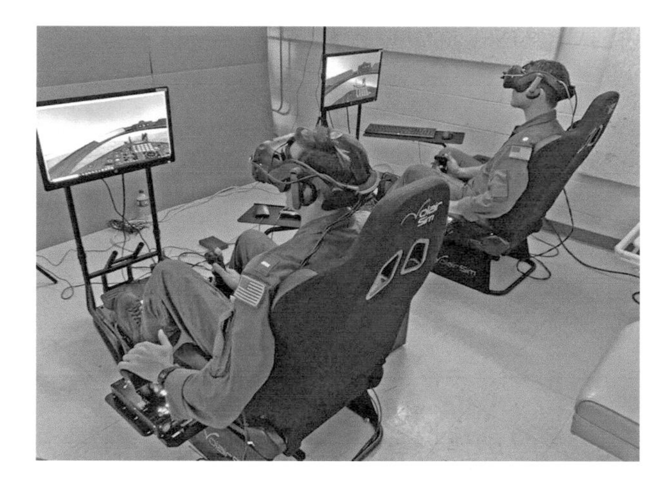

FIGURE 1.10 T-45C VR FTD at NAS Kingsville. (Courtesy of Defense Technical Information Center. U.S. Government Work, 17 USC §105.)

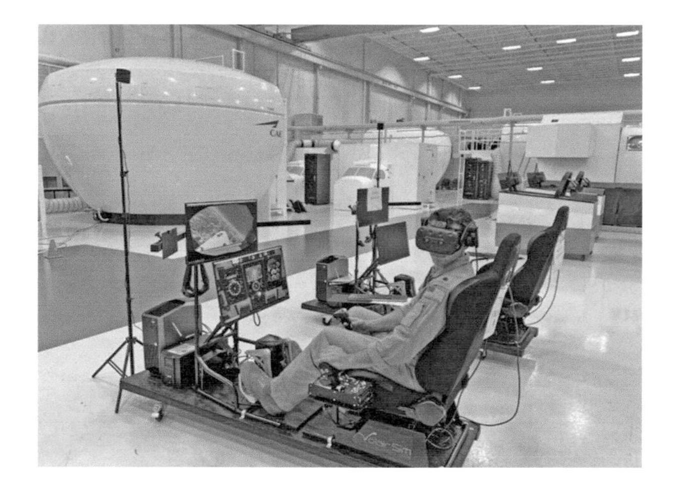

FIGURE 1.11 T-6B VR FTD at NAS Corpus Christi. (Courtesy of Defense Technical Information Center. U.S. Government Work, 17 USC §105.)

were best suited to support specific stages of training such as Contacts training (i.e., takeoff, aerobatics, and landing phases). The results also indicated that comfort with technology, prior use of VR devices, and general trust in automation were related to overall ratings of perceived utility of the technology. While VR sickness was an issue for some participants, their symptoms resolved within 30 minutes of completing the training evolution. Finally, the results indicated the importance of an intuitive user interface, the ability to interact with the virtual cockpit reliably in a natural manner, and VR HMD visual fidelity that approaches natural vision (e.g., field of view, visual acuity).

MR Flight Training Device

Traditionally, FTD rely on a large dome for their visual system. Figure 1.12 shows the US Army Aeromedical Research Lab (USAARL) Blackhawk helicopter full motion, visual simulator. Such devices provide a high-fidelity physical cockpit and a visual system that approximates the real environment; however, this drives a large device footprint and high costs. Recent advances in MR technology have the potential to change this paradigm. As discussed above, MR systems mix the RE with a VE on a common display. One notable limitation of VR systems is the inability to interact reliably with the real world or in this case a physical cockpit. MR systems use one of two approaches to merge the RE and VE in the user's sight picture (Rokhsaritalemi et al., 2020). Video passthrough integrates a video camera with a traditional VR HMD and blends the RE and VE on the user's screen, while optical passthrough (AKA See Through) uses a transparent lens to allow the user to view the RE and projects the VE onto the lens. Both have potential strengths and weaknesses and as in most cases, the appropriate solution depends on the problem to be solved. In conjunction with the VR FTD evaluation detailed above, the US Navy also conducted an initial evaluation of an MR FTD (McCoy-Fisher et al., 2019). In this case the MR device was used with an existing T-45C Operational Flight Trainer (OFT) (Figure 1.13). MR solutions that allow video passthrough rely on a pair of video cameras that are

FIGURE 1.12 USAARL Blackhawk Helicopter full-motion, full-visual flight simulator (the large footprint is evident when contrasted with the office chair in the lower-left corner). (https://www.dvidshub.net/image/6407391/usaarls-nuh-60fs-black-hawk-simulator-upgrade.) The appearance of U.S. Department of Defense (DoD) visual information does not imply or constitute DoD endorsement.

FIGURE 1.13 T-45C MR FTD at NAS Kingsville—An MR HMD integrated with existing T-45 Operational Flight Trainer that relies on traditional dome-based technology. (Courtesy of Defense Technical Information Center. U.S. Government Work, 17 USC §105.)

integrated into the external face of the HMD and provide stereoscopic video data that can be viewed and mixed with the VE on the interior display of the HMD. This allows the user to see the interior of the cockpit and interact naturally with it while displaying a virtual representation of the out-the-window view. The results were mixed but promising enough to encourage future development to address noted deficiencies.

AR Trainer for H-60R Preflight Procedures

Prior to flight, aircrew preflight the aircraft to ensure there are no discrepancies that could interfere with safe flight operations. Traditionally, training aircrew to conduct a preflight requires a qualified pilot and a significant amount of time to fully orient trainees on how to recognize different aircraft components and verify that they are safe for flight. The Navy MH-60R helicopter community recently completed a study to develop and evaluate the effectiveness of an AR for preflight training solution leveraging an iPad (NAWCTSD PAO, 2018). Figure 1.14 shows an MH-60R helicopter and Figure 1.15 shows the AR training device in use. This device allowed student aviators to practice the preflight on their own, with or without the aircraft, prior to the formal training evolution. Using the iPad's camera to view different aircraft components the system augmented the RE by overlaying labels, descriptions, and checklist items of components. The technology was viewed as very promising, but specific shortcomings with the initial design were as follows:

- The initial prototype is too cumbersome to be used comfortably by all trainees for preflight tasks performed on the ground.
- Important portions of the preflight checklists are performed by climbing on top of the aircraft, necessitating a hands-free solution for future implementation.

FIGURE 1.14 A MH-60R helicopter on the flight deck of the guided-missile cruiser USS Mobile Bay (CG 53). https://www.dvidshub.net/image/4371029/mh-60r. The appearance of U.S. Department of Defense (DoD) visual information does not imply or constitute DoD endorsement.

FIGURE 1.15 H-60R Augmented Reality Preflight Checklist Trainer: A tablet-based AR preflight checklist trainer for aircrew and student pilots. NAWCTSD PAO, 2018. The appearance of U.S. Department of Defense (DoD) visual information does not imply or constitute DoD endorsement.

XR Applications in Operational Support

Applications of XR for post-training operational support continue to expand across a variety of venues (Bond et al., 2018). Notable examples include the incorporation of XR into future Army tactical operations, medical procedures, and equipment maintenance operations.

AR in Army Tactical Operations

The Army is in the process of developing and evaluating a new tool for building and maintaining tactical situational awareness through a system called the integrated visual augmentation system (IVAS), based on HoloLens 2 hardware technology. The IVAS is designed for use in the operational environment as well as for training. IVAS is expected to incorporate low light and thermal vision, as well as sensors distributed across squad or platoon members to generate an integrated display, increasing engagement, awareness, unit effectiveness, and improved decision-making (Lang, 2021). While it must be acknowledged that this is purely a system in development, and not a fielded capability with documented measures of effectiveness, the Army is investing heavily in this technology.

VR in Operational Medicine

The use of VR to treat post-traumatic stress disorder (PTSD) has an extensive history (Rizzo & Shilling, 2017). The visual immersion yielded by a VR environment is a critical aspect of such treatment and helps PTSD and phobia sufferers gradually mitigate their fear and physiological responses to the environments that trigger these symptoms (Rizzo et al., 2019).

AR in Operational Medicine

AR tools have been used in operational medical procedures to directly improve patient outcomes. What these examples all have in common is the use of AR to help the practitioner better understand the spatial relationships between a surgical display and the patient's body, or to provide a visual aid to help the practitioner develop and maintain more accurate understanding of the locations of anatomical structures under the skin. Gregory et al. (2018) described the use of AR in operational surgical practice using a Microsoft HoloLens during a scapula replacement to help the provider visualize the location of specific bones and anatomical structures not visible to the naked eye during the procedure. Others have used AR overlays as operational aids for supplementing needle insertions (Heinrich et al., 2020). Liu et al. (2021) reported that the incorporation of an AR helmet-mounted display in a spinal surgery procedure limited radiation exposure and procedure time, while yielding similar patient outcomes to conventional surgery. In addition, Dias et al. (2021) and Qian et al. (2019) described the development, use, and effectiveness assessment of an AR trainer for endotracheal intubation training for providers learning to perform laryngoscopy. This device projected a depiction of the patient's airway onto the visual field of an HMD, giving the provider improved visualization of relevant anatomy without eliminating direct line of sight view of the point of insertion, and was found to provide improved training efficacy relative to unassisted rehearsal.

AR for Aircraft Maintainers

The development of job aids for maintainers has been of significant interest to the military for decades. For example, Kancler et al. (1998) investigated the use of HMDs to display maintenance procedures on maintainer performance and found that task completion was quicker and error rate was lower when using HMDs with augmented information. To illustrate military's continued interest in this area, consider that Class C mishaps between 2008 and 2017 increased from 7.5 mishaps per 100,000 flight hours to about 22 mishaps per 100,000 flight hours and reductions in the experience of maintainers was identified as a contributor to this increase (Eckstein, JUN 13, 2018). Class C mishaps include those mishaps that result in $50,000–$500,000 in damage and/or a non-fatal injury. The Navy is actively engaged in steps to address the experience level of maintainers (Eckstein, JUN 22, 2018). Developing AR-enabled maintainer solutions to support refresher and deployed training needs may be part of the solution (N193-D01, n.d.). The Navy is also interested in leveraging AR technology to support all maintenance levels, including complex or unusual maintenance activities (N201-024, n.d.). Appropriately, this effort specifies the need to ruggedize AR HMDs to operate in diverse military environments ranging from direct sunlight to moonless night.

Future Directions

A review of available resources provides insight into how industry and the US military are looking to leverage XR technologies in the future. For example, the US Army recently awarded Microsoft a contract to develop AR headsets to enable soldiers to map the battlespace and integrate intelligence to maintain situational awareness (Gregg & Greene, 2021). This section provides an overview of relevant innovations on the horizon for XR technologies.

XR HMD Enhancements

The DoD is encouraging the private sector to innovate XR HMD enhancements beyond commercially available capabilities and to address the fidelity requirements of military training applications (N192.087, n.d.). For example, the Headset Equivalent of Advanced Display Systems (HEADS) SBIR project specifies capabilities that Naval Aviation Training desires: (1) allows at least two pilots to interact with one another during any mission in any simulator, (2) full motion tracking of the HMD, (3) real-time imagery and/or accurate virtual representation of the cockpit, pilot's hands, and other pilots, (4) enhancements to instantaneous horizontal field of view (FOV), vertical FOV, binocular overlap, frame rate, refresh rate, and static spatial resolution, and (5) compatibility with existing Navy simulators. Such enhancements have the potential to address the limitations of existing XR HMDs described above.

AR for Red Air

Military strike pilot training requires that pilots be able to train against adversary aircraft typically referred to as Adversary Air or Red Air. Adversary aircraft are difficult to procure and expensive to maintain and operate. The USAF is investigating the use of AR to help solve this problem (Underwood, 2020). Specifically, the Air Force recently funded an effort to integrate AR technology into pilot helmets that will enable aircrew to reliably see the RE when virtual aircraft are injected onto their display. While there are several technical challenges such as the need to safely integrate the technology with various aircraft systems, this technology offers the potential to significantly reduce the requirement for adversary aircraft and to support a range of training needs, including airborne refueling and aerial combat.

VR for Spatial Disorientation Training

Spatial disorientation (SD) has long been recognized as a significant risk to aviation. Gibb et al. (2011) argue that nearly one-third of all military aviation mishaps are due to spatial disorientation. Bellenkes et al. (1992) found that SD was a causal factor in 33 Class A Naval Aviation mishaps between 1980 and 1989. Class A mishaps include those mishaps that result in $1 million or more in damage and/or an injury that results in a fatality or permanent total disability. Poisson and Miller (2014) found that over a 21-year period, SD was a causal factor in 72 US Air Force mishaps, resulting in 101 lost lives and an equipment cost of $2.32 billion. Given the cost both in lives and material it makes sense that there is significant interest in identifying methods to mitigate the risk of SD. Historically, training aircrew to recognize the symptoms of SD has relied on slide presentations, computer-based training, and/or the use of simulators. All have limitations. Recent advancements in VR technology provide a potential avenue to provide more immersive training on various visual illusions that can result in SD allowing student pilots to practice recognizing and responding in a timely fashion to these illusions. The US Navy is conducting an effort to explore VR solutions to address the need for more immersive SD training capabilities (N172-117, n.d.).

Pilot Training Next (PTN) and Naval Aviation Training Next (NATN)

In part motivated by recent pilot shortfalls, the USAF and USN are embracing the idea that new technology can revolutionize the training of military pilots. The USAF PTN program emphasizes the adoption of new VR FTDs, tailorable student-focused training, and VR-enabled self-study technologies (Tadjdeh, 2020). A recent Cost-Benefit Analysis (CBA) of the USAF PTN program estimated that over ten years the program can save a total of $12.92 billion and yield a 77% increase in the number of students trained (Pope, 2019). While this is quite promising, there are questions regarding the impact on pilot performance and quality. Similarly, the USN NATN program seeks to use new technologies in its aviation training programs (Shelbourne, 2020). These new technologies range from the use of VR FTDs (adapted from the USAF PTN program) to the incorporation of VR-enabled 360– videos that support self-study to MR solutions integrated with existing Instrument Flight Trainers (IFT) to provide a capability like that described as an MR FTD above. To illustrate how far training technology has evolved, consider the SNJ-5 Cockpit Checkout Recordings Device 12-AR-9a (Figure 1.16) that was developed by the Office of Naval Research in 1950. The SNJ-5 was a training aircraft used by many services, including the US Air Force and the US Navy. Student aviators could listen to these recordings to get an explanation of the location of all SNJ-5 instruments/controls and learn a variety of aircraft checklists such as start-up, takeoff, and landing. Today students have access to VR-enabled 360– videos.

FIGURE 1.16 SNJ-5 Cockpit checkout recordings student training device developed by the Office of Naval Research (1950).

The Navy is investigating additional technologies to support these innovations. For example, a recent effort is developing the Virtual Instructor Pilot Referee (VIPER), which uses artificial intelligence (AI) to provide adaptive, student-specific feedback on flight maneuvers that can be practiced without instructor involvement (Natali et al., 2020; SB052-009, n.d.). This technology can be integrated with the VR FTDs discussed above (McCoy-Fisher et al., 2019). Another tool is being used by the US Air Force to develop VR-enabled immersive training content, which can be used to support training in a range of areas, including how to respond to emergency situations (AF191-004, n.d.). A key to successful integration will be the use of open architecture.

While XR applications in military aviation and medicine were the primary focus of this discussion, numerous other markets are likely to benefit from recent advances in XR technology. The retail industry can offer online shoppers the opportunity to "see" their purchases in their home prior to purchase. The automotive repair industry can benefit from advances like those described for aircraft maintainers. The tourism industry can offer potential customers immersive VR experiences before their trip to decide what they want to see. Educators can provide students shared VR experiences to increase student engagement or give students access to additional information using AR technology. As an additional illustration of the applications of MR, consider the use of AR in decision-making discussed later in this chapter.

Challenges to the Adoption of XR Technology

There are numerous barriers to the adoption of XR technology for training and other applications. To illustrate this, examples in the following areas will be briefly discussed: XR device fidelity, user fatigue, integration with existing and new operational and training technologies, user resistance, safety, and regulations.

XR Device Visual Fidelity

The appropriateness of XR solutions for a given use case depends on the capability of the XR device(s) to meet the visual fidelity requirements of the proposed application. While current XR devices do a fairly good job of meeting the needs of the gaming community, current-generation XR devices still have a way to go to meet the needs of commercial, medical, and military users. For example, current generation XR HMDs cannot replicate the human horizontal or vertical field of view, which limits applications which rely heavily on use of peripheral vision and can induce unrealistic head movements in the training environment that can lead to negative training. Further, continued improvements to visual fidelity are necessary as numerous applications require the ability to replicate 20/20 vision. Current HMDs can lead to users "leaning in" to read displays. The integration of video (RE) and VE to create an MR environment presents its own challenges. For example, the visual fidelity of the RE and VE needs to be consistent and the transition between the two areas should be seamless and needs to be responsive to head movement such that the demarcation between the RE and VE is imperceptible.

User Fatigue

Using XR devices for an extended length of time can lead to user fatigue. The impacts of long-term use are still being investigated. A factor contributing to user fatigue and slowing the wider adoption of XR for sustained training operations is the vergence–accommodation conflict. The vergence–accommodation conflict results from the HMD wearer focusing for long periods on a focal point extremely close to the eye to track objects that appear to be at much greater distances in the visual display, in the case of VR. XR HMDs, including those that use AR instead, can also require the wearer's eyes to adapt to highly conflicting visual cues. This results in greater difficulty in resolving binocular images as well as visual fatigue resulting from extended use (Kramida, 2016). Progress is being made on these issues, largely through the development of HMDs that present visual cues at different optical distances (Lu et al., 2019).

Integration

XR devices may be stand-alone or integrated into an existing training system. Integration of XR devices with stand-alone systems can be reasonably expected to be simpler than integration with existing training systems as the stand-alone system is built with a specific XR device in mind. However, it is important that such systems be built recognizing that XR HMDs are being updated at a rapid pace. To illustrate this point, consider that the Oculus Quest was released in November 2019 and the Oculus Quest 2 was released in October 2020. Similarly, the Varjo VR-1 was released in February 2019, Varjo VR-2 was released in October 2019, and the Varjo VR-3 was released in early 2021.

Integrating XR devices with existing training systems presents a more formidable challenge. To illustrate this, consider two key issues. The XR device must be able to communicate with the training system. Depending on the age of the system, it may use a programming language or standard different from that commonly used today. Further, rendering a high-fidelity image in an XR HMD requires a significant amount of visual data. The legacy training system must both have the data at the fidelity required and be able to pass it to the XR HMD quickly enough to deliver the information to the user's display. McCoy-Fisher et al. (2019) described some of the challenges associated with integrating the MR HMD with the existing OFT (Operational Flight Trainer) that was not built to accommodate such technology. MR displays have the added challenge of having to use both virtual and real-world images on a single display. For example, they noted a recurring problem with aligning the display of VE and RE images on the display and indicated that in most instances this calibration had to occur prior to each event (McCoy-Fisher et al., 2019).

User Resistance

User resistance to new technology may be a barrier to adoption and widespread use of XR technology. Research has demonstrated that perceived usefulness and perceived ease of use predict intent to use technology (Davis et al., 1989; Venkatesh et al., 2002). That is, if users perceive the technology as instrumental to some desired outcome (e.g., improved performance) and the technology is intuitive and easy to use, then they are more likely to want to use it. Therefore, it is important to consider the perspective user community's experience with XR technology and attitudes toward it. It is equally important to develop a good user interface and test it prior to implementation. A negative first impression of the XR system will likely be difficult to overcome.

Safety

The use of XR technology also needs to consider the potential safety risks. In the medical domain this would refer to patient safety while in the aviation domain this would refer to safety of flight. Similarly, other domains must consider safety as they consider utilization of XR technology. For example, research has demonstrated that prolonged wear of head-mounted displays with a forward center of gravity (CG) rather than neutral is more likely to cause fatigue and injury, which highlights the need to consider the weight and CG of the HMD (Albery et al., 2008). Further, research has demonstrated that some individuals are susceptible to simulator sickness. McCoy-Fisher et al. (2019) found that mean oculomotor and disorientation scores increased from baseline following the use of an XR device though symptoms resolved shortly after the conclusion of the training event. This highlights the importance of research investigating characteristics of XR HMDs and individuals that might increase the likelihood and severity of simulator sickness. As a final example, consider the AR use case described above in which synthetic entities are displayed on a pilot's HMD. Integrating AR into a live flight environment drives the need to carefully consider risks to safe flight such as the need to distinguish between real and synthetic entities and to be able to quickly exit "training mode."

Regulations

Various agencies govern the certification and ongoing training requirements of professional communities. For example, the Federal Aviation Administration (FAA) regulates the initial and recurring training requirements of pilots. It also maintains standards for flight simulation training devices. Currently, XR devices are not certified to support mandatory training requirements of aircrew, which limits adoption and necessitates reliance on traditional simulators and live flight. Fully realizing the benefits of XR technology in pilot training will require establishing XR training device requirements and a certification process.

XR Is Not Always the Optimal Training Solution

Finally, it is important to note that extended reality technologies are not appropriate for every application (at least not with today's technology). Consideration of any instructional medium's potential as a learning aid must be grounded in a theoretical model that addresses the differences among learning outcomes, cognitive processes, and the conditions that distinguish among them. Gagne and Medsker's (1996) model of instruction is widely recognized, and holds that learning is based on three elements: a taxonomy of learning outcomes, conditions of learning, and nine possible events of instruction. Driscoll (2000) defined instructional theory as "identifying methods that will best provide the conditions under which learning goals will most likely be attained (p. 344)." This model holds that instructional methods should be matched to required conditions, to yield desired learning outcomes. Any evaluation of the effectiveness of different instructional media for achieving different learning outcomes must consider the nature of the knowledge, skills, and attitudes being taught, and match the strengths of the media under consideration to the targeted curricular tasks. To illustrate this point, consider the following evolution. The US Navy has used a VR-enabled parachute descent training device to teach aircrew how to recognize and resolve problems encountered during their parachute descent. IROK, the mnemonic memory aid taught to the students, stands for: Inflate the life preserver, Release the raft, Options, and Koch connector release. The decision-making training presents the trainee with simulated malfunctions and depending on the situation which options he/she should take such as when to jettison the attached seat kit. Execution of these tasks requires both visual and tactile access to the body-worn equipment (parachute release, life preserver, seat pan). Unfortunately, the VR device shown in Figure 1.17 occludes visual access to this equipment.

To address this issue, the US Navy worked with small businesses to develop a solution that supports both aviators' capability to execute the IROK procedures, identify and resolve parachute canopy issues, and navigate to a safe landing spot (N161-007, n.d.). This upgraded device is already in use at the Navy Aviation Survival Training Centers in Pensacola. Like its predecessor, this training device is designed to teach naval aviation students how to respond to parachute malfunctions, get and maintain control of the parachute, and perform operations needed to prepare for landing. Students hang suspended from a parachute harness, handle control lines, and interact with their safety equipment. The parachute canopy and physical environment are

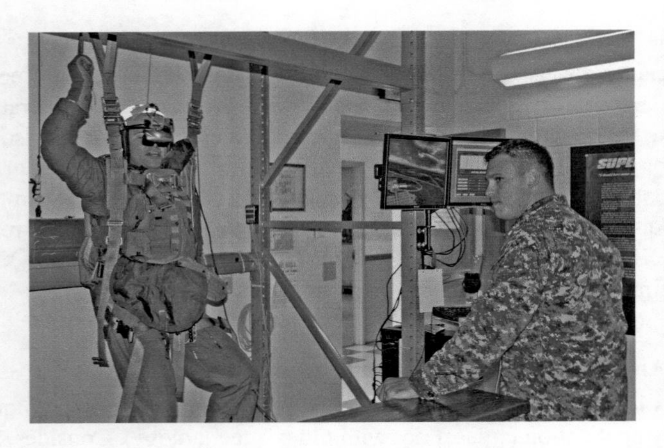

FIGURE 1.17 Legacy parachute descent trainer using VR goggles that inhibited trainees' ability to see and interact with flight gear such as parachute release, life preserver, and seat pan. (https://www.dvidshub.net/image/653875/aviation-survival-training-center-jacksonville.) The appearance of U.S. Department of Defense (DoD) visual information does not imply or constitute DoD endorsement.

rendered on the four monitors in the Unity game environment. The incorporation of monitors in lieu of VR goggles as depicted (Figure 1.18) remedied some mismatches between the training tasks and the visual medium. Specifically, using the previous VR system:

- Students could not see their hands, their gear, or the risers they had to grasp to control their direction during descent.
- Inflation of the survival vest during descent occluded the trainee's field of view, eliminating any training advantage otherwise yielded by the 360–FOV provided by VR headset.
- The time required to don/doff and calibrate VR headsets significantly reduced the throughput (number of students trained per unit time).

These mismatches reduced the students' ability to master the intellectual and motor skills required for proper parachute decent. Replacement of the VR headset with the flat panel screens remedied these limitations and increased student throughput. Additional situations in which AR tools are likely to yield training efficacy are discussed in Bond et al. (2018) and Militello et al. (2023).

This section of the chapter provided an overview of XR technology and highlighted applications for both training and operational use with an emphasis on aviation and medical domains. To be clear, this is not intended to be an exhaustive discussion of potential future applications of XR technology, but instead to provide an overview of opportunities on the horizon for XR to change the way we work. The next section will extend this discussion to the use of AR technology in decision-making.

FIGURE 1.18 Upgraded Immersive Parachute Descent Procedure, Malfunction, & Decision-Making Training System in use at ASTC Pensacola (PAO Clearance 21-ORL023). The appearance of U.S. Department of Defense (DoD) visual information does not imply or constitute DoD endorsement.

AUGMENTED REALITY IN DECISION-MAKING

In this section, we describe how augmented reality can facilitate the development of recognition skills for improved decision-making. Simulations of cognitive processes emerged in the 1980s. Kahneman et al. (1982) developed a simulation heuristic describing how decision-makers develop mental simulations. In fact, Daniel Kahneman, a psychologist, won the Nobel Prize in Economics in 2002 for his work in decision-making. He and his colleagues described how individuals use rules of thumb as opposed to a more rational analysis and the importance of properly framing the question.

Subsequently, Klein and Crandall (1995, p. 324) defined mental simulation as "the process of consciously enacting a sequence of events." They describe how urban fire ground commanders (FGCs) use mental simulation to allocate resources and direct their firefighters. According to Klein and Crandall, 72% of the accounts of mental simulation in their database referred explicitly to visual imagery.

Recognition skills require the participant to rapidly size up the situation and know what actions to take. Examples include organizational commanders and individuals making decisions under stress, firefighters locating individuals trapped in burning buildings, individuals learning how to escape from flooding compartments, and

SWAT personnel engaged in hostage rescue events. Sushereba et al. (2019) describe how an experienced US Army Apache helicopter pilot recognized a deviant wind pattern in the desert as a precursor to a larger sandstorm and diverted to a safer landing site, as well as how, during a routine assessment of a premature infant, a neonatal intensive care nurse recognized a potentially deadly infection based on a combination of nonthreatening symptoms. Both are examples of individuals whose prior experiences allowed them to identify relevant cues and meaningfully integrate them prior to making decisions. Recognition skills training is perhaps best accomplished by providing immersive and engaging scenario-based learning experiences. Militello et al. (2023, p. 49) specify the following three components of recognition skills:

1. Knowing "what to attend to, including which cues are relevant in which contexts."
2. Recognizing a situation and knowing how to act, "learners must be able to create meaning from the cues available."
3. Having "sophisticated mental models allows learners to evaluate potential interventions before acting by mentally playing them out, also known as mental simulation."

Based on their experience in applying recognition skills in multiple domains, Militello et al. (2023, p. 6) developed a handbook to "support training technology developers, training designers, and trainers in leveraging the strengths of augmented reality for training recognition skills." The nine design principles detailed in their handbook and listed in Table 1.1 are intended to facilitate the learning of relevant cues and clusters of cues, interpreting the situation correctly, and anticipating problems using mental simulation.

Space limitations preclude expansion on each of the principles listed in Table 1.1, so we refer the reader to the author's handbook (Militello et al., 2023) for a detailed discussion of the pedological principles involved in leveraging AR in designing training systems.

THE PERFECT STORM REVISITED: THE FUTURE OF HUMAN FACTORS IN SIMULATION AND TRAINING

We return to the perfect storm analogy presented at the beginning of this chapter to forecast the major challenges and opportunities as perceived by the authors.

Technological Trends

- The variety and quality of training devices will continue to increase, as will the hype. Therefore, we need to constantly verify that the technology is appropriate for the training objectives, as or more effective than traditional training, safe, and cost effective.

TABLE 1.1
Principles for Designing AR-based Recognition Skills Training

Principle	Definition
1. Fidelity and Realism Principles	
Sensory fidelity principle	Realistic cues are needed to support perceptual skill development
Scaling fidelity principle	Virtual props should be at a scale close to the real world
Assessment-action pairing principle	It is important to create a learning experience that allows the learner to both assess the situation and act
2. Scenario-Building Principles	
Periphery principle	Effective scenarios should include critical cues that are not obvious; rather, the learner must know to look for them and correctly interpret them
Perturbation principle	Training scenarios should expose trainees to novel conditions requiring adaptation and performance under non-routine conditions
3. Building Mental Models Principles	
Mental model articulation principle	Training techniques that require the learners to articulate what they are noticing, how they are assessing the situation, and predictions about how the situation will evolve and aid the learner in developing coherent mental models
Many variations principle	Training techniques that expose the learner to many variations with different levels of difficulty support development of robust mental models
4. Scaffolding and Feedback Principles	
Scaffolding principle	Scaffolding can be used to promote recognition skill development at different developmental stages
Reflection principle	Learners who reflect on the training experience are better able to extract insights from the experience and apply them to future performance

Source: Adapted with permission from Militello et al. (2023)

- Trainees will be equipped with wearable computing technologies which will record their physiological data as they train and provide tactile and kinesthetic cues.
- The use of artificial neural networks and AI will become more common, as will the use of avatars as intelligent personalized tutors.
- The availability of commercial off-the-shelf (COTS) computers, displays, and networking technologies will increase.
- Simulations using models of individual, team, and organizational behavior will become more common. These models will facilitate the understanding of cross-cultural differences.

- Training devices will rely more heavily on open architecture, making it easier to integrate new training technologies.
- Content development and integration/evaluation of XR technologies will continue to be the key cost drivers for XR training applications.
- Technology costs will continue to decrease.
- Market-driven innovations in XR technology will resolve the device fidelity and user fatigue concerns discussed in the XR section of this chapter.
- Integrating XR devices into existing, legacy, closed-architecture training systems will be much more challenging than either integrating with an open-architecture training system or building a stand-alone training system with a specific XR device in mind.
- As with all new technology, managing the users' initial experience to ensure it is positive will continue to be essential to user acceptance of new XR technologies.

Computation Power

- The fusion of inexpensive computation power, affordable broadband, and wireless networks will continue to increase, as costs, size, and power requirements decrease.
- The Department of Energy's exascale computing project (www.exascaleproject.org) has as the next milestone 10^{18} operations per second. This will be 1,000 times faster than the current fastest petascale (10^{15}) supercomputers enabling more realistic simulation of real-world processes and interactions. Udoh (2011) provides a point of reference, today's desktop personal computer's processor (e.g., i7 core) can execute less than 300 gigaflops per second. A gigaflop is 10^{9} operations. Current affordable supercomputers are over one million times faster per second (10^{6}).
- The continued development of high-performance computing (HPC)/supercomputers will make it easier to design, access, and integrate simulations into virtual environments.
- The proliferation of cloud computing, maturation of artificial intelligence, and open-source data analytics tools are rapidly changing both the scope and nature of data analyses performed in applied settings and in organizations. Use and interpretation of "big data" analyses requires more sophisticated tools and an ever-expanding combination of coding and visualization skillsets.
- DOE initiative to make AI models and data that adhere to FAIR data principles (Findable, Accessible, Interoperable, and Reusable) to produce data reusable both by researchers and machines with little human intervention. Exascale high-performance computing will accelerate the development of AI (Huerta et. al., 2020).
- Agent-based simulations and avatars will become more accurate computational representations of "self," "group," "organization," "city," and

"society." These simulations can alter their behavior through learning and their interactions to meet programmed goals. Their continued development will make it easier to design, access, and integrate simulations into virtual environments. This interactive environment will improve the education of individuals and teams that in the future will have access to patient, intelligent, adaptive tutors and coaches.

- Students and trainees will have 24/7 access to virtual training ranges to learn how to partner with each other and their AI team mates.
- The UAV commercial and military applications market will increase. The commercial drone market is expected to grow from $14 billion in 2018 to over $43 billion by 2024 (Research and Markets, March, 2019). The demand for simulators, trained operators, and maintainers will increase proportionately.
- Managers, educators, and learners will expect bigger and better simulations in the future.

Innovations in Education and Training

- As working from home becomes more commonplace, so will the expectancy to be educated and use simulators at home.
- Training anyplace, anytime will increase, particularly as "working from home" becomes more acceptable. The use of simulators for training by businesses will increase.
- As was demonstrated during the COVID-19 pandemic, hybrid learning is feasible but, while it is a viable mode of learning/training, significant investigation is needed to identify and remove the barriers to the effective use of hybrid learning in educating future learners.
- Regulatory agencies will need to establish processes for specifying XR training device requirements and certification.

The Changing Nature of Education

- Simulation will continue to blend training, mission planning, rehearsal, and execution into one continuum.
- Educators in high-risk environments such as the medical, chemical, nuclear, and aerospace industries will expect more tools that better enable them to deliver realistic training in safe environments which increase training transfer while reducing overall risk.
- Cohort training will decrease, while AI will facilitate the accommodation of individual differences in learning styles and rates.
- The criteria for advancement will be demonstrated performance to a criterion rather than the number of class hours completed.
- More individuals will be certified as qualified based on their demonstrated performance on simulators.

Acceptance of Simulation/Gaming

- Cloud gaming is the new reality. Cloud gaming stores everything about the game in the cloud and streams all the data to the gaming device. Thus, image quality and response lag are expected to be minimized.
- Not only has gaming been accepted by the public but it is also expected to grow exponentially. In 2019, the global gaming market was valued at $151.55 billion; and it's expected to increase to $206 billion by 2025 (https://www.gameindustry.com/news-industry-happenings/gaming-industry-growth-trends-and-forecast-2020-2025/).
- While we are a long way from the virtual environment matching the fidelity of the real environment, the fidelity of the virtual environment will continue to increase. This increase needs to be matched with a commensurate increase in the consideration of the safety of users immersed in these rich virtual environments. Strategies for combating simulator sickness, ensuring the physical safety of those using AR devices, and distinguishing the real and virtual environments quickly and reliably in an MR application must be developed and validated.

When considering the challenges/opportunities for human factors in simulation and training, we should keep John Miles's system design question in mind. Miles asked: "Can this *person* with this *training* perform these *tasks* to these *standards* under these *conditions*?" To answer that question, the simulation/training designer needs to adequately describe each of the italicized nouns which are expanded on briefly below:

- What are the characteristics of the trainee/trainees?
- What training have they received?
- What task(s) do they need to be trained for and when do they need to be trained?
- Exactly what are the standards to which they need to perform?
- Under what conditions is/are the trainee/trainees expected to perform?

ACKNOWLEDGMENTS

The authors acknowledge the assistance of friends in the modeling and simulation community, who not only shared material from their archives with us but reviewed portions of the chapter. The authors also thank their wives Kathy Moroney, Darlene Lilienthal, Shannon Foster, and their families for their patience and support.

DISCLAIMER

The views presented are those of the authors and do not necessarily represent the views of the DoD or its components.

REFERENCES

A16-076. (n.d.). Combat casualty care augmented reality intelligent training systems (C3ARESYS). https://www.sbir.gov/sbirsearch/detail/1514421

Adorian, P., Staynes, W. N., & Bolton, M. (1979). The evolution of the flight simulator. *Proceedings of Conference: 50 Years of Flight Simulation*, Vol. 1. London: Royal Aeronautical Society, pp. 1–23.

Aegis Research Corporation. (2002). War games bibliography. In *International Conference on Grand Challenges for Modeling and Simulation*, Lunceford, W. H. & Page, E. H., Eds. Society for Modeling and Simulation (www.scs.org).

AF191-004. (n.d.). *A Virtual Reality Content Creation, Training and Assessment Tool for the Department of Defense*. https://www.sbir.gov/sbirsearch/detail/1861567

Albery, C. B., Gallagher, H. L., & Caldwell, E. (2008). Neck muscle fatigue resulting from prolonged wear of weighted helmets. DTIC. https://apps.dtic.mil/dtic/tr/fulltext/u2/a491626.pdf

Alexis. (2020, June 26). Fixed base versus full-motion simulators. Simulator review. Retrieved June 23, 2021, from https://simulatorreview.com/fixed-base-vs-full-motion-simulators/

Allen, T. B. (1989). *War Games: The Secret World of the Creators, Players, and Policy Makers Rehearsing World War III Today*. Berkley Publishing Group.

AN/USQ-T46 Battle Force Tactical Training (BFTT). Retrieved June 23, 2021, from https://www.navy.mil/Resources/Fact-Files/Display-FactFiles/Article/2166789/anusq-t46-battle-force-tactical-training-bftt/no/

Andersen, S. A. W., Konge, L., & Sorensen, M. S. (2018). The effect of distributed virtual reality simulation training on cognitive load during subsequent dissection training. *Medical Teacher*, 40, 684–689.

Balci, O. (1998). Verification, validation, and testing. *Handbook of Simulation*, 10(8), 335–393.

Balci, O., Nance, R. E., Arthur, J. D., & Ormsby, W. F. (2002, December). Expanding our horizons in verification, validation, and accreditation research and practice. In *Proceedings of the Winter Simulation Conference* (Vol. 1, pp. 653–663). IEEE.

Barfield, W., & Furness, T. A. III (Eds.). (1995). *Virtual Environments and Advanced Interface Design*. Oxford University Press.

Basu, S., Dickes, A., Kinnebrew, J. S., Sengupta, P., & Biswas, G. (2013, May). CTSiM: A computational thinking environment for learning science through simulation and modeling. In *CSEDU* (pp. 369–378).

Baudhuin, E. S. (1987). The design of industrial and flight simulators. In *Transfer of Learning*, Cormier, S. M. & Hagman, J. D., Eds. San Diego, CA: Academic Press, pp. 217–237.

Beal, M. D., Kinnear, J., Anderson, C. R., Martin, T. D., Wamboldt, R., & Hooper, L. (2017). The effectiveness of medical simulation in teaching medical students critical care medicine: A systematic review and meta-analysis. *Simulation in Healthcare*, 12(2), 104–116.

Bellenkes, A., Bason, R., & Yacavone, D. W. (1992). Spatial disorientation in naval aviation mishaps: A review of class A incidents from 1980 through 1989. *Aviation, Space, and Environmental Medicine*, 63(2), 128–131.

Beringer, D. B. (1994). Issues in using off-the-shelf PC-based flight simulation for research and training: Historical perspective, current solutions and emerging technologies. *Proceedings of the Human Factors and Ergonomics Society 38th Annual Meeting*. Santa Monica, CA: Human Factors and Ergonomics Society, pp. 90–94.

Bernstein, M. (2000). *Grand Eccentrics: Turning the Century: Dayton and the Inventing of America*. Wilmington, OH: Orange Frazer Press.

Blacker, K. J., Pettijohn, K. A., Roush, G., & Biggs, A. T. (2020). Measuring lethal force performance in the lab: The effects of simulator realism and participant experience. *Human Factors*. DOI: 10.1177/0018720820916975

Boldovici, J. A. (1987). Measuring transfer in military settings. In *Transfer of Learning*, Cormier, S. M. & Hagman, J. D., Eds. San Diego, CA: Academic Press, pp. 239–260.

Bolling, E. (2020, September 25). Beyond reality: New AUGMED tool pushes limits of medical simulation. https://www.dvidshub.net/news/378712/beyond-reality-new-augmed-tool-pushes-limits-medical-simulation

Bond, A., Neville, K., Mercado, J., Massey, L., Wearne, A., & Ogreten, S. (2018). Evaluating training efficacy and return on investment for augmented reality: A theoretical framework. *Proceedings of the 2018 Interservice/Interagency Training, Simulation, and Education Conference*.

Boothe, E. M. (1994). *A Regulatory View of Flight Simulator Qualification, Flight Simulation Update*, 10th ed. Binghamton, NY: SUNY Watson School of Engineering.

Bork, F., Lehner, A., Eck, U., Navab, N., Waschke, J., & Kugelmann, D. (2020). The effectiveness of collaborative augmented reality in gross anatomy teaching: A quantitative and qualitative pilot study. *Anatomical Sciences Education*. https://doi-org.proxy.lib.fsu.edu/10.1002/ase.2016

Boud, A. C., Haniff, D. J., Baber, C., & Steiner, S. J. (1999). Virtual reality and augmented reality as a training tool for assembly tasks [Conference session]. *IEEE International Conference on Information Visualization* (Cat. No. PR00210). London, England (pp. 32–36).

Burki-Cohen, J., Sparko, A., & Go, T. (2007, August). Training value of a fixed-base flight simulator with a dynamic seat. In *AIAA Modeling and Simulation Technologies Conference and Exhibit* (p. 6564).

Caffrey, M. B. (2019). *On Wargaming: How Wargames Have Shaped History and How They May Shape the Future* (Vol. 43). Naval War College Press.

Carnahan, C. D. (2012). *The Effects of Learning in an Online Virtual Environment on K-12 Students*. Indiana University of Pennsylvania.

Caro, C.W. (1979). Development of simulator instructional feature design guides.In *Proceedings of Conference: Fifty Years of Flight Simulation* (p. 75). London: Royal Aeronautical Society.

Caro, P. W. (1988). Flight training and simulation. In *Human Factors in Aviation* (pp. 229–261). New York: Academic Press.

Cisco, U. (2020). *Cisco Annual Internet Report (2018–2023) White Paper*. https://www.cisco.com/c/en/us/solutions/collateral/executive-perspectives/annual-internet-report/white-paper-c11-741490.html

Cook, D. A., Brydges, R., Hamstra, S. J., Zendejas, B., Szostek, J. H., Wang, A. T., Erwin, P. J., & Hatala, R. (2012). Comparative effectiveness of technology-enhanced simulation versus other instructional methods: A systematic review and meta-analysis. *Simulation in Healthcare*, 7(5), 308–320. doi: 10.1097/SIH.0b013e3182614f95. https://journals.lww.com/simulationinhealthcare/Fulltext/2012/10000/Comparative_Effectiveness_of_Technology_Enhanced.6.aspx

Custer, L. L. (1930). Aviation training machine, U. S. Patent No. 481,831. Washington, DC: United States Patent Office. New York: Academic Press, pp. 229–261.

D'Angelo, C., Harris, C., & Rutstein, D. (2013). *Systematic Review and Meta-Analysis of STEM Simulations*. Menlo Park, CA: SRI International.

Davis, F. D., Bagozzi, R. P., & Warshaw, P. R. (1989). User acceptance of computer technology: A comparison of two theoretical models. *Management Science*, 35, 982–1002.

Department of Defense. (2009). DoD Instruction on DoD Modeling and Simulation VV&A (DoDI 5000.61 December 9, 2009, Incorporating Change 1, October 15, 2018) DoDI

5000.61, December 9, 2009, Incorporating Change 1, October 15, 2018. Retrieved from https://www.esd.whs.mil/Portals/54/Documents/DD/issuances/dodi/500061p.pdf
Dias, P. L., Greenberg, R. G., Goldberg, R. N., Fisher, N., & Tanaka, D. T. (2021). Augmented reality-assisted video laryngoscopy and simulated neonatal intubations: A pilot study. *Pediatrics*, 147(3), 1–11. doi: https://doi.org/10.1542/peds.2020-005009
DoD Enterprise DevSecOps Reference Design Version 1.0 12 August 2019. https://dodcio.defense.gov/Portals/0/Documents/DoD%20Enterprise%20DevSecOps%20Reference%20Design%
DoD Directive 5000.59 August 8, 2017 Incorporating Change 1, October 15, 2018, DoD Modeling and Simulation (M&S) Management, Retrieved from DoDD 5000.59 – Executive Services Directorate.
Doolani, S., Wessels, C., Kanal, V., Sevastopoulos, C., Jaiswal, A., Nambiappan, H., & Makedon, F. (2020). A review of Extended Reality (XR) technologies for manufacturing training. *Technologies*, 8, 77.
Drascic, D., & Milgram, P. (1996, April). Perceptual issues in augmented reality. *Stereoscopic Displays and Virtual Reality Systems III*, 2653, 123–135.
Driscoll, M. P. (2000). Gagne's theory of instruction. In *Psychology of Learning for Instruction*, Allyn & Bacon (Eds.). Florida State University Press, pp. 341–372.
Eckstein, B. M. (2018, June 13). Naval Safety Center Standing Up Data Analytics Office Amid Surface, Aviation Mishap Increases. USNI News. https://news.usni.org/2018/06/13/naval-safety-center-standing-data-analytics-office-amid-surface-aviation-mishap-increases
Eckstein, B. M. (2018, June 22). Less Experienced Maintainers Contribute to Rise in Naval Aviation Mishaps. USNI News. https://news.usni.org/2018/06/22/less-experienced-maintainers-contribute-rise-naval-aviation-mishaps
Electronics Technician Training. (2018, November 16). *How Simulation Tools are Transforming Education and Training*. Retrieved from https://www.etcourse.com/simulation-tools-transform-education-and-training.html
Eshet-Alkalai, Y., Caspi, A., Eden, S., Geri, N., Tal-Elhasid, E., & Yair, Y. (2010). Challenges of integrating technologies for learning: Introduction to the IJELLO special series of Chais Conference 2010 best papers. *Interdisciplinary Journal of E-Learning and Learning Objects*, 6(1), 239–244.
Esteves, M., Fonseca, B., Morgado, L., & Martins, P. (2011). Improving teaching and learning of computer programming with the Second Life virtual world. *British Journal of Educational Technology*, 42(4), 624–637.
Fischetti, M. A. & Truxal, C. (1985). Simulating "The right stuff." *IEEE Spectrum*, 22(3), 38–47.
Flavian, C., Ibanez-Sanchez, S., & Orus, C. (2019). The impact of virtual, augmented, and mixed reality technologies on the customer experience. *Journal of Business Research*, 100(July), 547–560.
Flexman, R. E., Roscoe, S. N., Williams, A. C., & Williges, B. H. (1972). Studies in pilot training: The anatomy of transfer. *Aviation Research Monographs*, 2(1). Champaign, IL: University of Illinois Aviation Research Laboratory.
Gagne, R. M., & Medsker, K. L. (1996). *The conditions of learning: Training applications*. Fort Worth: Harcourt Brace College Publishers.
Gawron, V. J. (13 May 2019). Simulation applications in training. Presented at the Flight and Ground Simulation Update, Binghamton University.
Gawron, V. J. (2000). *Human Performance Measures Handbook*. Hillsdale, NJ: Lawrence Erlbaum Associates.
Geiselman, E. E., Johnson, C. M., Buck, D. R., & Patrick, T. (2013). Flight deck automation: A call for context-aware logic to improve safety. *Ergonomics in Design*, 21(4), 13–18.

Gibb, R., Ercoline, B., & Scharff, L. (2011). Spatial disorientation: Decades of pilot fatalities. *Aviation, Space, and Environmental Medicine*, 82(7), 717–724. https://doi.org/10.3357/asem.3048.2011

Gray, A. (2017, June 8). New realistic submarine bridge trainer opens at Trident Training Facility Bangor. https://www.militarynews.com/norfolk-navy-flagship/news/quarterdeck/new-realistic-submarine-bridge-trainer-opens-at-trident-training-facility-bangor/article_88eba425-df65-5003-8cc6-d50df80e0a51.html

Gregg, A., & Greene, J. (2021). Microsoft wins $21 billion Army contract for augmented reality headsets. *The Washington Post*. https://www.washingtonpost.com/business/2021/03/31/microsoft-army-augmented-reality/

Gregory, T. M., Gregory, J., Sledge, J., Allard, R., & Mir, O. (2018) Surgery guided by mixed reality: Presentation of a proof of concept. *Acta Orthopaedica*, 89(5), 480–483.

Gross, B., Rusin, L., Kiesewetter, J., Zottmann, J. M., Fischer, M. R., Prückner, S., & Zech, A. (2019). Crew resource management training in healthcare: A systematic review of intervention design, training conditions and evaluation. *BMJ Open*, 9(2), e025247.

Hays, R. T., Vincenzi, D. A., Seamon, A. G., & Bradley, S. K. (1998) Training effectiveness evaluation of the VESUB technology demonstration system. Technical Report 98-003, Naval Air Warfare Center Training Systems Division. https://apps.dtic.mil/sti/pdfs/ADA349219.pdf

Heim, M. (1998). *Virtual Realism*. Oxford: Oxford University Press.

Heinrich, F., Schwenderling, L., Joeres, F., Lawonn, K., & Hansen, C. (2020) Comparison of augmented reality display techniques to support medical needle insertion. *IEEE Transactions on Visualization and Computer Graphics, Visualization and Computer Graphics*, 26(12), 3568–3575. doi:10.1109/TVCG.2020.3023637

Helmreich, R. L., & Schaefer, H. G. (1998). Team performance in the operating room. In *Human Error in Medicine*, Bogner, M. S., Ed. Hillsdale, NJ: Lawrence Erlbaum Associates.

Helmreich, R. L., Wilhelm, J. A., Gregorich, S. E., & Chidester, T. R. (1990). Preliminary results from the evaluation of cockpit resource management training: Performance ratings of flight crews, *Aviation, Space, and Environmental Medicine*, 61, 576–579.

Hiskens, I. A., Peng, H., & Fathy, H. K. (2011, July). Transportation electrification education for k-12 students. In *2011 IEEE Power and Energy Society General Meeting* (pp. 1–5). IEEE.

Hofstadter, D. R. (1979). *Godel, Escher, Bach: An Eternal Golden Braid*. New York: Vintage Books.

Huerta, E. A., Khan, A., Davis, E., Bushell, C., Gropp, W. D., Katz, D. S., ... Saxton, A. (2020). Convergence of artificial intelligence and high-performance computing on NSF-supported cyberinfrastructure. *Journal of Big Data*, 7(1), 1–12.

Insinna, V. (2019, May 2). Defense news. One of the F-35′s cost goals may be unattainable. https://www.defensenews.com/air/2019/05/02/one-of-the-f-35s-cost-goals-may-be-unattainable/

Jones, E. R., Hennessy, R. T., & Deutsch, S. (1985). *Human Factors Aspects of Simulation: Report of the Working Group on Simulation*. National Research Council Washington DC Committee on Human Factors.

Kahneman, D., Slovic, P., & Tversky, A. (1982). *Judgment under Uncertainty: Heuristics and Biases*. Cambridge: Cambridge University Press.

Kancler, D. E., Quill, L. L., Revels, A. R., Webb, R. R., & Masquelier, B. L. (1998). Reducing cannon plug connector pin selection time and errors through enhanced data presentation methods. *Proceedings of the Human Factors and Ergonomics Society Annual Meeting*, 42(18), 1286–1290. https://doi.org/10.1177/154193129804201802

Kaplan, A. D., Cruit, J., Endsley, M., Beers, S. M., Sawyer, B. D., & Hancock, P. A. (2019, November). Transfer of training from virtual reality and augmented reality: A meta-analysis extended abstract. In *Proceedings of the Human Factors and Ergonomics Society Annual Meeting* (Vol. 63, No. 1, pp. 2142–2143). Los Angeles, CA: SAGE Publications.

Kaplan, A. D., Cruit, J., Endsley, M., Beers, S. M., Sawyer, B. D., & Hancock, P. A. (2020). The effects of virtual reality, augmented reality, and mixed reality as training enhancement methods: A meta-analysis. *Human Factors*. https://doi.org/10.1177/0018720820904229

Kiourexidou, M., Natsis, K., Bamidis, P., Antonopoulos, N., Papathanasiou, E., Sgantzos, M., & Veglis, A. (2015). Augmented reality for the study of human heart anatomy. *International Journal of Electronics Communication and Computer Engineering*, 6(6), 658–662.

Kitroeff, N., & Gelles, D. (2020, January 8). In reversal, Boeing recommends 737 Max simulator training for pilots. *New York Times*. https://www.nytimes.com/2020/01/07/business/boeing-737-max-simulator-training.html

Klein, G. A., & Crandall, B. W. (1995). The role of mental simulation in naturalistic decision making. In *Local Applications of the Ecological Approach to Human-Machine Systems*, Hancock, P., Flach, J., Caird, J., & Vicente, K., Eds. Mahwah, NJ: Lawrence Erlbaum Associates, Vol. 2, pp. 324–358.

Kramida, G. (2016). Resolving the Vergence-accommodation conflict in head-mounted displays. *IEEE Transactions on Visualization and Computer Graphics*, 22(7), 1912–1931. https://ieeexplore.ieee.org/document/7226865

Lane, N. E. (1986). *Issues in Performance Measurement for Military Aviation with Applications to Air Combat Maneuvering*. Orlando, FL: ESSEX CORP.

Lang, B. (2021, March 31). Microsoft Signs $22B Contract with US Army to Bring HoloLens 2 Tech to the Battlefield. U.S. Naval Institute. https://www.roadtovr.com/microsoft-hololens-2-us-army-contract-production-phase/

Langley, A., Lawson, G., Hermawati, S., D'cruz, M., Apold, J., Arlt, F., & Mura, K. (2016). Establishing the usability of a virtual training system for assembly operations within the automotive industry. *Human Factors and Ergonomics in Manufacturing & Service Industries*, 26(6), 667–679.

Lindenberger Group. (2017). 8 top benefits of training simulations in the workplace. Retrieved from https://lindenbergergroup.com/8-top-benefits-training-simulations-workplace/

Lintern, G. (1991). An informational perspective on skill transfer in human-machine systems. *Human Factors*, 33(3), 251–266.

Lintern, G., Roscoe, S. N., & Sivier, J. E. (1990). Display principles, control dynamics, and environmental factors in augmentation of simulated visual scenes for teaching air-to-ground attack, Human. *Factors*, 32, 299–371.

Liu, X., Sun, J., Zheng, M., & Cui, X. (2021). Application of mixed reality using optical see-through head-mounted displays in transforaminal percutaneous endoscopic lumbar discectomy. *BioMed Research International*, 1–8. https://www.ncbi.nlm.nih.gov/pmc/articles/PMC7902133/

Loesch, R. L., & Waddell, J. (1979). The importance of stability and control fidelity in simulation. *Proceedings of Conference: 50 Years of Flight Simulation*. London: Royal Aeronautical Society, pp. 90–94.

Losey, S. (2018, September 30). The Air Force is revolutionizing the way airmen learn to be aviators. *Air Force Times*. https://www.airforcetimes.com/news/your-air-force/2018/09/30/the-air-force-is-revolutionizing-the-way-airmen-learn-to-be-aviators/

Lu, Y., Deng, B., Yan, Y., Qie, Z., Li, J., & Gui, J. (2019, December 9). Vergence-*Accommodation Conflict Potential Solutions* in *Augmented Reality Head Mounted Displays*. AIP Publishing. https://aip.scitation.org/doi/abs/10.1063/1.5137845?journalCode=apc.

Lutz, R. R., Frederick, P. S., Walsh, P. M., Wasson, K. S., & Fenlason, Nl. L. (2017). Integration of unmanned aircraft systems into complex airspace environments. *Johns Hopkins APL Technical Digest*, 33(4), 291–302. www.jhuapl.edu/techdigest

Magnuson, S. (2019, Jan 2). Services declare breakthrough in LVC training. National Defense: The Business and Technology Magazine of NIDA, 103(782), 12. Retrieved from http://www.nationaldefensemagazine.org/articles/2019/1/2/services-declare-breakthrough-in-lvc-training

Mazuryk, T., & Gervautz, M. (1996). *Virtual Reality: History, Applications, Technology, and Future*. Vienna: Institute of Computer Graphics Vienna University of Technology.

McCoy-Fisher, C., Mishler, A., Bush, D., Severe-Valsaint, G., Natali, M., Riner, B., & Naval Air Warfare Center Training Systems Division. (2019). Student Naval Aviation Extended Reality Device Capability Evaluation (NAWCTSD-TR-2019-001). DTIC. https://apps.dtic.mil/sti/citations/AD1103227

Mearian, L. (2017). CW@ 50: Data storage goes from $1 M to 2 cents per gigabyte. *Computerworld*. https://www.computerworld.com/article/3182207/cw50-data-storage-goes-from-1m-to-2-cents-per-gigabyte.html)

Milgram, P., Takemura, H., Utsumi, A., & Kishino, F. (1995). Augmented reality: A class of displays on the reality virtuality continuum. *SPIE: Telemanipulator and Telepresence Technologies*, 2351, 282–292. http://etclab.mie.utoronto.ca/publication/1994/Milgram_Takemura_SPIE1994.pdf

Milgram, P., & Kishino, F. (1994). A taxonomy of mixed reality visual displays. *IEICE Transactions on Information and Systems*, 77, 1321–1329.

Militello, L. G., Sushereba, C. E., & Ramachandran, S. (2023). *Handbook of Augmented Reality Training Design Principles*. Cambridge: Cambridge University Press. https://unveilsystems.com/

Miller, R. B. (1954). *Psychological Considerations in the Design of Training Equipment* (Tech. Rep. No. 54-563). Wright Patterson AFB, OH: Wright Air Development Center.

Mixon, T. R., & Moroney, W. F. (1982). *An Annotated Bibliography of Objective Pilot Performance Measures*. Orlando, FL: Naval Training Equipment Center.

Moreno-Ger, P., Burgos, D., & Torrente, J. (2009). Digital games in eLearning environments: Current uses and emerging trends. *Simulation & Gaming*, 40(5), 669–687.

Moroney, W. F., & Moroney, B. W. (2010). Flight simulation. In *Handbook of Aviation Human Factors*, Wise, J., Hopkin, V. D., & Garland, D., Eds. Boca Raton, FL: Taylor and Francis Group.

Morse, K. L., Coolahan, J., Lutz, R., Horner, N, Vick, S., & Syring, R. (2010). *Best Practices for the Development of Models and Simulations*. John Hopkins University – Applied Physics Laboratory, NSAD-R-2010-037. Final Report, June 2010. https://www.acqnotes.com/Attachments/JHU%20APL%20Best%20Practices%20for%20the%20Development%20of%20M&A,%20June%2010.pdf

Mueller, S. (2021, April). Our Sims. *Ourselves Wired*, 20–21.

Mulenburg, J. (2011). Crew resource management improves decision making. *ASK Magazine is published by NASA*, Washington DC, 11–13.

N161-007. (n.d.). SkyFall. https://www.navysbir.com/16_1/12.htm

N172-117. (n.d.). Mishap awareness scenarios and training for operational readiness responses. https://www.sbir.gov/node/1254605

N192.087. (n.d.). Headset Equivalent of Advanced Display Systems (HEADS). https://www.navysbir.com/n19_2/N192-087.htm

N193-D01. (n.d.). On demand training solutions for maintenance technicians. https://www.navysbir.com/n19_3/N193-D01.htm

N193-D01. (n.d.). On demand training solutions for maintenance technicians. https://www.navysbir.com/n19_3/N193-D01.htm

N201-024. (n.d.). Augmented reality headset for maintainers. https://www.navysbir.com/n20_1/n201-024.htm

Natali, M., Mercado, J., & Foster, C. (2020, Aug, p 19–22). Integrating AI with aviation training: A preview of the virtual instructor pilot referee. *Call Signs*. https://navyaep.com/wp-content/uploads/2020/08/Call-Signs-9.1.pdf

Nausheen, F., & Bhupathy, R. (2020). Disrupting clinical education: Point-of-care ultrasound merged with HoloLens augmented reality during early medical training. *Clinical Teacher*, 17(2), 146–147. https://doi-org.proxy.lib.fsu.edu/10.1111/tct.13155

Naval Air Warfare Center Training Systems Division (NAWCTSD) Public Affairs Office (PAO). (2018). H-60R AR preflight checklist trainer: A tablet-based Augmented Reality (AR) preflight checklist trainer for aircrew and student pilots. *Project Summary Presented at the 2018 Interservice/Industry Training, Simulation, and Education Conference (I/ITSEC)*. https://www.navair.navy.mil/nawctsd/sites/g/files/jejdrs596/files/2019-01/iitsec_2018_fact_sheet_h-60r_ar_dist_statement_a.pdf

Naval Technology. (2021). MQ-4C Triton Broad Area Maritime Surveillance (BAMS) UAS: MQ-4C Triton is a new broad area maritime surveillance (BAMS) unmanned aircraft system (UAS) unveiled by Northrop Grumman for the US Navy. https://www.naval-technology.com/projects/mq-4c-triton-bams-uas-us/

Nestler, V., Moore, E. L., Huang, K. Y. C., & Bose, D. (2013). The use of second life® to teach physical security across different teaching modes. In *Information Assurance and Security Education and Training* (pp. 188–195). Berlin, Heidelberg: Springer. https://www.youtube.com/watch?v=I5VXf5Ky5ms

Office of the Under Secretary of Defense for Personnel and Readiness. (2018, March 16 with Change 1 Effective February 15, 2022). *Department of Defense Instruction 1322.24: Medical Readiness Training (MRT)*. Retrieved from: https://www.esd.whs.mil/Portals/54/Documents/DD/issuances/dodi/132224p.pdf?ver=pDae3SN8brdnRhNUZFrztw%3D%3D

Orlansky, J., & String, J. (1977). *Cost-Effectiveness of Flight Simulator for Military Training* (Rep. No. IDA NO. HQ 77-19470). Arlington, VA: Institute for Defense Analysis.

Perla, P. P. (1990). *The Art of Wargaming: A Guide for Professionals and Hobbyists*. Naval Institute Press.

Petty, M. D. (2010). Verification, validation, and accreditation. *Modeling and Simulation Fundamentals: Theoretical Underpinnings and Practical Domains*, 325–372.

Picciano, A. G., Seaman, J., Shea, P., & Swan, K. (2012). Examining the extent and nature of online learning in American K-12 education: The research initiatives of the Alfred P. Sloan Foundation. *The Internet and Higher Education*, 15(2), 127–135.

Pohman, D. L., & Fletcher, J. D. (1999). Aviation personnel selection and training. In *Handbook of Aviation Human Factors*, Daniel, T. J., Garland, J. W., & Hopkin, V. D., Eds. Mahwah, NJ: Lawrence Erlbaum Associates.

Poisson, R. J., & Miller, M. E. (2014). Spatial disorientation mishap trends in the U.S. Air Force 1993–2013. *Aviation, Space, and Environmental Medicine*, 85(9), 919–924. https://doi.org/10.3357/ASEM.3971.2014

Pope, T. M. (2019). *A cost-benefit analysis of pilot training next*. Theses and Dissertations. 2314. https://scholar.afit.edu/etd/2314

Qian, M., Nicholson, J., Tanaka, D., Dias, P., Wang, E., & Qiu, L. (2019). Augmented reality (AR) assisted laryngoscopy for endotracheal intubation training. In *Proceedings from the International Conference on Human-Computer Interaction*; July 26–31, Orlando, FL.

Raser, J. R. (1969). *Simulation and Society: An Exploration of Scientific Gaming*. Boston: Allyn & Bacon. https://eric.ed.gov/?id=ED043220

Research and Markets. (2019, March). The drone market report 2019: Commercial drone market size and forecast (2019–2024). https://www.researchandmarkets.com/reports/4764173/the-drone-market-report-2019-commercial-drone

Ribeiro, R., Ramos, J., Safadinho, D., Reis, A., Rabadão, C., Barroso, J., Pereira, A., & Web, A. R. (2021). Solution for UAV pilot training and usability testing. *Sensors*, 21, 1456 https://doi.org/10.3390/s21041456

Ricci, K. E., Salas, E., & Cannon-Bowers, J. A. (1996). Do computer-based games facilitate knowledge acquisition and retention? *Military Psychology*, 8(4), 295–307.

Rizzo, A. A., & Shilling, R. (2017). Clinical virtual reality tools to advance the prevention, assessment, and treatment of PTSD. *European Journal of Psychotraumatology*, 8(Suppl 5), 1414560.

Rizzo, A., Koenig, S. T., & Talbot, T. B. (2019). Clinical results using virtual reality. *Journal of Technology in Human Services*, 37(1), 51–74. https://doi.org/10.1080/15228835.2019.1604292

Roese, N. J., & Olson, J. M. (1995). *What Might Have Been: The Social Psychology of Counterfactual Thinking*. Mahwah, NJ: Lawrence Erlbaum Associates.

Rokhsaritalemi, S., Sadeghi-Niaraki, A., & Choi, S.-M. (2020). A review on mixed reality: Current trends, challenges and prospects. *Applied Sciences*, 10(2), 636. MDPI AG. https://doi.org/10.3390/app10020636

Roscoe, S. N. (1991). Simulator qualification: Just as phony as it can be. *The International Journal of Aviation Psychology*, 1(4), 335–339.

Roscoe, S. N. & Williges, B. H. (1980). Measurement of transfer of training. In *Aviation Psychology*, Roscoe, S. N., Ed. Ames, IA: Iowa State University Press, p. 182.

Royal Aeronautical Society. (1979). *Proceedings of Conference: 50 Years of Flight Simulation, I, II, and III*. London: Royal Aeronautical Society.

Ruben, P., & Gray, J. (2020). The WIRED guide to virtual reality. https://www.wired.com/story/wired-guide-to-virtual-reality/

Salas, E., Bowers, C. A., & Rhodenizer, L. (1998). It is not how much you have but how you use it: Toward a rational use of simulation to support aviation training. *The International Journal of Aviation Psychology*, 8(3), 197–208.

Salas, E., Rhodenizer, L., & Bowers, C. A. (2000). The design and delivery of crew resource management trainee: Exploiting available resources. *Human Factors*, 42, 490–511.

SB052-009. (n.d.). Virtual instructor pilot exercise referee for CNATRA. https://www.sbir.gov/sbirsearch/detail/1851489

Schrage, M. (2000). *Serious Play*. Boston: Harvard Business School Press.

Schultz, C. (2014). A military contractor just went ahead and used an Xbox controller for their new giant laser cannon. *Smithsonian Magazine*. Retrieved June 23, 2021, from https://www.smithsonianmag.com/smart-news/military-contractor-just-went-ahead-and-used-xbox-controller-their-new-giant-laser-cannon-180952647

Selvander, M., & Asman, P. (2012). Virtual reality cataract surgery training: Learning curves and concurrent validity. *Acta Ophthalmologica*, 90, 412–417.

Shelbourne, M. (2020). Navy harnessing new technology to restructure aviation training. *USNI News*, September 14, 2020, https://news.usni.org/2020/09/14/navy-harnessing-new-technology-to-restructure-aviation-training

Spannaus, T. W. (1978). What is a simulation? *Audiovisual Instruction*, 235, 16–17.

Sparaco, P. (June, 1994). Simulation acquisition nears completion. *Av. Week Space Technol.*, 71.

Sparko, A., Burki-Cohen, J., & Go, T. (2010). Transfer of training from a full-flight simulator vs. a high-level flight-training device with a dynamic seat. In *AIAA Modeling and Simulation Technologies Conference* (p. 8218)

Stark, E. A. (1994). *Training and Human Factors in Flight Simulation, Flight Simulation Update*, Vol. 10. Binghamton, NY: SUNY Watson School of Engineering.

Sternberg, R. J., & Gastel, J. (1989a). Coping with novelty in human intelligence: An empirical investigation. *Intelligence*, 13(2), 187–197.

Sternberg, R. J., & Gastel, J. (1989b). If dancers ate their shoes: Inductive reasoning with factual and counterfactual premises. *Memory & Cognition*, 17, 1–10.

Sushereba, C., Militello, L., & Patterson, E. (2019, June). The potential of augmented reality for training macrocognitive skills in combat medics. In *Proceedings from The International Conference on Naturalistic Decision Making 2019*, San Francisco, CA.

Sushereba, C. E., & Militello, L. G. (2020, December). Virtual patient immersive trainer to train perceptual skills using augmented reality. In *Proceedings of the Human Factors and Ergonomics Society Annual Meeting* (Vol. 64, No. 1, pp. 470–472). Los Angeles, CA: SAGE Publications.

Sweigart, L., & Hodson-Carlton, K. (2013). Improving student interview skills: The virtual avatar as client. *Nurse Educator*, 38(1), 11–15.

Tadjdeh, Y. (2020, November 25). Air Force embracing new tech to solve pilot shortage. https://www.nationaldefensemagazine.org/articles/2020/11/25/air-force-embracing-new-tech-to-solve-pilot-shortage

Tanaka, A., Craighead, J., Taylor, G., & Sottilare, R. (2019, July). Adaptive learning technology for AR training: Possibilities and challenges. In *International Conference on Human-Computer Interaction* (pp. 142–150). Springer, Cham.

Tanaka, A., Craighead, J., & Taylor, G. (2019, December). The application of augmented reality for immersive TC3 training. *Proceeding at the Interservice/Industry Training, Simulation, and Education Conference.*

Tang, Y. M., Ng, G. W. Y., Chia, N. H., So, E. H. K., Wu, C. H., & Ip, W. H. (2021). Application of virtual reality (VR) technology for medical practitioners in type and screen (T&S) training. *Journal of Computer Assisted Learning*, 37(2), 359–369. https://onlinelibrary.wiley.com/doi/abs/10.1111/jcal.12494

Tang, Y., Shetty, S., Jahan, K., Henry, J., & Hargrove, S. K. (2012). Sustain city – A cyber-infrastructure-enabled game system for science and engineering design. *Journal of Computational Science Education*, 3(1), 57–65.

Taylor, G., Deschamps, A., Tanaka, A., Nicholson, D., Bruder, G., Welch, G., & Guido-Sanz, F. (2018, July). Augmented reality for tactical combat casualty care training. In *International Conference on Augmented Cognition* (pp. 227–239). Springer, Cham.

Tetlock, P. E., & Belkin, A. (1996). *Counter-Factual Thought Experiments in World Politics: Logical, Methodological and Psychological Perspectives*. Princeton, NJ: Princeton University Press.

Udoh, E. (Ed.). (2011). *Cloud, Grid and High-Performance Computing: Emerging Applications: Emerging Applications*. Hershey, PA: IGI Global.

Ulrich, T. A., Lew, R., Werner, S., & Boring, R. L. (2017, September). Rancor: A gamified microworld nuclear power plant simulation for engineering psychology research and process control applications. In *Proceedings of the Human Factors and Ergonomics Society Annual Meeting* (Vol. 61, No. 1, pp. 398–402). Los Angeles, CA: SAGE Publications.

Underwood, K. (2020, April). Augmented reality goes airborne. *Signal AFCEAs International Journal*. https://www.afcea.org/content/augmented-reality-goes-airborne

Venkatesh, V., Speier, C., & Morris, M. G. (2002). User acceptance enablers in individual decision making about technology: Toward an integrated model. *Decision Sciences*, 33(2), 297–316.

Voices of VR Podcast. (2015, November 17). *No.245: 50 Years of VR with Tom Furness: The Super Cockpit, Virtual Retinal Display, HIT Lab, & Virtual World Society* [Audio

Podcast]. http://voicesofvr.com/245-50-years-of-vr-with-tom-furness-the-super-cockpit-virtual-retinal-display-hit-lab-virtual-world-society/

VR Headset Authority. (2021). Why is refresh rate and field-of-view important for VR headset? Retrieved June 23, 2021, from https://vrheadsetauthority.com/why-is-refresh-rate-and-fov-important-for-a-vr-headset/

Waldrop, M. M. (2001). The origins of personal computing. *Scientific American*, 285(6), 84–91. https://www.scientificamerican.com/article/the-origins-of-personal-computing/

Weigmann, D. A., & Shappell, S. A. (1999). Human error and crew resource management failures in naval aviation mishaps: A review of U. S. Naval Safety Center data, 1990–96. *Aerospace Medicine and Human Performance*, 70, 147–1151.

Weiss, I., Kramarski, B., & Talis, S. (2006). Effects of multimedia environments on kindergarten children's mathematical achievements and style of learning. *Educational Media International*, 43(1), 3–17.

Wendland, W., & Honold, P. (2006). *Virtual Reality (VR) Research and Demonstration Project: Virtual Environment for Submarine Ship Handling Training (VESUB).* Unpublished technical report.

Wiener, E. L., & Nagel, D. C. (Eds.). (1988). *Human Factors in Aviation*. Gulf Professional Publishing.

Wiener, E. L., Kanki, B. G., & Helmreich, R. L. (1993). *Cockpit Resource Management*. San Diego, CA: Academic Press.

Williams, A. C., & Flexman, R. E. (1949). *An Evaluation of the Link SNJ Operational Trainer as an Aid in Contact Flight Training* (SDC-71-16-3). Navy Special Devices Center, Port Washington, New York.

Wilson, J. R. (2000). Technology brings change to simulation industry. *Interavia Bus Technology*, 55(643), 19–21.

Wilson, J. R. (2018, September 1). The increasing role of COTS in high-fidelity simulation. *Military Aerospace Electronics Magazine*. Retrieved June 23, 2021, from https://www.militaryaerospace.com/home/article/16707147/the-increasing-role-of-cots-in-highfidelity-simulation

Wong, Y. H., Bae, S. J., Bartels, E. M., & Smith, B. (2019). Next Generation Wargaming for the US Marine Corps. Santa Monica, CA: Rand National Defense Research Inst.

Young, M. F., Slota, S., Cutter, A. B., Jalette, G., Mullin, G., Lai, B., ... Yukhymenko, M. (2012). Our princess is in another castle: A review of trends in serious gaming for education. *Review of Educational Research*, 82(1), 61–89.

Youngblood, S. M., Pace, D. K., Eirich, P. L., Gregg, D. M., & Coolahan, J. E. (2000). Simulation verification, validation, and accreditation. *Johns Hopkins APL Technical Digest*, 21(3), 359–367.

2 Justification for Use of Simulation

Meredith Carroll, Summer Rebensky, Maria Chaparro Osman, and John Deaton

CONTENTS

INTRODUCTION

Simulation is used to provide an effective means of conducting training, system evaluations, and research, as well as for enjoyment and connection, while balancing elements of capability, cost-efficiency, and safety to achieve these goals. To sufficiently justify and validate simulator usage and applicability within these contexts,

DOI: 10.1201/9781003401360-2

advantages and disadvantages must be considered. Although disadvantages exist, they are surpassed by the magnitude of overall benefits simulation can provide in achieving a range of goals.

Simulation is an interactive system that represents the operational system through artificial duplication or replication of the system and its equipment, environment, and capabilities (Jones, Hennessy, & Deutsch, 1985). Advancements in computer hardware and software technology allow for the creation, manipulation, and control of complex, realistic situations and environments (McGrath et al., 2017). Advancement in virtual reality (VR), internet connectivity, and artificial intelligence have expanded the range of applicability of simulation usage and associated benefits (Fletcher et al., 2017). For example, today an individual interested in buying a home may take a simulated, virtual tour of a prospective home on a real estate website and even evaluate the fit of new furniture in the rooms through the use of simulation and augmented reality (AR) applications.

Traditionally, most simulators were classified as training devices; however, their use now extends to engineering simulators used for design, development, and evaluation processes (Lee et al., 2013); research simulators (Webster et al., 2014); and simulators used for recreational purposes such as enjoyment and social connection (Hromek & Roffey, 2009). Simulation is widely accepted in a vast array of domains, in part due to the opportunities provided to the user, which may not be safe or feasible in a real-world or operational environment. The value of this safety feature is priceless, with respect to cost-efficiency and human safety, both illustrated in terms of the ability to safely practice emergency procedures, and avoiding equipment or system damage (Moroney & Moroney, 2009). Acceptance of the simulation as a training device is demonstrated through the reliance and confidence of the transfer of training to the real-world environment. This acceptance is such that licenses, certifications, and qualifications may be received from simulator training alone. Simulation used for systems engineering and research purposes often facilitates analysis related to system design, development, testing, and evaluation, and the analysis and evaluation of standards and procedures for licensing and certification (Webster et al., 2014). Simulation is also used for research across domains to recreate task environments, collect rich data, and improve performance (Deterding et al., 2015). The use of simulation for enjoyment has benefited organizations such as the military by allowing them to adopt the simulators once issues have been resolved by the companies in the entertainment industry (Balinova, 2020). The following sections will provide a discussion of the primary purposes for which simulation is currently utilized, domains in which their use is most prominent, and achievable outcomes which justify their widespread use.

PURPOSES

There are four primary purposes for which simulation is typically utilized: (a) training, (b) systems engineering evaluation, (c) research, and (d) recreation.

Training

Training reshapes behavior using practice, direction, measurement, and feedback to teach task performance at a level of skill not previously possible (Caro, 1988, p. 229). As a training device, simulators make their contribution in skill development, maintenance, and assessment while allowing for enjoyment (Morgan et al., 2002; Rolfe & Staples, 1986). Serious games, educational simulations, and edutainment have all emerged as a way to facilitate trainees and learners developing required skills in a more interactive and stimulating environment (Breuer & Bente, 2010; Djaouti et al., 2011). Serious games and educational gaming developed for educational purposes engage learners and have commonly been found to lead to learning gains and improved performance (Vlachopoulos & Makri, 2017). For example, a virtual board game originally designed for fun is now utilized for military training to teach strategy and principles of tactical warfare (Breuer & Bente, 2010). Simulations also allow trainees to practice their skills while allowing for more autonomy and involvement during multiple phases of training (Landon-Hays et al., 2020; Crea, 2011). For instance, flight simulations allow trainees to practice specific aspects of flight for which they need more training (e.g., landing), without spending the time to conduct an entire flight. Simulated training can more flexibly facilitate learning across different problem domains and skills compared to traditional training (Hoffman et al., 2014). For example, low-fidelity desktop simulators can be used in initial training to familiarize a trainee with a system, low-medium fidelity simulators that provide realism in switches and controls can facilitate procedural training, and high-fidelity trainers that integrate full motion and affective cues can be used in the final stage of training in which skills are consolidated and proficiency is assessed.

Research has shown that training tasks requiring procedural knowledge and skills are better trained within a simulated environment than through lectures (Nestel et al., 2011). Simulation is also effective for facilitating the consolidation of skills under realistic and stressful situations (Andreatta et al., 2010). Simulations can additionally be used to put trainees through multi-player, team-building exercises, and cooperation-building challenges in a more enjoyable and realistic environment (De Gloria et al., 2014). However, the utilization or simulation is not ideal for all training tasks. For example, utilizing simulators to train declarative knowledge may not be as effective as learners can grasp how to obtain a desired outcome such as a passing score rather than the knowledge the system is intended to teach (Douglas et al., 2019). This can make it difficult to ascertain whether the user is learning the intended knowledge or how to "game the system." The enjoyment experienced with simulations can lead to issues such as recall of irrelevant material, if the user is focusing on aspects which are not the focus of the training (Chowdhury et al., 2017).

Systems Engineering Evaluation

Although training is the principal application of simulators, simulation is also extremely valuable as a systems engineering tool. Simulations are utilized for system design and development processes involving a multitude of applications. Modeling can be used to analyze complex systems in use cases for manufacturing, transportation, and healthcare to determine the impact of different designs and aid in selecting the best design (Lee et al., 2013). Simulations are also used for test and evaluation purposes of system and subsystem performance, and operational capabilities and limitations (Jones et al., 1985). Simulators for these purposes, and even to determine system maintenance scheduling, staffing, inventory, cost, and policies, have been largely used for manufacturing systems, airline hubs, and highway design and maintenance (Alabdulkarim et al., 2013; Liu et al., 2019). Simulation can also be used to predict and model human behavior such as energy consumption in a household to facilitate the design of houses in an energy-efficient manner (Hong et al., 2016), and human performance in aircraft to evaluate flight deck technologies for the Next-Generation Air Transport System (NextGen; Gore, 2011). The average consumer can even use simulation and AR technology to determine if new furniture fits in their home (Tang et al., 2014). In the automotive industry, simulation can be used to recreate crashes and replicate similar scenarios with newer safety settings to evaluate effectiveness (Alvarez et al., 2017). In aviation and aeronautics, simulation can be used to test and certify aircraft before physical development and even explore the potential impacts on vehicle structures in the event of a water landing (Hughes et al., 2013). It can also be used to test, model, and validate features, including methods for vertical takeoff and landing (VTOL) for emerging unmanned aircraft vehicle concepts (Yuksek et al., 2016), and collision avoidance technology for FAA certification (Webster et al., 2014). Modeling simulation can be used to assess the fit of humans in various systems such as pilots in aircraft seats to ensure proper reach and spine support (Lindsey et al., 2019). However, conclusions drawn from simulation-based analyses are only as accurate as the simulation itself. Many simulations assume systematic operations, such as in the case of human energy consumption in the home, but such an approximation can sometimes be off by as much as 30% from actual life consumption (Hong et al., 2016). As a result, modern advances in these fields have looked at methods to improve model accuracy, including ways to quantify uncertainty, novel or different modeling methods, combining multiple modeling methods, and incorporating more data for consideration into the model (Wang & Zhai, 2016; Lee et al., 2013).

Research

Simulators are often used as research instruments and within research settings as they make data collection and manipulation easier, and make study replication with minimal changes between participants possible (Deterding et al., 2015; Lukosch et al., 2018). Due to their prevalence, both researchers and participants often know how to interact with this technology, making their use in research straightforward (Alexander et al., 2005). Further, the addition of supporting internet capability/connection allows for the potential to gain access to a larger pool of participants, as participants who are not local can connect via the internet from their location. The use of simulations in research applications often involves investigation of human–computer interactions, visual/motion systems, and human performance assessments, such as measurements of workload, decision-making skills, psychomotor abilities, multi-tasking abilities, situation awareness, stressor effects, and spatial abilities (Blodgett et al., 2018; Stroosma et al., 2003; Woda et al., 2017; Vlakveld et al., 2018; Zhang et al., 2017). The ability of simulators to examine human performance phenomena is unmatched as they afford researchers control over the environment while limiting ambient influences (Baubien & Baker, 2004). Researchers are able to examine a wide range of conditions and characteristics (e.g., performer reactions to different types of decision events) while holding other aspects of the environment constant (e.g., information available, time to make the decision), thereby increasing the validity of the research (Beaubien & Baker, 2004). Additionally, modernization of simulators has led to the ability to pair emerging technology such as heads-up displays (HUDs), AR, and VR technology. The use of simulators in conjunction with VR and AR has led to advances in research across disciplines such as healthcare, education, and training (Cook et al., 2011; Radianti et al., 2020). For example, research involving simulated surgery software utilizing VR technology has shown support for improved training performance (Kim et al., 2017). These advancements allow for a more immersive environment and can increase the authenticity of participants' responses (Radianti et al., 2020; Kronqvist et al., 2016). When an individual uses VR, they are presented with fewer visual distractions from their physical location as well as a 3D view of the virtual environment, making the experience much more realistic (Cecil et al., 2014; Kim et al., 2017).

Simulators are also widely used in studies exploring team process and team performance measures. More recently, the use of simulators to examine human–agent collaboration has gained traction as simulators can be used to simulate interaction with the autonomous agent. The use of simulators in this area has provided insight into how multiple virtual agents affect operator performance, requirements, and training needs (Fraune et al., 2021; Mairaj et al., 2019). Simulation has also been used as a means of evaluating situation awareness associated with certain system displays. There are drawbacks to the use of simulation for research purposes. Human performance simulators, although similar to the intended task, do not always elicit the same responses present in the natural domain. For example, a flight simulator may have vibration and utilize VR to make the experience immersive, but the operator is aware that a simulated crash will not lead to the same outcome as a true crash. Emotional and environmental stressors are not easily replicable in simulators (Patterson et al., 2008). Additionally, data and results from research simulators can be difficult to interpret, at times requiring additional analysis software (Sena et al., 2012).

Recreation

A somewhat new and still emerging use of simulation is for recreational purposes such as enjoyment and social connection. Simulations within the entertainment industry are used for gaming, arts, theme park rides, sports, and film, among other reasons (Bouyer et al., 2017; Smith, 2010). Given that simulators provide immersion and enjoyment, now at an affordable cost, their popularity has increased dramatically, leading them to be purchased for home entertainment (Granic et al., 2014; Klimmt, 2003; Rui, 2020). For example, golfing simulators have become a popular item for golfers to enjoy the sport from the convenience of their homes. Operational simulations can also be used in a more informal context for fun. For example, aviation enthusiasts often invest in commercial, desktop flight simulators for recreational purposes. Enjoyment associated with simulator use has been tied to the use of suspense to cultivate curiosity, control leading to a sense of empowerment, and escapism through immersion, among other factors (Granic et al., 2014; Klimmt, 2003; Ritterfeld et al., 2009). For example, simulators which provide a storyline cultivate curiosity, while the ability to immediately observe the results of their own decisions within a game has also been found to provide players with enjoyment (Klimmt et al., 2007).

Simulations can also foster communication between individuals as they can mimic complex social systems (Lukosch et al., 2018). For example, many games provide chat rooms which allow users to communicate using personalized avatars. Users are able to both join and create groups based on similar interests and goals, as well as work in teams in certain simulators like the Eurogamer Flight Simulator for PC. Simulations which enable connection have been linked to counteracting loneliness, promoting positive feelings and physical activity in older adults (Kahlbaugh et al., 2011). Simulations commonly make use of avatars which can be personalized, virtual environments and worlds where users can interact, and are even used in amusement parks (Cocking & Matthews, 2000). The use of simulations in amusement parks ranges from utilization on rides to improving user experience while waiting in lines. Unfortunately, the novelty effect can be an issue with simulation for recreational purposes, as over time the simulator becomes less engaging to the user and their initial response is only due to initial engagement levels, which can dwindle (Arıcı & Yılmaz, 2020).

DOMAINS OF APPLICATION

The past several decades have seen an explosion in the use of simulation across multiple domains. Around the turn of the century, simulation was limited primarily to military, aviation, and systems engineering domains; however, simulation use is now prevalent in domains ranging from maintenance and emergency services to education and entertainment. Here we provide a brief discussion of simulation applications within both traditional and emerging domains.

AVIATION

"Within the aviation community, the effectiveness of simulators is accepted as an article of faith. Indeed, the aviation industry could not function without simulators and flight training devices (FTDs), whose existence is mandated by Federal Aviation Administration (FAA) regulations" (Hancock et al., 2008, p. 46). Even decades later, this statement remains true, as many aviation sectors have heavily invested in simulator training, with 30–50% of training conducted in simulators (Yoon et al., 2019).

Within the aviation industry, manufacturers, users, and regulators rely on simulators for training and testing purposes across manned, unmanned, and urban air mobility (UAM) contexts. The FAA establishes standards and policies for the certification of civil aviation pilots, and Federal Aviation Regulations (FARs) permit simulators to be used for various training objectives, including crew resource management (CRM), initial training, recurrent training, emergency operations, and stall procedures (FAA, 2020). Currently, many flight schools use a range of simulators from motion-based to fixed-base to personal computer-based flight simulators (Reweti et al., 2017). Simulators are also used to test and train future concepts and issues that have not yet reached the commercial aviation sector, such as EFB integration into the flight deck panel (Carroll et al., 2021; Pittorie et al., 2019), as well as emergency or novel situations that would be difficult to replicate otherwise, such cybersecurity hacking of systems or UAS search and rescue (Carroll et al., 2020; Rebensky et al., 2020). Simulation is widely used in CRM training in which aircrews are taught how to utilize every resource available to them, including communication. Simulated scenarios can include line-oriented flight training (LOFT) in which an aircrew flies a full simulated mission, incorporating hands-on learning, practice, and feedback available to the crew afterward in the form of a video recording of the flight. This technique is well accepted by the aviation industry and believed to provide an excellent training platform (Kanki, 2019). Additionally, simulators like the Virtual Drone Search Game have been utilized to research search and rescue, drone-teaming practices (Fraune et al., 2021).

Military

Simulators have been used in the military realm for gunnery, driver, infantry, pilot, commander, and maintenance training (Pinheiro et al., 2012; Straus et al., 2019). Other uses of simulation in the military include support for development of tactics and combat management skills, as well as to simulate medical emergencies and evaluate operational systems (Straus et al., 2019; Fletcher et al., 2017; Hall et al., 2014). Modern simulators can be used for military purposes by leveraging video game technology to address training needs, including synthetic task environments that appear realistic and allow for simulation of crews, vehicles, and communication, as well as assessment and scoring capabilities for ensuring warfighter readiness (Kumm & Burwell, 2017). For example, Virtual Battlespace 3 (VBS3), an Army game-based training software, was developed from a popular commercial first-person shooter game, Arma 3, and is used to train both individual- and team-level skills for army personnel (Straus et al., 2019).

Medical

Simulation within the medical industry primarily focuses on surgical procedures as surgical skills require constant practice (Tan & Sarker, 2011; van de Ven et al., 2017). Simulations have also been widely used as a way to roleplay, and to practice and cultivate communication skills for physicians with future patients (Cegala & Broz, 2003;

Hardoff & Schonmann, 2001). Simulation can also benefit nurses with emergency response procedures such as surgical resuscitation, by increasing knowledge, feelings of confidence, and satisfaction, through realistic practice (McRae et al., 2017; Warren et al., 2016). Across various studies, simulation, particularly high-fidelity simulation in the critical care medical domain, has been found more effective than other methods of teaching for both skill and knowledge acquisition (Beal et al., 2017).

Driving

Driving simulators are used in human factors research, including systems engineering research, and research examining human behaviors, perceptions, and operational features. Research demonstrates that simulators are as effective as on-road studies at uncovering and studying performance outcomes like mirror checking, maintaining speed, monitoring behaviors, and following the rules of the road (Meuleners & Fraser, 2015; Bella, 2008; Underwood et al., 2011). Simulator-based driving research has shown mixed effectiveness in using simulation to explore other driving behaviors and situations such as lane changing, speeding, driving performance and errors, braking, steering, mental workload, and eye fixations as high-fidelity and low-fidelity simulators vary in effectiveness depending on the constructs explored (Wynne et al., 2019). Driving simulations can also be used to explore environmental and design areas such as autonomous driving assistance and environmental hazardous conditions, as well as at-risk-population driving behaviors (Abdelgawad et al., 2017; Campos et al., 2017). Simulation has been utilized in all of these contexts to conduct studies aimed at reducing accident rates and human fatalities in a safe and efficient manner.

Emergency Response

Simulations are used to train the police force, military, firefighters, and medical personnel for emergency situations that require dynamic decision-making, in stressful and unpredictable environments, in an interactive way (Taber, 2008). Some simulators have been developed for police training, utilizing VR systems that, through motion tracking technology, can help police officers learn how to read body language and emotions of suspects, in order to train effective responses (Himona et al., 2011). Additional simulators that can be used by police, medical teams, and firefighters have been developed for training personnel on response to crowd control in escalated situations (Kamiński et al., 2020). Early simulations allowed emergency personnel to practice different tactics for controlling a situation such as house fires using desktop computers, but these simulators required large developmental effort to simulate realistic fire and animations and still had limited realism (St. Julien & Shaw, 2003). However, modern simulators can replicate a range of features with little to no development expertise, including visual features, and even physical features such as haptic feedback to simulate the force of using a fire hose (Rebensky et al., 2020; Nahavandi et al., 2019). These high-fidelity environments replicate the difficulty associated with emergency situations and allow repeated practice of necessary skills that are beneficial for high-risk, low-frequency emergency events (Kamiński et al., 2020; Nahavandi et al., 2019; Jerald et al., 2020; Marler et al., 2020).

Education and STEM

Simulation can be utilized to teach difficult science, technology, engineering, and math (STEM) concepts that may otherwise be challenging to convey—such as allowing students to explore space in virtual reality to learn about planetary motion (Lindgren et al., 2016). A substantial amount of support has been found for the use of simulators to promote learner discovery of important concepts, motivation to learn, and conceptual understandings in science (Lindgren et al., 2016; Honey & Hilton, 2011; Rutten et al., 2012). Simulations in the form of virtual learning environments have gained interest from STEM educators for the support they provide to instructors, the wealth of resources and information to students, and their ability to engage students (Mpu & Adu, 2020). For example, the use of the virtual environment "SimScientists," which allows students to examine the interactions between organisms and their roles in an interactive environment, has been linked to improved science process skills and learning goals (Honey & Hilton, 2011; Quellmalz et al., 2020). STEM fields commonly require hands-on experiments and experiences, in which learning by doing is essential (Pellas et al., 2017). The ability of simulators to rapidly simulate consequences and feedback that may naturally take weeks allows for a "visual" understanding that would normally take longer (Skinner et al., 2020).

Entertainment

The use of simulation in entertainment spans from motion simulators such as simulated rides in amusement parks, and viewing simulators such as online simulators which allow people to virtually visit stores, to full human simulators used in some museums to bring a flair of interactivity to the exhibit. The use of simulators for entertainment is important, as their lessons learned and innovative advancements are commonly instantiated in other areas such as the military (Alexander et al., 2005; Balinova, 2020). The collaboration between entertainment and other industries for simulations is common. For example, the first medical mannequin designed to simulate mouth-to-mouth simulation was created by a toy manufacturer asked to create a doll for CPR training (Cooper & Taqueti, 2008). Additionally, the Marine Corps Infantry Immersion Trainer, a highly realistic, live, training environment, was designed with help from Hollywood set designers, who helped bring the environment to life with the sights, sounds, and smells of an actual battlefield (Dean et al., 2008). Additionally, their use can lead to sparked interest and solutions in other domains. The increased engagement found in gaming simulators used for entertainment has led to their adoption in training and learning areas, such as college engineering courses (Coller & Scott, 2009).

Maintenance

Simulators for maintenance have been widely used in the manufacturing domain in order to optimize scheduling of maintenance, cost estimation, policy development, and system reliability, and have also been used for maintenance training (Alabdulkarim et al., 2013; Pinheiro et al., 2012; Winther et al., 2020; Neges et al., 2017; Liu et al., 2019). Research conducted by Winther et al. (2020) using simulation for training maintenance tasks has demonstrated that using VR simulation can be an effective medium for training procedural knowledge. However, tasks that require interacting with physical objects still pose limitations as the sensations of touch, motion, and force cannot currently be replicated entirely in virtual reality. Hands-on, in-person experiences are still superior. For maintenance tasks, training must extend beyond procedural knowledge to include physical aspects such as how much torque to use, how it feels to turn a part, or the right grip to utilize (Neges et al., 2017; Winther et al., 2020). Without haptic or physical objects to interact with, the maintainer will not be able to experience key signs, such as a part clicking into place, that indicate a maintenance task was completed correctly. These skills are necessary to accurately perform maintenance on various technologies in a wide array of domains. Newer technology allows for the integration of physical objects, into a virtual environment, with the virtual environment dynamically responding to user interactions with the physical object in the real world (Neges et al., 2017). An example of this is turning valves on a low-fidelity panel of physical levers with sensors that are presented in virtual reality that update the simulated hydraulic system allowing users to practice resolving emergency situations. Advances in this space will continue to enhance the applicability and transfer of training of simulation for maintenance tasks.

ACHIEVABLE OUTCOMES

The benefits of using simulation for the above-specified purposes provide several beneficial outcomes that allow practitioners to achieve their goals. Simulators can often provide more in-depth, safer, less expensive, and more effective training, evaluation, and experiential opportunities compared to those provided by the actual system being simulated. This is due, in part, to simulation environments allowing the incorporation of instructional features that facilitate instructor intervention, sensors and measurement algorithms that provide precise assessments, and multimodal and high-fidelity environments that facilitate realistic and enjoyable interaction and immersion. As technology has advanced, more benefits of simulation use for a range of purposes have emerged. For education and training alone, a meta-analysis, across a wide range of applications that utilize AR and VR technology, demonstrated positive outcomes, including enhanced immersion, increased learning, and reductions in learning time and skill decay (Fletcher et al., 2017). The following sections outline several positive outcomes that can be achieved through simulation, including associated advantages and disadvantages.

Cost Benefit

Simulation has shown to be a cost-effective training device by providing the required training at lower costs than the actual system, or live training events, which often have high operating and maintenance costs (Champney et al., 2017; Moroney & Moroney, 2009). When simulators first gained popularity, military simulators operated at 5–20% of the cost of operating the actual aircraft (12% being the median cost; Orlansky & String, 1977). Yoon et al. (2019) conducted an analysis demonstrating that simulator cost has reduced to 3.3–14% of actual aircraft cost with an average of 5.9%. They also demonstrated that, across various aviation industries, an average of 3 simulator hours can replace up to 1 hour of live flight training and that as training time in the simulator grows to 30–50% of total training, costs for training can be reduced by 22.8% to upward of 76%. In medical domains, simulation training can also reduce costs that could be incurred by adverse outcomes such as infections caused by catheter placements, which can result in $82,000 in costs per occurrence, making the $112,000 investment in simulator training a worthy investment as it could ultimately result in over $700,000 in savings annually by reducing catheter placement infections by ten patients in a year (Cohen et al., 2010). Simulation training can also lead to reductions in time to complete training, which can have positive financial impacts as well (Lubner et al., 2019).

Further reductions in costs can be achieved by utilizing personal computer aviation training devices (PCATD), which have demonstrated comparable performance to full flight simulators in instrument procedures, air crew coordination skills, and reductions in training times (Taylor et al., 1999; Jentsch & Bowers, 1998; Taylor et al., 2004). PCATDs also appeal as low-cost tools in education courses and research as these systems can cost as little as $6,000 and can include modern technology such as electronic flight bags, touchscreens, and dual cockpit configurations (Taylor et al., 2004; Jentsch & Bowers, 1998; Pittorie et al., 2019).

Development of specialized simulation-based trainers could cost anywhere between $100,000 and $800,000; however, leveraging game development software can reduce initial cost significantly, with preliminary analysis indicating development costs as low as $45,000 (Rebensky et al., 2020). Further, newer technologies such as COTS game developer software allow lower development and maintenance costs, often with reusable assets for future use cases (McGrath et al., 2017). Emerging areas, at the time of this chapter, are now expanding into VR-based aviation training that could be utilized for cockpit familiarization and checklist training (Sikorski et al., 2017). VR can greatly reduce the cost of the physical components of a simulator as these can be virtually represented and configured to a range of applications.

Many simulators, even in the military sector, now utilize commercial-off-the-shelf (COTS) technologies and tools, as they can reduce both initial cost as well as potential future cost due to increased reusability compared to made-from-scratch simulator technology with proprietary technology (Kumm & Burwell, 2017). The military is also utilizing low-fidelity game technology for training, which has been shown to exhibit significantly lower procurement, maintenance, and training costs per soldier and with increased collective mission training capabilities (Straus et al.,

2019). For example, the costs of using a traditional high-fidelity combat training system can range from $750 to upwards of $7,000 per solider, per day, for air-combat simulators, compared to $200 per solider, per day, utilizing games for training (Straus et al., 2019).

High costs associated with live training, which often utilizes real people, actual systems, and, in some cases, live ammunition for military training, often far surpass the cost of even the highest-fidelity simulators (Straus et al., 2019). For example, simulator-based hospital evacuation training can result in cost reduction over the years, compared to live evacuation training, even with extremely high initial acquisition costs (Farra et al., 2019). Such examples clearly demonstrate the cost-benefit of using simulators.

On the other hand, simulators can come with extreme costs that need to be weighed against the training benefit they provide, including the cost of development, testing, and certification, which can be well into the millions for high-fidelity simulators (Straus et al., 2019). Additionally, different simulated tasks have different transfer of training effectiveness ratios, which could greatly impact the justification for using simulation over live training (Moroney & Moroney, 2009). Special attention should also be given to the adjustment time of the trainee when transitioning from the simulator to the live environment, as the real environment can differ from required inputs and physical demands and does not allow the same features in simulators, such as the ability to pause (Moroney & Moroney, 2009). Another consideration is that certain simulators, specifically dome and motion-based simulators, require large spaces, air conditioning, and specialized maintenance personnel (Moroney & Moroney, 2009; Straus et al., 2019). Further, each simulator comes with its own set of hardware and software maintenance issues, often requiring specialized technicians and additional upkeep costs not required for live training. For example, unique simulators may require specialized technical support when updates are instantiated, and parts becoming obsolete may be an issue. Due to this, many organizations have shifted from utilizing self-developed simulator software to COTS technologies that allow the off-loading of software update and maintenance responsibilities to the manufacturers (Kumm & Burwell, 2017).

In some cases, simulator technology cannot keep up with the rapidly advancing technology and therefore quickly becomes obsolete (Straus et al., 2019). For example, in the four years that the Oculus Rift's first developer's kit was released, it had been superseded by two new versions of the virtual reality headset, a new controller set, and multiple software updates and revised hardware requirements (Kumm & Burwel, 2017). The breakneck pace of technology innovation in recent years had made it difficult for developers of software to keep simulators current and compatible. Because of this, simulators may need "futureproofing," or going above and beyond the needed hardware requirements, in hopes of an extended simulator life cycle that can handle the demands of future technology innovations. However, even with these drawbacks, the cost-benefit advantages of simulation clearly outweigh the disadvantages, as demonstrated by acceptance in various domains and regulatory agencies (Straus et al., 2019; Yoon et al., 2019).

Safety

Simulators also provide a safe environment for evaluation and training. Before the introduction and utilization of simulators, more accidents resulted from the practice of emergency situations than from actual emergencies (Rolfe & Staples, 1986). Simulators provide hazardous situations without worry of risk or threat to humans and the system (Sharma & Scribner, 2017) and can therefore be used to provide training to inexperienced and novice individuals in a system or situation that may pose a serious threat.

Simulators can provide opportunities for the introduction of unexpected, low-probability events or system errors that require the operator to respond quickly during a stressful situation (Neges et al., 2017). Hence, simulators can be used to expose operators to conditions or experiences that they would otherwise be unlikely to experience in actual environments, increasing their ability to respond in such situations if encountered. Example tasks include a hydraulic failure during a maintenance task, emergency resuscitation during surgery, and emergency evacuation of a building (Neges et al., 2017; McRae et al., 2017; Sharma & Scribner, 2017). Simulators can also be used to demonstrate operations without concern for safety requirements or violation of regulations. For example, an instructor can deliberately permit a trainee to make a mistake or error to demonstrate associated results or consequences, which may not be feasible in the actual system due to being operationally illegal (Moroney & Moroney, 1999).

The utilization of simulators reduces hours of operation of the operational system, which in turn reduces wear and tear on the system and its equipment. It also reduces health and environmental impacts associated with pollution and noise (e.g., smoke and fire cannot be safely replicated in a live training environment without potential health threats to the trainee) (Yoon et al., 2019; Rebensky et al., 2020; Straus et al., 2019; Sharma & Scribner, 2017). As the military uses simulation to develop tactics, combat management skills, and to evaluate operational systems, use of simulation to achieve this can also reduce the impact on resources (e.g., damage to land associated with military missile and firing training; Straus et al., 2019).

However, there are disadvantages to training in "safe" and "sanitized" environments. Trainees may not feel willing or able to practice within simulators. When simulators began dominating the aviation training space, simulators were associated with negative effects on morale and retention, as trainees were anxious to be in the operational environment: "I'm here to fly airplanes, not simulators" (Moroney & Moroney, 1999). As simulation begins to enter a new potential space of VR-based aviation training, research is examining the factors that impact one's attitudes toward VR-based training, including perceptions of potential impacts on performance, and subsequent likelihood to engage in VR-based training—which may result in limited acceptance as seen with early use of simulators in aviation due to lack of experience (Fussell, 2020). However, as simulation technology enters the entertainment space, many individuals are obtaining experience with simulated environments prior to simulation-based training. Research shows that this may lead to a potentially more positive acceptance of VR-based, simulation training (Luiser et al., 2019; Fussell, 2020; Fussell & Truong, 2020).

Data

Simulators also allow for data collection to facilitate (a) performance comparisons (e.g., to a performance criteria or performance of other trainees), (b) performance and learning diagnosis (e.g., learning progress and problem areas), and (c) performance evaluation (measures of performance effectiveness; Moroney & Moroney, 1999). Modern simulators can track user decision—actions and performance—to facilitate providing feedback and assessment to the trainee and trainers (Marler et al., 2020). In medical domains, VR-based surgery simulators can be used as performance assessment tools during surgery, providing the assessor with information related to the accuracy of surgical instruments or the number of times a trainee touched unintended areas with surgical instruments (Pfandler et al., 2017). This type of information allows for the unobtrusive and objective measurement of performance and errors that can be utilized to provide feedback. For example, in emergency scenarios, simulators can be used to objectively assess the possibility of human error in off-nominal events, which would otherwise be deduced from subjective expert opinion (Musharraf et al., 2019). In the automotive domain, simulators can collect data to assess event recognition and driving errors utilizing steering wheel position, brake and acceleration, and head-position data collected in VR simulators (Taheri et al., 2017). In the system engineering and evaluation space, specific user-defined values can be entered, and simulators can output the specific impacts to the system such as particular stress levels experienced by aircraft structures (Hughes et al., 2013). The data provided by simulation allows for unique opportunities to remove biases from assessment strategies and objectively collect data previously unobtainable in live training. The benefit of data collection by simulation is equally beneficial for research purposes in which the goal is often to study response to the phenomenon being simulated.

However, there are drawbacks to using simulators for data collection. With any technology, technical issues arise that can lead to data loss, such as power outages or issues with connection to the internet, although this can often be rectified through the use of generators and stable network connections (Mehta et al., 2009). Additionally, the fact that some simulators are connected to the internet can raise ethical concerns related to the privacy of the data, and the degree to which the data being collected is protected (Mugunthan et al., 2020). Finally, the output of the data from the software can sometimes be difficult and time-consuming to handle, leading to errors in interpretation. However, the ubiquity of technology has given rise to programs that facilitate the transformation and interpretation of data. Furthermore, the use of proper documentation and tutorials can help reduce the time it takes to perform data cleaning and setup (Mehta et al., 2009). There can also be a learning curve for users, requiring additional time to get to a point where they are comfortable and understand how to operate the simulator (Colaco et al., 2017; Howells et al., 2009). These hurdles can be mitigated through the use of tutorials, practice sessions, and feedback.

Intervention

Simulators support the delivery of a range of different interventions. Simulation-based training can be enhanced by incorporating automated assessment and feedback to improve skill levels. Simulators that do not provide feedback have been found to yield little to no benefits for users without instructor intervention (Mahmood & Darzi, 2004; Van Heukelom et al., 2010). Simulators have the capability to provide learning interventions that not only alleviate instructor requirements, but also provide consistency in instructional strategy delivery. For example, the Metacognitive Scaffolding Service (MSS) software has been utilized in medical simulators to facilitate self-regulated learning during their training (Berthold et al., 2012). Another advantage is the capacity for real-time performance measurement and feedback, adaptive training, programmed demonstrations and malfunctions, and the immediate placement of position and situation (Waag, 1978; Rebensky et al., 2020). Trainees have the ability to receive immediate feedback, go back and reflect on their performance, and receive individualized remediation based on their performance (McGrath et al., 2017).

Increasing the challenge of the learning task to match the learner's abilities can be accomplished via simulators. In a simulator, this can be achieved without affecting other students who may not be ready to move to that level of challenge, resulting in positive impacts on motivation and performance (Bauer et al., 2012; Orvis et al., 2008). Determining how the human operator can work together with automated systems is a critical issue that can best be duplicated in a simulated environment. Unlike real-world scenarios, simulators can be paused to work through potential solutions in risky situations and assess operator states (Moroney & Moroney, 2009; Rebensky et al., 2020, Endsley, 1995)

Keep in mind that simulation-based training is only as valid and valuable as the instructor utilizing the simulator and the associated assessments and interventions being utilized. The performance evaluations, training content, level of acceptance, and actual usage are subject to opinions and attitudes of the operators, instructors, and management (Moroney & Moroney, 1999; Green, 2000). Modern simulations help ensure effectiveness by allowing for standardized learning content delivery and objective performance assessment (McGrath et al., 2017).

Flexibility and Availability

The flexibility of simulators also adds to their effectiveness. For example, traffic and other distractions can be eliminated from the training task. Simulators also provide standardized scenarios to allow various operators to practice functions, procedures, and dynamics within identical environmental conditions. The scenarios can be repeated until required performance criteria are met, or to promote efficiency or even automated responses (e.g., over learning of a task; Moroney & Moroney, 2009). Further, modern PCATDs can be easily moved and reconfigured into different buildings and different cockpit configurations, as 3D printing of controls and novel touch-based displays allow for inexpensive and rapid alterations to the system (Pittorie et al., 2019).

Simulators often increase the availability of training and operations. Simulators can have 24-hour accessibility, void of dependency on actual system availability and usability requirements, such as suitable weather conditions for flights or test mannequins for surgeries (McGrath et al., 2017; Moroney & Moroney, 2009). Simulators also provide immediate access to operating areas (e.g., a flight simulator allows students to practice multiple landings in succession without time or fuel costs; Moroney & Moroney, 2009). Users can repeat tactics in various scenarios, allowing for a safe environment to apply knowledge to novel situations (Straus et al., 2019). This approach can allow users to achieve mastery of skills as well as refresh skills that might otherwise decay. Additionally, simulators can allow for distance learning during times when travel to training sites is restricted, such as during deployments or unsafe travel conditions from a remote location (McGrath et al., 2017). It should be noted however, this may require trainees to be tech-savvy enough to load, control, and troubleshoot any issues with the simulator.

VR continues to improve the portability and availability of these systems in various domains. For example, cockpit familiarization can move to VR environments as opposed to classrooms or more space- or hardware-intensive simulators (Sikorski et al., 2017). VR simulators can also use game-based software that allows for quick changes of training scenarios and interaction methods such as controller-based or speech-based (Marler et al., 2020; Kamiński et al., 2020). VR headsets are also becoming more compact and portable, allowing for a portable, fully immersive, simulation (Oculus, 2020).

Realism

Many skills require training in realistic environments, which for emergency situations and complex tasks can be difficult to replicate in the real world. However, modern simulation technology now makes the re-creation of these environments possible. For effective training, simulators must replicate accurate cognitive loads, allow realistic communication, and facilitate the practice of requisite motor skills (Bennett et al., 2017). Realism is typically achieved through simulator fidelity, which at a high level can be broken into two main elements: task fidelity and instructional fidelity. Task fidelity refers to the similarity between the operational system or environment and the simulator and its missions, whereas instructional fidelity refers to the system's ability to transfer new skills to the pilot (Macfarlane, 1997). High-task fidelity is usually associated with physical fidelity and therefore high cost. Instructional fidelity is found in lower physical fidelity simulators such as low-cost, part-task, and procedural trainers.

Physical fidelity is attributed to two main factors: high levels of visual detail and motion. Each of these factors has been explored by researchers, including the extent to which the factors affect transfer of training. With the advancement in simulator technology, it is often assumed that increases in fidelity are accompanied by increases in training effectiveness. This is not necessarily the case. Often the addition of motion or haptics can add significant development time and costs, with little impact on training effectiveness (Jerald et al., 2020). Often, simulators with lower levels of physical fidelity (e.g., vibration features instead of a full motion simulator) can achieve equivalent training gains (Jerald et al., 2020; Nahavandi et al., 2019). However, for complex and mission-based training, visual and physical fidelity may be necessary (Straus et al., 2019). For example, to train psychomotor skills such as appropriate maneuvers during turbulence or understanding the appropriate force to apply for a scalpel during surgery (Yoon et al., 2019; McGrath et al., 2017).

If the simulator aims to train a cognitive task, the simulator may only need to replicate the cognitive aspects to effectively train these tasks (Hochmitz & Yuviler-Gavish, 2011). As a result, military branches have shifted to lower-cost, game-based training systems instead of high-fidelity combat trainers, as the game-based systems are effective with much lower operation and maintenance costs (Straus et al., 2019). For example, using low-fidelity virtual environments to train room-clearing tactics can be just as effective as training with high-fidelity environments for some aspects of the task (Champney et al., 2017). Some research has even demonstrated that game-based training that has high cognitive fidelity is more effective at improving aspects of mission performance when compared to high-fidelity combat trainers (Straus et al., 2019). As a result, many domains have shifted to lower fidelity, lower-cost simulators to train procedural skills and processes.

Realistic, high-fidelity simulators also come with additional disadvantages, for example, simulator sickness, which poses problems for some users. Simulator sickness results from discrepancies between visual and vestibular cues encountered in simulators in which the visual system (quality and field of view) is more technologically advanced and therefore disproportionate to the motion system, which can lead to various symptoms such as nausea, headaches, and eye strain (Brooks et al., 2010). Pilots may devise strategies to compensate for, or avoid, simulator sickness, which may in turn have negative impacts on performance when transferred to the operational system. Some participants in driving simulators have to stop participation altogether as simulator sickness can become too intense (Brooks et al., 2010; Fletcher et al., 2017). Furthermore, any discrepancies, deviations, and misrepresented information depicted in terms of the motion, visual, and auditory cueing systems can impact learning and overall performance in the operational environment as trainees learn to respond to inaccurate information (Ray, 2000; Fletcher et al., 2017).

One last consideration with respect to realism is that performance in a simulator may not necessarily reflect an individual's operational performance. Never assume that proficiency in a simulator equates to proficiency in the operational system or environment. Increased performance outcomes may be due to reduced stress levels in simulation compared to an actual situation and operational performance may not be adequately reflected, as emergencies, system malfunctions, and unscheduled events are expected during training. On the other hand, added stress may occur in simulation evaluations due to socio-evaluative stress. Further, trainees might not perform as they would in the real world due to reduced levels of risk. For example, in a simulated driving experiment, some drivers may be more likely to speed in simulators due to the reduced risk perception in this environment (Bella, 2008). However, other research suggests that experienced drivers continue to engage in safe and hazard monitoring behavior regardless of the simulated nature, and driving speeds do not differ in simulated studies (Underwood et al., 2011; Bella, 2005). Other factors that can influence performance in a simulator compared to the real world could include: (a) above-average performance due to preparation before simulation evaluations (Moroney & Moroney, 2009), (b) lack of fatigue, complacency, or boredom compared to the operational environment (Moroney & Moroney, 2009), (c) the learning curve required to become proficient in the simulator (Ronen & Yair, 2013), and (d) shift in responsibilities and expectations when transferring from the simulator to the operational environment (Green, 2000). As such, there will always be concerns when comparing performance in the simulator to performance in the real world.

CONCLUSION

Simulation has many purposes, including the ability to train in safe environments, practice novel situations, evaluate futuristic concepts, research difficult-to-observe constructs, connect people across the globe, and experience new fantasy worlds. Simulation has various applications and is widely accepted in numerous domains, including aviation, military, medical, driving, emergency response, education, entertainment, and maintenance domains. As with any technology, disadvantages exist; however, they are outweighed by the benefits derived from their benefits in terms of cost reduction, safety, data access, intervention capabilities, flexibility, and realism. This chapter aimed to illustrate the advances in simulation technology, range of uses, and the associated beneficial outcomes across a range of domains and uses. As technology continues to advance, use of simulation will continue to increase—slowly overtaking live applications altogether (Yoon et al., 2019). As we continue to find ways to innovate, cost reductions, benefits, and gains from simulation will continue to broaden, leading to increased adoption across domains.

REFERENCES

Abdelgawad, K., Gausemeier, J., Stöcklein, J., Grafe, M., Berssenbrügge, J., & Dumitrescu, R. (2017). A platform with multiple head-mounted displays for advanced training in modern driving schools. *Designs*, *1*(2), 8.

Alabdulkarim, A. A., Ball, P. D., & Tiwari, A. (2013). Applications of simulation in maintenance research. *World Journal of Modelling and Simulation*, *9*(1), 14–37.

Alexander, A. L., Brunyé, T., Sidman, J., & Weil, S. A. (2005). From gaming to training: A review of studies on fidelity, immersion, presence, and buy-in and their effects on transfer in pc-based simulations and games. *DARWARS Training Impact Group*, *5*, 1–14.

Alvarez, S., Page, Y., Sander, U., Fahrenkrog, F., Helmer, T., Jung, O., Hermitte, T., Düering, M., Döering, S., & Op den Camp, O. (2017). Prospective effectiveness assessment of ADAS and Active safety systems via virtual simulation: A review of the current practices. *25th International Technical Conference on the Enhanced Safety of Vehicles (ESV)*.

Andreatta, P. B., Hillard, M., & Krain, L. P. (2010). The impact of stress factors in simulation-based laproscopic training. *Surgery*, *147*(5), 631–639.

Arıcı, F., & Yılmaz, R. M. (2020). The effect of laboratory experiment and interactive simulation use on academic achievement in teaching secondary school force and movement unit. *Elementary Education Online*, *19*(2), 465–476.

Balinova, D. (2020). Military-Entertainment Complex: The myth of the War on Terror. Chicago.

Bauer, K., Brusso, R., & Orvis, K. (2012). Using adaptive difficulty to optimize video-game-based training performance: The moderating variable of personality. *Military Psychology*, *24*(2), 148–165.

Beal, M. D., Kinnear, J., Anderson, C. R., Martin, T. D., Wamboldt, R., & Hooper, L. (2017). The effectiveness of medical simulation in teaching medical students critical care medicine: A systematic review and meta-analysis. *Simulation in Healthcare: The Journal of the Society for Simulation in Healthcare*, *12*(2), 104–116.

Beaubien, J. M., & Baker, D. P. (2004). The use of simulation for training teamwork skills in health care: How low can you go? *BMJ Quality & Safety*, *13*(suppl 1), i51–i56. Chicago.

Bella, F. (2005). Validation of a driving simulator for work zone design. *Transportation Research Record: Journal of the Transportation Research Board*, *1937*(1), 136–144.

Bella, F. (2008). Driving simulator for speed research on two-lane rural roads. *Accident Analysis & Prevention*, *40*(3), 1078–1087.

Bennett, W. B., Rowe, L. J., Craig, S. D., & Poole, H. M. (2017). Training issues for remotely piloted aircraft systems from a human systems integration perspective. In N. J. Cooke, L. J. Rowe, W. Bennett, & D. Q. Joralmon (Eds.), *Remotely Piloted Aircraft Systems A Human Systems Integration Perspective* (pp. 15–39). West Sussex, UK: John Wiley & Sons Ltd.

Berthold, M., Moore, A., Steiner, C. M., Gaffney, C., Dagger, D., Albert, D., ... & Conlan, O. (2012, September). An initial evaluation of metacognitive scaffolding for experiential training simulators. In *European Conference on Technology Enhanced Learning* (pp. 23–36). Springer, Berlin, Heidelberg.

Blodgett, N. P., Blodgett, T., & Kardong-Edgren, S. E. (2018). A proposed model for simulation faculty workload determination. *Clinical Simulation in Nursing*, *18*, 20–27.

Bouyer, G., Chellali, A., & Lécuyer, A. (2017, March). Inducing self-motion sensations in driving simulators using force-feedback and haptic motion. In *2017 IEEE Virtual Reality (VR)* (pp. 84–90). IEEE.

Breuer, J., & Bente, G. (2010). Why so serious? On the relation of serious games and learning. *Journal for Computer Game Culture*, *4*, 7–24.

Brooks, J. O., Goodenough, R. R., Crisler, M. C., Klein, N. D., Alley, R. L., Koon, B. L., Logan, J. W. C., Ogle, J. H., Tyrrell, R. A., & Wills, R. F. (2010). Simulator sickness during driving simulation studies. *Accident Analysis and Prevention*, *42*(3), 788–796. https://doi-org.portal.lib.fit.edu/10.1016/j.aap.2009.04.013

Campos, J. L., Bédard, M., Classen, S., Delparte, J. J., Hebert, D. A., Hyde, N., Law, G., Naglie, G., & Yung, S. (2017). Guiding framework for driver assessment using driving simulators. *Frontiers in Psychology*, *8*, 1428.

Caro, P. W. (1988). Flight training and simulation. In E. L. Wiener & D. C. Nagel (Eds.), *Human Factors in Aviation*. San Diego, CA: Academic Press.

Carroll, M., Rebensky, S., & Sanchez, P. (2020). Examining pilot response to cybersecurity events on the flight deck. *National Training Aircraft Symposium*, March 2–4, Daytona Beach, FL.

Carroll, M., Rebensky, S., Wilt, D., Pittorie, W., Hunt, L., Chaparro, M., & Sanchez, P. (2021). Integrating uncertified information from the electronic flight bag into the aircraft panel: Impacts on pilot response. *International Journal of Human Computer Interaction*, *37*:7, 630–641. https://doi.org/10.1080/10447318.2020.1854001

Cecil, J., Ramanathan, P., Pirela-Cruz, M., & Kumar, M. B. R. (2014). A virtual reality based simulation environment for orthopedic surgery. In *OTM Confederated International Conferences "On the Move to Meaningful Internet Systems"* (pp. 275–285). Springer, Berlin, Heidelberg.

Cegala, D. J., & Broz, S. L. (2003). Provider and patient communication skills training. In T. L. Thompson, A. K. Dorsey, K. Miller, & R. Parrott (Eds.), *Handbook of Health Communication*, 95–119. Mahwah, NJ: Erlbaum.

Champney, R. K., Stanney, K. M., Milham, L., Carroll, M. B., & Cohn, J. V. (2017). An examination of virtual environment training fidelity on training effectiveness. *International Journal of Learning Technologies*, *12*(1), 42–65.

Chowdhury, T. I., Ferdous, S. M. S., & Quarles, J. (2017). Information recall in a virtual reality disability simulation. In *Proceedings of the 23rd ACM Symposium on Virtual Reality Software and Technology* (pp. 1–10).

Cocking, D., & Matthews, S. (2000). Unreal friends. *Ethics and Information Technology*, *2*(4), 223–231.

Cohen, E. R., Feinglass, J., Barsuk, J. H., Barnard, C., O'Donnell, A., McGaghie, W. C., & Wayne, D. B. (2010). Cost savings from reduced catheter-related bloodstream infection after simulation-based education for residents in a medical intensive care unit. *Simulation in Healthcare: The Journal of the Society for Simulation in Healthcare*, *5*(2), 98–102. 10.1097/SIH.0b013e3181bc8304

Colaco, H. B., Hughes, K., Pearse, E., Arnander, M., & Tennent, D. (2017). Construct validity, assessment of the learning curve, and experience of using a low-cost arthroscopic surgical simulator. *Journal of Surgical Education*, *74*(1), 47–54.

Coller, B. D., & Scott, M. J. (2009). Effectiveness of using a video game to teach a course in mechanical engineering. *Computers & Education*, *53*(3), 900–912.

Cook, D. A., Hatala, R., Brydges, R., Zendejas, B., Szostek, J. H., Wang, A. T., ... & Hamstra, S. J. (2011). Technology-enhanced simulation for health professions education: A systematic review and meta-analysis. *Jama*, *306*(9), 978–988.

Cooper, J. B., & Taqueti, V. (2008). A brief history of the development of mannequin simulators for clinical education and training. *Postgraduate Medical Journal*, *84*(997), 563–570. Chicago.

Crea, K. A. (2011). Practice skill development through the use of human patient simulation. *American Journal of Pharmaceutical Education*, *75*(9), 188.

Dean, S, Milham, L., Carroll, M., Schaeffer, R., Alker, M., & Buscemi, T. (2008). Challenges of scenario design in a mixed-reality environment. In *Proceedings of the Interservice/Industry Training, Simulation, and Education Conference (I/ITSEC) Annual Meeting*, Orlando, FL.

De Gloria, A., Bellotti, F., & Berta, R. (2014). Serious Games for education and training. *International Journal of Serious Games*, *1*(1).

Deterding, S., Canossa, A., Harteveld, C., Copper, S., Nacke, L. E., Whitson, J. R. (2015). Gamifying research: Strategies, opportunities, challenges, ethics. In *Proceedings of the 33rd Annual Acm Conference Extended Abstracts on Human Factors in Computing Systems* (pp. 2421–2424).

Djaouti, D., Alvarez, J., Jessel, J. P., & Rampnoux, O. (2011). Origins of serious games. In M. Ma, A. Oikonomou, & L. Jain (Eds.), *Serious Games and Edutainment Applications* (pp. 25–43). London: Springer.

Douglas, S., Hood, C., Overmans, T., & Scheepers, F. (2019). Gaming the system: Building an online management game to spread and gather insights into the dynamics of performance management systems. *Public Management Review*, *21*(10), 1560–1576.

Endsley, M. (1995). Measurement of situation awareness in dynamic systems. *Human Factors*, *37*(1), 65–84.

Farra, S. L., Gneuhs, M., Hodgson, E., Kawosa, B., Miller, E. T., Simon, A., Timm, N., & Hausfeld, J. (2019). Comparative cost of virtual reality training and live exercises for training hospital workers for evacuation. *Computers, Informatics, Nursing*, *37*(9), 446–454. 10.1097/CIN.0000000000000540

Federal Aviation Administration. (2020). FAR 121 Subpart N—Training program. Federal Aviation Administration.

Fletcher, J. D., Belanich, J., Moses, F., Fehr, A., & Moss, J. (2017). Effectiveness of augmented reality & augmented virtuality. In *MODSIM Modeling & Simulation of Systems and Applications*. World conference.

Fraune, M. R., Khalaf, A. S., Zemedie, M., Pianpak, P., NaminiMianji, Z., Alharthi, S. A., ... & Toups, Z. O. (2021). Developing future wearable interfaces for human-drone teams through a virtual drone search game. *International Journal of Human-Computer Studies, 147*, 102573.

Fussell, S. G. (2020). Determinants of aviation students' intentions to use virtual reality for flight training. Dissertation.

Fussell, S. G., & Truong, D. (2020). Preliminary results of a study investigating aviation students' intentions to use virtual reality for flight training. *International Journal of Aviation, Aeronautics, and Aerospace*, *7*(3), 2.

Gore, B. F. (2011). Man-machine integration design and analysis system (MIDAS) v5: Augmentations, motivations, and directions for aeronautics applications. In P. C. Cacciabu, M. Hjalmdahl, A. Luedtke, & C. Riccioli (Eds.), *Human Modelling in Assisted Transportation* (pp. 43–54). Heidelberg, Germany: Springer.

Granic, I., Lobel, A., & Engels, R. C. (2014). The benefits of playing video games. *American Psychologist*, *69*(1), 66. Chicago.

Green, M. F. (2000). Aviation instruction through flight simulation: Enhancing pilots' decision-making skills. *Flight Simulation—The Next Decade: Proceedings of the Royal Aeronautical Society*, London, May 10–12.

Hall, A. B., Riojas, R., & Sharon, D. (2014). Comparison of self-efficacy and its improvements after artificial simulator or live animal model emergency procedure training. *Military Medicine*, *179*(3), 320–323.

Hancock, P. A., Vincenzi, D. A., Wise, J. A., & Mouloua, M. (Eds.). (2008). *Human Factors in Simulation and Training*. Boca Raton, FL: CRC Press.

Hardoff, D., & Schonmann, S. (2001). Training physicians in communication skills with adolescents using teenage actors as simulated patients. *Medical Education*, *35*(3), 206–210.

Himona, S. L., Stavrakis, E., Loizides, A., Savva, A., & Chrysanthou, Y. (2011). SIMPOL VR – A virtual reality law enforcement training simulator. *MCIS 2011 Proceedings*.

Hochmitz, I., & Yuviler-Gavish, N. (2011). Physical fidelity versus cognitive fidelity training in procedural skills acquisition. *Human Factors*, *53*(5), 489–501.

Hoffman, R. R., Ward, P., Feltovich, P. J., DiBello, L., Fiore, S. M., & Andrews, D. H. (2014). *Expertise: Research and Applications: Accelerated Expertise: Training for High Proficiency in a Complex World*. Psychology Press.

Honey, M. A., & Hilton, M. L. (2011). *Learning Science Through Computer Games*. Washington, DC: National Academies Press.

Hong, T., Taylor-Lange, S. C., D'Oca, S., Yan, D., & Corgnati, S. P. (2016). Advances in research and applications of energy-related occupant behavior in buildings. *Energy and Buildings*, *116*, 694–702.

Howells, N. R., Auplish, S., Hand, G. C., Gill, H. S., Carr, A. J., & Rees, J. L. (2009). Retention of arthroscopic shoulder skills learned with use of a simulator: Demonstration of a learning curve and loss of performance level after a time delay. *JBJS*, *91*(5), 1207–1213.

Hromek, R., & Roffey, S. (2009). Promoting social and emotional learning with games: "It's fun and we learn things". *Simulation and Gaming*, *40*(5), 626–644.

Hughes, K., Vignjevic, R., Campbell, J., De Vuyst, T., Djordjevic, N., & Papagiannis, L. (2013). From aerospace to offshore: Bridging the numerical simulation gaps-simulation advancements for fluid structure interaction problems. *International Journal of Impact Engineering*, *61*, 48–63.

Kamiński, J., Jurczak, J., & Jakubczyk, R. (2020). Simulator of police actions in crisis situations as an application of an intelligent decision support system in the process of improving Polish police actions. *Internal Security*, Sp Issue, 137–145.

Jentsch, F., & Bowers, C. A. (1998). Evidence for the validity of PC-based simulation in studying aircrew coordination. *International Journal of Aviation Psychology*, *8*(3), 243–260.

Jerald, J., Haskins, J., Eadara, S., Gainer, S., Zhu, B., & Huse, W. (2020). Utilizing physical props to simulate equipment in immersive environments. *I/ITSEC 2020.*

Jones, E. R., Hennessy, R. T., & Deutsch, S. (1985). *Human Factors Aspects of Simulation.* Washington, DC: National Academy Press.

Kahlbaugh, P. E., Sperandio, A. J., Carlson, A. L., & Hauselt, J. (2011). Effects of playing Wii on well-being in the elderly: Physical activity, loneliness, and mood. *Activities, Adaptation & Aging*, *35*(4), 331–344.

Kanki, B. G. (2019). Communication and crew resource management. In B. G. Kanki, J. Anca, & T. R. Chidester (Eds.), *Crew Resource Management* (pp. 139–184). Elsevier Science & Technology.

Kim, Y., Kim, H., & Kim, Y. O. (2017). Virtual reality and augmented reality in plastic surgery: A review. *Archives of Plastic Surgery*, *44*(3), 179.

Klimmt, C. (2003). Dimensions and deteminants of the enjoyment of playing digital games: A three-level model. In M. Copier & J. Raessens (Eds.), *Level Up: Digital Games Research Conference* (pp. 246–257). Utrecht, The Netherlands: Faculty of Arts, Utrecht University.

Klimmt, C., Hartmann, T., & Frey, A. (2007). Effectance and control as determinants of video game enjoyment. *Cyberpsychology & Behavior*, *10*(6), 845–848.

Kronqvist, A., Jokinen, J., & Rousi, R. (2016). Evaluating the authenticity of virtual environments: Comparison of three devices. *Advances in Human-Computer Interaction*, *3*, 1–14.

Kumm, C., & Burwell, J. (2017). "Systems of systems" approach for the development of next generation modular simulation-based training systems. *MODSIM World*, *7*, 1–17.

Landon-Hays, M., Peterson-Ahmad, M. B., & Frazier, A. D. (2020). Learning to teach: how a simulated learning environment can connect theory to practice in general and special education educator preparation programs. *Education Sciences*, *10*(7), 184.

Lee, L. H., Chew, E. P., Frazier, P. I., Jia, Q., & Chen, C. (2013). Advances in simulation optimization and its applications. *IIE Transactions*, *45*(7), 683–684. https://doi.org/10.1080/0740817X.2013.778709

Lindgren, R., Tscholl, M., Wang, S., & Johnson, E. (2016). Enhancing learning and engagement through embodied interaction within a mixed reality simulation. *Computers & Education*, *95*, 174–187. https://doi.org/10.1016/j.compedu.2016.01.001

Lindsey, S., Ganey, H. N., & Carroll, M. (2019). Designing military cockpits to support a broad range of personnel body sizes. *20th International Symposium on Aviation Psychology*, 145–150.

Liu, Y., Wang, T., Zhang, H., Cheutet, V., & Shen, G. (2019). The design and simulation of an autonomous system for aircraft maintenance scheduling. *Computers & Industrial Engineering*, *137*, 106041.

Lubner, M., Dattel, A. R., Allen, E., Henneberry, D., & DeVivo, S. (2019). Six-year follow-up of intensive, simulator-based pilot training. *20th International Symposium on Aviation Psychology*, 450–455.

Luisier, J., Yooyen, S., & Deebhijarn, S. (2019). Perceptions of Thai aviation students on consumer grade VR flight experiences. *Multidisciplinary Digital Publishing Institute Proceedings*, *39*(1), 8.

Lukosch, H. K., Bekebrede, G., Kurapati, S., & Lukosch, S. G. (2018). A scientific foundation of simulation games for the analysis and design of complex systems. *Simulation & Gaming*, *49*(3), 279–314.

Mahmood, T., & Darzi, A. (2004). The learning curve for a colonoscopy simulator in the absence of any feedback: no feedback, no learning. *Surgical Endoscopy and Other Interventional Techniques*, *18*, 1224–1230.

Mairaj, A., Baba, A. I., & Javaid, A. Y. (2019). Application specific drone simulators: Recent advances and challenges. *Simulation Modelling Practice and Theory*, *94*, 100–117.

Marler, T., Straus, S. G., Mizel, M. L., Hollywood, J. S., Harrison, B., Yeung, D., Klima, K., Lewis, M. W., Rizzo, S., Hartholt, A., & Swain, C. (2020). Effective game-based training for police officer decision-making: Linking missions, skills, and virtual content. *I/ITSEC 2020.*

Macfarlane, R. (1997). Simulation as an instructional procedure. In G. J. F. Hunt (Ed.), *Designing Instruction for Human Factors Training in Aviation*. Aldershot, UK: Avebury Aviation.

McGrath, J. L., Taekman, J. M., Dev, P., Danforth, D. R., Mohan, D., Kman, N., Crichlow, A., & Bond, W. F. (2017). Using virtual reality simulation to assess competence for emergency medicine learners. *Academic Emergency Medicine*, *25*(2), 186–195. https://doi.org/10.1111/acem.13308

McRae, M. E., Chan, A., Hulett, R., Lee, A. J., & Coleman, B. (2017). The effectiveness of and satisfaction with high-fidelity simulation to teach cardiac surgical resuscitation skills to nurses. *Intensive and Critical Care Nursing*, *40*, 64–69.

Mehta, S., Ullah, N., Kabir, M. H., Sultana, M. N., & Kwak, K. S. (2009). A case study of networks simulation tools for wireless networks. In *2009 Third Asia International Conference on Modelling & Simulation* (pp. 661–666). IEEE.

Meuleners, L., & Fraser, M. (2015). A validation study of driving errors using a driving simulator. *Transportation Research Part F: Traffic Psychology and Behaviour*, *29*, 14–21.

Morgan, P. J., Cleave-Hogg, D., McIlroy, J., & Devitt, J. H. (2002). Simulation technology: A comparison of experiential and visual learning for undergraduate medical students. Anesthesiology. *The Journal of the American Society of Anesthesiologists*, *96*(1), 10–16.

Moroney, W. F., & Moroney, B. W. (1999). Flight simulation. In D. J. Garland J. A. Wise, & V. D. Hopkin (Eds.), *Handbook of Aviation Human Factors* (pp. 355–388). Mahwah, NJ: Lawrence Erlbaum Associates.

Moroney, W. F., & Moroney, B. W. (2009). Flight simulation. In D. J. Garland, J. A. Wise, & V. D. Hopkin (Eds.), *Handbook of Aviation Human Factors*. Boca Raton, FL: CRC Press.

Mpu, Y., & Adu, E. O. (2020). Collaborative Virtual Learning in Education in STEM Education. *Management*, *8*(4), 315–324.

Mugunthan, V., Peraire-Bueno, A., & Kagal, L. (2020, October). Privacyfl: A simulator for privacy-preserving and secure federated learning. In *Proceedings of the 29th ACM International Conference on Information & Knowledge Management* (pp. 3085–3092).

Musharraf, M., Moyle, A., Khan, F., & Veitch, B. (2019). Using simulator data to facilitate human reliability analysis. *Journal of Offshore Mechanics and Arctic Engineering*, *141*(2), 021607.

Nahavandi, S., Wei, L., Mullins, J., Fielding, M., Deshpande, S., Watson, M., Korany, S., Nahavandi, D., Hettiarachchi, I., Najdovski, Z., Jones, R., Mullins, A., & Carter, A. (2019). Haptically-enabled VR-based immersive fire fighting training simulator. *Intelligent Computing: Proceeding of the 2019 Computing Conference*, *1*, 11–21.

Neges, M., Adwernat, S., & Abramovici, M. (2017). Augmented virtuality for maintenance training simulation under various stress conditions. *6th Annual Conference on Through-life Engineering Services*, 7–8.

Nestel, D., Groom, J., Eikeland-Husebø, S., & O'Donnell, J. M. (2011). Simulation for learning and teaching procedural skills: The state of the science. *Simulation in Healthcare*, *6*(7) Supplement, S10–S13.

Oculus. Quest 2 details. https://www.oculus.com/quest-2/?locale=en_US

Orlansky, J., & String, J. (1977). Cost-effectiveness of flight simulator for military training (Rep. No. IDA NO. HQ 77-19470). Arlington, VA: Institute for Defense Analysis.

Orvis, K., Horn, D., & Belanich, J. (2008). The roles of task difficulty and prior videogame experience on performance and motivation in instructional videogames. *Computers in Human Behavior*, *24*, 2415–2433.

Patterson, M. D., Blike, G. T., & Nadkarni, V. M. (2008). In situ simulation: Challenges and results. Advances in patient safety: New directions and alternative approaches (Vol. 3: performance and tools). Chicago.

Pellas, N., Kazanidis, I., Konstantinou, N., & Georgiou, G. (2017). Exploring the educational potential of three-dimensional multi-user virtual worlds for STEM education: A mixed-method systematic literature review. *Education and Information Technologies*, *22*(5), 2235–2279.

Pfandler, M., Lazarovici, M., Stefan, P., Wucherer, P., & Weigl, M. (2017). Virtual reality-based simulators for spine surgery: A systematic review. *The Spine Journal*, *17*(9), 1352–1363. https://doi.org/10.1016/j.spinee.2017.05.016

Pinheiro, A., Fernandes, P., Maia, A., Cruz, G., Pedrosa, D., Fonseca, B., Paredes, H., Martins, P., Morgado, L., & Rafael, J. (2012). Development of a mechanical maintenance training simulator in opensimulator for F-16 aircraft engines. *Procedia Computer Science*, *15*, 248–255.

Pittorie, W., Lindsey, S., Wilt, D. F., & Carroll, M. (2019). Low-cost simulator for flight crew human-factors studies. In *Proceedings of the AIAA Aviation and Aeronautics Forum and Exposition*, June 17–21, Dallas, Texas.

Quellmalz, E. S., Silberglitt, M. D., Buckley, B. C., Loveland, M. T., & Brenner, D. G. (2020). Simulations for supporting and assessing science literacy. In I. Management Association (Ed.), *Learning and Performance Assessment: Concepts, Methodologies, Tools, and Applications* (pp. 760–799). Hershey, PA: IGI Global.

Radianti, J., Majchrzak, T. A., Fromm, J., & Wohlgenannt, I. (2020). A systematic review of immersive virtual reality applications for higher education: Design elements, lessons learned, and research agenda. *Computers & Education*, *147*, 103778.

Ray, P. A. (2000). Is today's flight simulator prepared for tomorrow's requirements? Flight simulation—The next decade. In *Proceedings of the Royal Aeronautical Society*, May 10–12.

Rebensky, S., Carroll, M., Bennett, W., & Hu, X. (2020). Collaborative development of a synthetic task environment by academia and military. *Interservice/Industry Training, Simulation, and Education Conference (I/ITSEC) 2020.*

Reweti, S., Gilbey, A., & Jerffery, L. (2017). Efficacy of low-cost PC-based aviation training devices. *Journal of Information Technology Education: Research*, *16*, 127–142.

Rutten, N., van Joolingen, W. R., & van der Veen, J. T. (2012). *The learning effects of computer simulations in science education Computers & Education*, *58*(1), 136–153.

Ritterfeld, U., Cody, M., & Vorderer, P. (2009). *Serious Games: Mechanisms and Effects.* London, UK: Routledge.

Rolfe, J. M., & Staples, K. J. (1986). *Flight Simulation.* New York: Cambridge University Press.

Ronen, A., & Yair, N. (2013). The adaptation period to a driving simulator. *Transportation Research Part F: Traffic Psychology and Behaviour*, *18*, 94–106.

Rui, L. (2020). *Behind the Popularity: Simulation Game in China*. [Published undergraduate honors thesis, UNC Digital Library]. https://doi.org/10.17615/h3a0-k802

Sena, P., Attianese, P., Carbone, F., Pellegrino, A., Pinto, A., & Villecco, F. (2012). A fuzzy model to interpret data of drive performances from patients with sleep deprivation. Chicago.

Sharma, S., & Scribner, D. (2017). Megacity: A collaborative virtual reality environment for emergency response, training, and decision making. *IS&T International Symposium on Electronic Imaging* (pp. 70–77).

Sikorski, E., Palla, A., & Brent, L. (2017). Developing an immersive virtual reality aircrew training capability. *2017 IITSEC.*

Skinner, H., Possignolo, R. T., Wang, S. H., & Renau, J. (2020, August). LiveSim: A fast hot reload simulator for HDLs. In *2020 IEEE International Symposium on Performance Analysis of Systems and Software (ISPASS)* (pp. 126–135). IEEE.

Smith, R. (2010). The long history of gaming in military training. *Simulation & Gaming, 41*(1), 6–19. Chicago.

St. Julien, T., & Shaw, C. D. (2003). Firefighter command training virtual environment. *TAPIA '03: Proceedings of the 2003 conference on Diversity in Computing.* https://doi-org.portal.lib.fit.edu/10.1145/948542.948549

Straus, S. G., Lewis, M. W., Connor, K., Eden, R., Boyer, M. E., Marler, T., ... Smigowski, H. (2019). *Collective Simulation-Based Training in the U.S. Army.* Santa Monica, CA: RAND Corporation.

Stroosma, O., Van Paassen, M. M., & Mulder, M. (2003, August). Using the SIMONA research simulator for human-machine interaction research. In *AIAA Modeling and Simulation Technologies Conference and Exhibit* (p. 5525).

Taber, N. (2008). Emergency response: Elearning for paramedics and firefighters. *Simulation and Gaming, 39*(4), 515–527.

Taheri, S. M., Matsushita, K., Sasaki, M. (2017). Development of a driving simulator with analyzing driver's characteristics based on a virtual reality head mounted display. *Journal of Transportation Technologies, 7*(3). 10.4236/jtts.2017.73023

Tan, S. S. Y., & Sarker, S. K. (2011). Simulation in surgery: A review. *Scottish Medical Journal, 56*(2), 104–109.

Tang, J., Lau, W., Chan, K., & To, K. (2014). AR interior designer: Automatic furniture arrangement using spatial and functional relationships. 2014 *International Conference on Virtual Systems & Multimedia (VSMM).* 10.1109/VSMM.2014.7136652

Taylor, H. L., Lintern, G., Hulin, C. L., Talleur, D. A., Emanuel, T. W., & Phillips, S. I. (1999). Transfer of training effectiveness of a personal computer aviation training device. *International Journal of Aviation Psychology, 9*(4), 319–335.

Taylor, H. L., Talleur, D. A., Rantanen, E. M., & Emanuel, T. W. (2004). The effectiveness of a personal computer aviation training device, a flight training device, and an airplane in conducting instrument proficiency checks. *Aviation Human Factors Division Institute of Aviation.*

Underwood, G., Crundall, D., & Chapman, P (2011). Driving simulator validation with hazard perception. *Transportation Research Part F: Traffic Psychology and Behaviour, 14*(6), 435–446.

Van Heukelom, J. N., Begaz, T., & Treat, R. (2010). Comparison of postsimulation debriefing versus in-simulation debriefing in medical simulation. *Simulation in Healthcare, 5*(2), 91–97.

van de Ven, J., van Baaren, G. J., Fransen, A. F., van Runnard Heimel, P. J., Mol, B. W., & Oei, S. G. (2017). Cost-effectiveness of simulation-based team training in obstetric emergencies (TOSTI study). *European Journal of Obstetrics & Gynecology and Reproductive Biology, 216*, 130–137.

Vlachopoulos, D., & Makri, A. (2017). The effect of games and simulations on higher education: A systematic literature review. *International Journal of Educational Technology in Higher Education*, *14*(1), 1–33.

Vlakveld, W., van Nes, N., de Bruin, J., Vissers, L., & van der Kroft, M. (2018). Situation awareness increases when drivers have more time to take over the wheel in a Level 3 automated car: A simulator study. *Transportation Research Part F: Traffic Psychology and Behaviour*, *58*, 917–929.

Waag, W. L. (1978). Recent studies of simulation training effectiveness. *Proceedings of the Society of Automotive Engineers: Town and County*, San Diego, November 27–30.

Wang, H., & Zhai, Z. (2016). Advances in building simulation and computational techniques: A review between 1987 and 2014. *Energy and Buildings*, *128*(15), 319–335.

Warren, J. N., Luctkar-Flude, M., Godfrey, C., & Lukewich, J. (2016). A systematic review of the effectiveness of simulation-based education on satisfaction and learning outcomes in nurse practitioner programs. *Nurse Education Today*, *46*, 99–108.

Webster, M., Cameron, N., Fisher, M., & Jump, M. (2014). Generating certification evidence for autonomous unmanned aircraft using model checking and simulation. *Journal of Aerospace Information Systems*, *11*(5), 258–278.

Winther, F., Ravindran, L., Svendsen, K. P., & Feuchtner, T. (2020). Design and evaluation of a VR training simulation for pump maintenance based on a use case at Grundfos. 2020 *IEEE Conference on Virtual Reality and 3D User Interfaces.*

Woda, A., Hansen, J., Paquette, M., & Topp, R. (2017). The impact of simulation sequencing on perceived clinical decision making. *Nurse Education in Practice*, *26*, 33–38.

Wynne, R. A., Beanland, V., & Salmon, P. M. (2019). Systematic review of driving simulation validation studies. *Safety Science*, *117*, 138–151.

Yoon, S., Park, T., Lee, J., & Kim, J. (2019). A study on transfer effectiveness and appropriate training hours. *ITEC 2019.*

Yuksek, B., Vuruskan, A., Ozdemir, U., Yukselen, M. A., & Inalhan, G. (2016). Transition flight modeling of a fixed-wing VTOL UAV. *Journal of Intelligent & Robotic Systems*, *84*, 83–105.

Zhang, X., Liu, X., Yuan, S. M., & Lin, S. F. (2017). Eye tracking based control system for natural human-computer interaction. *Computational Intelligence and Neuroscience*, *2017*, 1–9.

3 Simulation Fidelity

Dahai Liu, Jiahao Yu, Nikolas D. Macchiarella, and Dennis A. Vincenzi

CONTENTS

INTRODUCTION

With the increasing demand on training to function in highly complex situations, researchers and practitioners strive to build high-fidelity simulation devices as similar to real situations as possible. Each year, technological advances bring simulation closer and closer to duplicating precise, authentic environments. Simulation-based training has many benefits such as improving speed and technical skills (Bur et al., 2017), boosting knowledge, critical thinking, and decision-making (Akalin & Sahin, 2020), enhancing team communication (Chamberland et al., 2018), and increasing students' self-confidence (Haddeland et al., 2019). Unfortunately, a high financial cost is associated with these highly sophisticated devices.

Simulation quality and human capabilities are critical factors that determine training effectiveness and efficiency. The key issue related to simulation quality is the "degree to which the training devices must duplicate the actual equipment" or environment. This degree of similarity is called simulation fidelity (Allen, 1986). The issue of fidelity must be addressed not only because it is quite possibly the most important factor in the assessment of simulation quality, and a key factor of

DOI: 10.1201/9781003401360-3

simulation training transfer, but is also a critical factor related to cost-effective simulation device design. Often simulators are not utilized because the equipment is too costly to purchase (Garrison, 1985). It is widely accepted that training devices with excessive levels of fidelity may not be cost-effective (Allen, 1986; Fortin, 1989; Roza, 2000; Scott & Gartner, 2019). Medium-fidelity simulation had a significantly higher satisfaction rate than high-fidelity simulation in nursing students, and it was proved to be more cost-effective for the acquisition of basic skills at a lower cost (Alconero-Camarero et al., 2021).

Existing research in the area of flight simulation training, aviation psychology, military aviation, and many other domains has contributed greatly to the understanding of simulation. Despite this work, many questions remain unanswered. Research indicates that simulation is a valuable tool for training (Akalin & Sahin, 2020; Bur et al., 2017; Chamberland et al., 2018; Haddeland et al., 2019; Hays et al., 1992). What is not fully understood is how to make simulation more efficient, i.e., how to determine what level of fidelity is sufficient for effective transfer of training.

Despite this, fidelity, or "level of detail," continues to be a major issue in simulation development (Hughes & Rolek, 2003). Due to the complex nature of simulation tasks, large numbers of objects and attributes, and random human behaviors involved, quantification of simulation fidelity becomes the most challenging aspect of fidelity measurement. Schricker et al. (2001) concluded that the main issues regarding fidelity and how it is addressed in the literature are as follows:

> (1) No detailed, agreed-upon definition; (2) Rampant subjectivity; (3) No method of quantifying the assignment of fidelity; and (4) No detailed example of a referent.

Other major issues involve the accepted methods by which fidelity is measured. There are two major methods described in the literature for fidelity measurement. The first is through mathematical measurement that calculates the number of identical elements shared between the real world and the simulation; the greater the number of shared identical elements, the higher the simulation fidelity. This is referred to as the objective method (Gross & Freeman, 1997; Schricker et al., 2001). A second method to measure fidelity is through a trainees' performance matrix. By assessing a human's performance and then comparing it to real-world performance to measure the transfer of training, fidelity of a simulation can be measured indirectly (Parrish et al., 1983; Ferguson et al., 1985; Nemire et al., 1994; Field et al., 2002; Mania et al., 2003). We will address these methods in detail in a later section. First, a clear definition is needed to understand what is meant by "fidelity."

DEFINITION OF FIDELITY

Simulation fidelity is an umbrella term defined as the extent to which the simulation replicates the actual environment (Alessi, 1988; Gross et al., 1999). In aviation, simulation fidelity refers to the extent to which a flight training device looks, sounds, responds, and maneuvers like a real aircraft. Many simulation professionals attempt to define fidelity comprehensively, whereas others argue that fidelity is a far too nebulous idea than can even be defined. This implies that efforts of defining fidelity are currently unsuccessful (Schricker et al., 2001). Rehmann et al. (1995) investigated over 30 years of research on fidelity and found that there was no agreed-upon single definition; in fact, at least 22 different definitions can be drawn from the literature. For example, fidelity can be defined as simply as "how closely a simulation imitates reality" (Alessi, 1988), or "the levels of physical and visual similarity with real work settings" (Hontvedt & Øvergård, 2020), or more specifically, on different fidelity dimensions, including but not limited to equipment fidelity, environmental fidelity, psychological and cognitive fidelity, task fidelity, physical fidelity, and functional fidelity. Furthermore, Rehmann et al. (1995) found that none of these terms or definitions is applicable to overall aircraft simulation in general. This lack of a well-defined and widely accepted fidelity concept causes miscommunications between researchers and inconsistencies in their research (Roza, 2000).

Based on the previous research and a definition by Fidelity-ISG, Roza (2000) developed the following theorems regarding simulation fidelity:

1. Fidelity models are multidimensional; they involve and can be quantified using a variety of factors.
2. Fidelity is application-independent; it is an intrinsic and inherent property of a simulation model.
3. Fidelity must be quantified and qualified with respect to a referent; this means that metrics (i.e., size, weight, shape) should exist on how to determine if a simulation resembles its referent.
4. Fidelity quantification has a level of uncertainty.
5. Fidelity comparison should be based on a common referent in order to make sense. For example, comparing fidelity levels of an aircraft simulator should be drawn from the same or similar aircraft.

Seven descriptive concepts were defined to further understand and quantify fidelity: detail, resolution, error, precision, sensitivity, timing, and capacity. Specific metrics or measurements can be defined in depth for each of these concept factors.

A simulation referent is important not only to define fidelity, but also for its measurement. It can be simply described "in terms of the extent to which a representation reproduces the attributes and behaviors of a referent" (Hughes & Rolek, 2003). A referent is "an entity or collection of entities and/or conditions—together with their attributes and behaviors—present within a given operational domain" (Hughes & Rolek, 2003).

For example, in the small aircraft aviation industry, one is particularly interested in studying the single pilot performance in a Cessna 172. Thus, the cockpit of the Cessna 172 is the *reality*. The simulated cockpit would be the referent (display, radio, controls, pedals, chair, etc.); the models would be the computer-simulated models that produce this referent such as Microsoft Flight Simulator.

Roza (2000) defined a referent as "a formal specification of all knowledge about reality plus indicators to determine the uncertainty levels and quality of this knowledge to judge the confidence level of this referent data." In other words, a referent is the abstract model of the reality that is relevant to simulated tasks, and serves as a basis for measuring the simulated environment and tasks. According to Roza (2000), a referent structure consists of the following elements:

1. A referent identification section.
2. A referent applicability section.
3. A referent developer and validation agent.
4. A referent knowledge sources section.
5. A real-world structural properties section.
6. A real-world parametric and behavioral data section.

Roza (2000) claims that elements 2, 3, and 4 contain the indicators needed to assess the fidelity of the simulation reference, and elements 5 and 6 can be used to measure the fidelity.

By using the concept of a referent, fidelity measurement can be simplified. As Schricker et al. (2001) pointed out, if one tried to consider fidelity issues on a real-world system, it would become far too intricate. Simulations are developed to represent a certain object or group of objects in a certain domain, which can be regarded as the simulated models of a certain referent of the reality.

Fidelity can be further broken down into sub-definitions that describe detailed elements that are categorized based on the different aspects of the simulated tasks or environment. The categorization among researchers varies a great deal and no consensus has been reached. The different fidelity element categorizations depended largely on the different fidelity experiments and different simulated tasks. For example, in Zhang's (1993) study, simulation fidelity is broken down into six elements: hardware fidelity, software fidelity, fidelity for a whole tested system, fidelity of the pilot's subjective impression, simulation mission (task) fidelity, and simulation experience fidelity. Hontvedt and Øvergård (2020) introduced a framework with three central approaches: physical and functional accuracy which was conceptualized as technical fidelity, problem-solving strategies, mental models and feelings which were conceptualized as psychological fidelity, and precise coordination and collaborative patterns within a team which was conceptualized as interactional fidelity.

Hays and Singer (1989) identified the following three major types of variables that were believed to interact with fidelity:

1. Task-related variables, including task domain, task type, task difficulty, task frequency, task criticality, task learning difficulty, task practice requirement, and task skills, abilities, and knowledge.
2. Training environment and personal variables (e.g., purpose of training, instructional principles, and student population).
3. Device utilization variables that are the least understood and the most "potent" in determining training device effectiveness.

TABLE 3.1
Fidelity Definitions

Word	References	Definition
Simulation fidelity	Gross et al. (1999); Alessi (1988)	Degree to which device can replicate actual environment, or how "real" the simulation appears and Feds
Physical fidelity	Allen (1986)	Degree to which device looks, sounds, and feels like actual environment
Visual–audio fidelity	Rinalducci (1996)	Replication of visual and auditory stimulus
Equipment fidelity	Zhang (1993)	Replication of actual equipment hardware and software
Motion fidelity	Kaiser and Schroeder (2003)	Replication of motion cues felt in actual environment
Psychological–cognitive fidelity	Kaiser and Schroeder (2003)	Degree to which device replicates psychological and cognitive factors (i.e., communication, situational awareness)
Task fidelity	Zhang (1993); Roza (2000); Hughes and Rolek (2003)	Replication of tasks and maneuvers executed by user
Functional fidelity	Allen (1986)	How device functions, works, and provides actual stimuli as actual environment
Interactional fidelity	Hontvedt and Øvergård (2020)	The accuracy and relevance of participant collaboration and enactment of work tasks in simulator training

Table 3.1 lists a number of different aspects of fidelity and how they relate to each other as well as how researchers have attempted to describe each aspect. These definitions have been compiled from research that has focused on structuring and defining the various components of fidelity. Definitions of fidelity mainly fall within two categories: those that describe the physical experience and those that describe the psychological or cognitive experience. These two categories are briefly described in Table 3.1, and will be explored further.

PHYSICAL FIDELITY

The most commonly discussed fidelity categorization is physical fidelity (Allen, 1986; Hays & Singer, 1989; Andrews et al. 1995). Not to be confused with the broader term of simulation fidelity, this term specifically deals with the physical properties of the simulation experience. To consider high physical fidelity, the simulator must have high visual–audio fidelity, or the look, sound, feel, and, in some cases, smell of the real aircraft (Allen, 1986). Physical fidelity encompasses other definitions of fidelity such as visual–audio fidelity, equipment fidelity, and motion fidelity. Hontvedt and

Øvergård (2020) named it technical fidelity after adding the functional accuracy, to describe "the degree of accuracy to which the technical and environmental cues are recreated by the simulator technology."

VISUAL–AUDIO FIDELITY

Visual–audio fidelity is the most frequently studied aspect of fidelity in the available literature (Rinalducci, 1996). It can be thought of as the level of visual and aural detail that the simulator displays. For example, a visually simulated airport can include several elements or artifacts that could be found when directly viewing a real-world airport. The runways, lights, hangers, control towers, ground vehicles, natural surroundings, and other airplanes could be included in the simulation. Also, communication between the control tower and pilot can also be simulated.

A low-fidelity simulation would include only a few of these artifacts mentioned. For example, the low fidelity simulator may contain simply a runway and landing threshold lights, and no aural detail at all. As the simulation includes more artifacts, the level of fidelity will increase.

Technology has advanced visual fidelity, specifically, to a very high level. Although a perfect copy of the real world is beyond reach at this point, advances in satellite and aerial mapping for the purpose of simulation are producing increasingly accurate representations of the real environment and geography. For example, it is now possible to simulate and practice military operations in a computer-generated city that displays the buildings, streets, obstacles, and other physical features that may be encountered in the actual battle environment. This mitigates the chance of surprises and can greatly aid in the planning and rehearsal of military operations without risking lives.

Visual–audio fidelity in flight simulation can be decomposed into two basic parts: the "what" and the "where." The "what" refers to the pilot's central vision, which is referred to as the foveal and parafoveal regions. This area includes the items which the pilot is directly viewing (e.g., instruments and the windscreen). The "where" refers to visual stimulus that is in the pilot's peripheral vision. The stimuli in the peripheral vision provide the pilot with a sense of speed, motion, situational awareness, and attitude of the aircraft. Both regions of vision have been shown to be important to the control of an aircraft (Kaiser & Schroeder, 2003). Unfortunately, computer-generated graphics often lack the power to generate visual stimulus in both regions at the same time to the same degree. Thus, a trade-off exists, and displays must decide what stimulus to show at what time.

As technology advances and the power of image generators improve, this confound may become a nonissue. Additionally, instrument flight requires pilots to engage and view only their instruments. Thus, there is no need for "outside" visual stimuli. In this situation, an outside visual scene may even hinder the transfer of training by distracting the pilot.

Bradley and Abelson (1995) investigated visual fidelity issues for desktop flight simulators. Although improvements in computing power and software advances have enabled high-animation in visual display, the most significant factor limiting visual fidelity (quality of performance) is still the speed of frame refresh, or "frame rate." In highly detailed visual scenes, the computational demand for visual frame rate is still the bottleneck for fast visual feedback.

EQUIPMENT FIDELITY

To consider high physical fidelity, proper equipment fidelity must also be present. Equipment fidelity refers to the extent to which a simulator can emulate or replicate the equipment being used, which includes all software and hardware components of the system (Zhang, 1993). Occasionally, it may become unfeasible or too costly to use actual equipment for the simulation, in which case a replica or substitute may be used in its place.

It is important, however, to maintain a certain degree of equipment fidelity for training purposes. Reed (1996) studied equipment fidelity when trying to simulate an aerial gunner station of a helicopter. During the simulation, the actual weapons system appeared to interfere with the simulation equipment, and therefore had to be replaced by a replica. Using similarly shaped and weighted equipment in combination with parts of the actual weapons system, Reed was able to preserve some equipment fidelity without endangering the training effectiveness.

MOTION FIDELITY

The role of motion fidelity is another important component of physical fidelity. It can be defined as the degree to which a simulator can reproduce the sense of motion felt by humans in the operational environment. The use of motion does increase the physical fidelity and realism of the simulation, but the benefits realized toward measurable transfer of training are minimal and insignificant. At best, empirical research on motion simulation has indicated that the addition of motion provides very limited benefits. For example, reproduction of the movement a pilot feels when banking for a standard turn adds to the realism experienced by the pilot, but has not been shown to improve the ability of the pilot to perform that turn in the real world.

However, the motion component of simulation has been found to be quite critical for the training of pilots for certain types of aircraft. For example, military fighter pilots often rely heavily on motion cues to perform complicated maneuvers in jet airplanes (Thomas, 2004).

Motion also has its limitations. The brain, for example, can often be tricked into sensing motion where motion does not exist. Motion, it seems, provides very little to overall training effectiveness (Garrison, 1985; Ray, 1996). It also does not reduce the real flight time necessary to reach proficiency when performing many tasks in the real world such as the turning maneuver mentioned earlier (Ray, 1996). Furthermore, Advani and Mulder (1995) argue that reproduction of motion cues with ground-based flight simulators is principally impossible due to the kinetic limitations inherent in the motion system. Simulators may be capable of reproducing the banking angle that a real aircraft may encounter in an operational setting, but they have not been able to sufficiently produce and maintain the centrifugal forces experienced when actually executing the banking maneuver in the real world.

PSYCHOLOGICAL–COGNITIVE FIDELITY

Beyond the look and feel of a simulation, there exists another component of fidelity that results in the robustness (or lack thereof) of the psychological and cognitive experience that a person receives from being in the simulator. This component is known as psychological–cognitive fidelity.

Psychological–cognitive fidelity is the extent to which psychological and cognitive factors are replicated within the simulation (Kaiser & Schroeder, 2003), and is related to an individual's problem-solving strategies and the establishment and use of mental models (Hontvedt & Øvergård, 2020). The result of the degree of psychological–cognitive fidelity present within a simulation is the degree to which the user is psychologically and cognitively engaged in the same manner when compared to the degree to which the actual equipment would engage the user. This kind of fidelity involves humans' ability to perform the cognitive aspects of tasks, including decision-making, situation awareness, problem-solving, and sense-making (Hontvedt & Øvergård, 2020).

The cockpit environment is a demanding environment, requiring the pilot to constantly monitor flight systems and instrumentation, looking for failures, and maintaining a flight plan (Wickens et al., 2004). As the learning situation is important for psychological–cognitive fidelity, aspects like simulator immersiveness, the designed task, guidance, and social affordances need to be noted (Hontvedt & Øvergård, 2020). To consider a simulation with high psychological and cognitive fidelity, the simulator must require the same attentional resources from the pilot and produce similar psychological effects such as stress and workload.

Research concerned with cognitive and psychological fidelity focuses on those and other factors that affect performance. It is known that too much stress may cause critical performance decrement in flight. If, when immersed in a simulation environment, a user experiences symptoms of stress similar to those felt in the operational setting, it is generally accepted that some level of psychological–cognitive fidelity has been achieved.

OTHER FIDELITY

Task fidelity is the degree to which a simulator replicates the tasks involved in the actual environment (Zhang, 1993; Roza, 2000; Hughes & Rolek, 2003). Flight training devices must act like actual airplanes; therefore, a pilot must be able to "fly" the simulator just as he or she would fly an aircraft. This means that all tasks that need to be executed in an aircraft must be done in the same fashion when in the simulator. The extent to which these tasks are simulated may be an issue for future research.

Not all tasks may need to be replicated for all training exercises, and some may be isolated to investigate specific problems or issues. When training for cockpit communication, for example, the tasks involving communication must be exactly the same as those in the operational setting. Other tasks (e.g., stick and rudder tasks) do not need to be the same, or can be eliminated altogether, if the objective of the training does not include these tasks.

Much like task fidelity, functional fidelity is also important. Functional fidelity can be described as how the simulator reacts to the tasks and commands being executed by the pilot (Allen, 1986). Not only does the pilot have to react and fly as he or she would in a real aircraft but the simulator must also react and maneuver as a real aircraft would in an operational setting. This ensures operational correctness and accuracy, as well as adding realism and believability to the simulation. When a trainee pulls back on the yoke, for example, the simulator must react as if the aircraft were pulling up and climbing in altitude. This fidelity, along with task fidelity is essential for training effectiveness and positive transfer of training to occur.

Interactional fidelity is kind of different as it focused on the collaborative and coordinating patterns of a socio-technical system (Hontvedt & Øvergård, 2020). Interactional fidelity was described as "the accuracy and relevance of participant collaboration and enactment of work tasks in simulator training" (Hontvedt & Øvergård, 2020). Hutchins and Klausen (1996) used a flight simulator to discover the interactions between pilots in the flight deck which showed a high interactional fidelity as pilots' cognitive efforts to collaborate. Pilots need to learn how to work with others in the same way that they can expect in a real flight. Simulations with high interactional fidelity can also provide the recreation of social interaction as aiding communication between different entities in the system (Hontvedt & Øvergård, 2020).

Simulator fidelity is a complex subject that includes a number of factors. It is important to understand that these dimensions are not mutually exclusive; there is a large degree of overlap. Although expensive, high-fidelity devices are often used for advanced pilot training. However, research has shown that high fidelity may not be necessary to produce effective training results (Connolly et al., 1989; Hays et al., 1992; Duncan & Feterle, 2000). Future research should focus on the specific knowledge to determine exact fidelity requirements. A significant related topic is that of fidelity measurement.

MEASURING FIDELITY

Although it is well-accepted that fidelity is the degree of similarity between the simulation and reality, it is critical to have a detailed and precise capability to measure that fidelity. Due to the complex nature of simulated tasks, different definitions of fidelity can lead to different measures of fidelity, and thus cause inconsistency among research results. Yu (1997) points out that simulation fidelity needs to be defined on the basis of "application purpose and technical possibility." Each aspect of fidelity must be verified in a way that is both consistent and specific.

Equipment fidelity can be verified by checking whether the simulator performance is within some predetermined accuracy compared with real aircraft. In other words, visual fidelity can be measured by some consistent and specific visualization standard such as resolution.

In terms of fidelity quantification, there are several types of metrics available. Roza (2000) found all these metrics either subjective or qualitative in nature, which is not good enough for current simulation requirements. Existing fidelity metrics can be classified either as a singular metric or a set of metrics that can be statistically combined to create a meaningful multidimensional metric.

Objective measurement of simulation fidelity attempts to compare the simulated objects with the corresponding referent or real-world environment. Due to the level of complexity involved, especially with complex simulation setup and tasking, it is nearly impossible to count and compare every single element of the simulation. To better illustrate this picture, take a look at your surroundings, either an office or a room. Imagine that your task is to count every object around you to compare every feature and detail with the simulated environment. You will have some idea of the difficulties involved in accomplishing that task.

Thus, exact measure of realism is not feasible at this time, and is considered by some to be "a goal which can never be accomplished" (Roza et al., 2001). It is practically impossible to count everything, or know everything about the reality or referent due to (1) the high degree of uncertainty, (2) the overwhelming information involved, (3) complicated attributes and behaviors associated with the reality or referent, and (4) human limitations needed to observe and explain real-world information. A recent report on Fidelity Definition and Metrics (FDM-ISG) attempts to specify the fidelity requirements in a formal way (Gross and Freeman, 1997; Roza et al., 2001). Simulation fidelity is defined as "The degree to which a model or simulation reproduces the state and behavior of a real world object or perception of a real world object, feature, condition or standard in a measurable or perceivable manner."

The importance of fidelity measurement is also addressed by the FDM-ISG (Gross et al., 1999) as "what aspect should be simulated and how to observe the simulation purpose and objectives best."

These fidelity requirements are essential for simulation system design because the fidelity requirements will ultimately affect the simulation context, purpose, and hardware and software requirements, and thus affect the trade-off results between cost and achieved transfer of training. Research on simulation has been primarily focused on hardware and software development, which is targeted at "the ultimate display" to produce the real-time simulated environment.

What is the minimum fidelity that is required to achieve the required level of transfer of training? To find the most appropriate level of fidelity needed for the simulation tasks, one still needs to be able to accurately assess the existing or proposed level of fidelity. The most common methods of fidelity measurement are mathematical modeling, research experiments, and rating methods (Schricker et al., 2001; Kaiser & Schroeder, 2003).

THE MATHEMATICAL MODEL

For years, research on fidelity quantification focused mostly on the objective mathematical formulation. Schricker et al. (2001) offer a thorough review of mathematical models on fidelity measurements.

SUBJECTIVE METHODS

Subjective methods use expert opinions (including developers and users) to determine the degree of fidelity. Clark and Duncan proposed one of the simplest methods (Schricker et al., 2001). In this model, fidelity is measured by a binary scoring system. Simulation conditions are evaluated by either a "0" or "1," with "0" meaning simulation does not duplicate the real-world conditions, and "1" indicating that the simulation does reproduce the real-world conditions. Averaging those ratings together provides an assessment of the overall fidelity. Although the simplicity is appealing, this method is subject to the same rating issues as any other subjective form of rating. An arbitrary judgment, or guess, can significantly affect the fidelity score.

A more advanced mathematical model is proposed by Gross and Freeman (Schricker et al., 2001). The model is based on four theorems:

I. $0 \leq F(A) \leq 1$
II. If $F(A) = 1$, then $A \equiv R$
III. $F(A) \geq F(Meta\ A)$
IV. $F(A) + F(B) = \min(F(A), F(B))$

where A and B are models of interest, $F(A)$ is the fidelity of A, *Meta A* is a model of referent (including A itself), and R is the referent of A.

The simple formula for determining the overall value of fidelity of a simulation system is as follows: *Fs* 5o*FiWi*, where F_i is the fidelity of each referent characteristics and F_s is the fidelity of the entire simulation system, W_i is the relative importance rate of characteristics i. This formula, however, contradicts Theorem IV (Schricker et al., 2001). The main reason for this contradiction is because it did not clearly define the set, operation on the set, and function.

Liu and Vincenzi (2004) modified this model definition as follows. Let set S be defined as the following: $S = \{A : A = meta_i\ R\}$ (where R is the referent of the real-world group of objects and S is the group of all possible simulation models of referent R). Then, the function of fidelity is defined on set S as $F: S \rightarrow [0, 1]$. Furthermore, we define the following operations for $A \in S$ and $B \in S$, $A \oplus B = A \cup B$ (one can easily prove that $A + B$ is commutative and associative). Based on these notations, Gross and Freeman's (1997) model can be modified as follows:

I, II, and III are still true and IV becomes

$$\min(F(A), F(B)) \leq (A \oplus B) \leq \max(F(A), F(B))$$

Gross and Freeman stated that the fidelity of any simulated system is equal to the fidelity of the individual component of the simulation of the lowest fidelity. It can be argued that this might not be the case. If we add low-fidelity components to a high-simulation model, it will certainly affect the high-fidelity components, but it should also improve the low-fidelity components if this add-on interacts with the low-fidelity component. Readers who are interested in this model should refer to Schricker et al. (2001) for more details.

With models like this, we have a more comprehensive mathematical framework in place to facilitate the assessment of fidelity. Several other attempts to measure fidelity mathematically currently exist. Interested readers should refer again to Schricker et al. (2001) for more details.

FIDELITY EVALUATION FRAMEWORKS

It can be argued that although mathematical modeling is beneficial to an understanding of the concept of fidelity, it has little practical implication for actual measurements due to the complexity and uncertainty involved. Researchers are attempting to develop other alternatives to measure fidelity in the field. One way of doing this is to utilize evaluation framework (Roza et al., 2001; Schricker et al., 2001). Figure 3.1 illustrates a generic model of measurement framework modified from Schricker et al. (2001).

By using this framework, all simulation task-critical objects can be identified, as well as their associated behaviors and attributes for the referent. By comparing the level (percentage) of the corresponding objects from a simulation model, the fidelity can be estimated quantitatively. This framework is based on the assumption of Perato's law (20% of the elements contribute 80% of the training effects in simulation).

Roza (2000) summarized existing research on simulation fidelity, especially on fidelity characterization and quantification. Based on his study, a preliminary fidelity theory and a practical tool called Fidelity Management Process Overlay Model (FiMO) was proposed to assess and quantify simulation fidelity. After investigating the distributed simulation fidelity requirement for the U.S. Department of Defense's Defense Modeling and Simulation Office (DMSO) High-Level Architecture (HLA), Roza (2000) found that although HLA is well-defined by the Federation Development

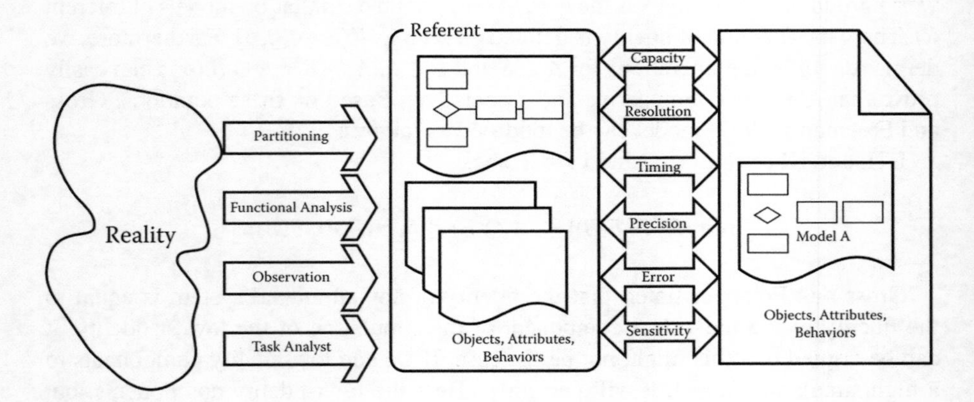

FIGURE 3.1 A conceptual illustration of measurement procedure. (Adapted from Schricker, B., Franceschini, R., and Johnson, T., "Fidelity evaluation framework," *Proceedings of the 34th Annual Simulation Symposium,* 2001.)

and Execution Process (FEDEP), it focuses primarily on technological aspects and cannot answer many questions that arise regarding fidelity or fidelity quantification. A systematic and structured way is needed to efficiently characterize aspects of fidelity. FiMO is one model employed for this purpose. This approach maps to the FEDEP framework and provides a process view for characterization of fidelity issues along simulation development stages and activities. The basic framework consists of five major, iterative activities. For detailed information on this approach, readers can refer to Roza's (2000) paper. This framework is believed to have the ability to handle large amounts of fidelity data in a progressive manner.

Bell and Freeman (1995) developed a draft Fidelity Description Requirement (FDR) they believed can be used to quantify simulation fidelity. An assessment process was proposed, and taxonomy was developed, in a hierarchical format. The top level is the simulation resource consisting of a combination of hardware and software solutions (i.e., a Cessna 172 Flight Simulator). Level 2 is the fidelity domain (i.e., physical fidelity or visual fidelity). Level 3 is the capability level, level 4 is the implementation specific instantiations of the capability, level 5 consists of individual characteristics of an implementation, and level 6 defines the measurements for level 5. Bell and Freeman (1995) also discussed the possibility of using Fuzzy logic as a means to help quantify fidelity as accurately as possible, taking into account unknowns and assumptions associated with characteristics of fidelity. This approach may be appropriate as there are a number of uncertainties involved.

Fidelity can also be measured by other indirect measurements such as by the evaluation of human performance. It is assumed that the function of fidelity is objective, i.e., for one referent; two simulation models have the same transfer of training effect if these two simulated models have the same fidelity. (Please note the opposite might not be true.) As the ultimate goal of simulation is to transfer the skills gained in training to the real-world situation and the objective measurement of fidelity is far more intriguing, the measurement of human task performance would be a good metric for any application that mainly targets transfer of training in the real world (Mania et al., 2003). Although human performance assessment alone cannot provide a quantitative assessment of simulation fidelity, it can provide a measure or indication of the relative efficiency of different simulated models for the same referent. It is more intuitive and hands-on than many other methods, and this measurement is widely accepted and used (Mania et al., 2003).

Lehmer and Chung (1999) applied Image Dynamic Measurement System (IDMS-2) to verify simulation fidelity. Measurement of delay in the visual system response time was used to assess the fidelity.

FIDELITY AND TRANSFER OF TRAINING

It is natural to assume that the higher the level of fidelity, the higher degree of transfer of training will occur. Based on the "identical elements" theory by Thorndike (1903), this notion is still strongly held by simulator designers and industries today. Thorndike argued that there would be transfer between the first task (simulation) and the second task (real world) if the first task contained specific component activities that were held by the second task. There are a number of theoretical studies (Noble, 2002) that support the notion that higher fidelity will produce higher degrees of training transfer.

The "Alessi Hypothesis" states that there is a certain point at which adding more fidelity does not transfer training at the same rate as early or beginning training (Alessi, 1988). Some proposed a U-shaped curve, whereas some theories proposed a normal curve. But according to Alessi (1988), experimental research has not provided sufficient evidence to support this high-fidelity notion as theory predicted. Some results even indicate that lower fidelity has an advantage. Alessi (1988) believed that the explanation for failing to show the high-fidelity advantage is that (1) high fidelity also means high complexity, which will require more cognitive skills, thus increasing trainee's workload, which will, in turn, impede learning; and (2) proven instructional techniques, which improve initial learning, do not depend upon high-fidelity components which, in turn, tend to lower the overall fidelity of the simulation.

Alessi (1988) proposed that the relationship between learning and fidelity is nonlinear and also dependent on other factors such as the trainee experience. For different trainees, there are different learning curves, and all these curves are nonlinear and different.

According to Alessi (1988), when the level of fidelity is increased, the corresponding change in transfer of training depends largely on the trainee's characteristics and ability to respond to this increase in fidelity. For novices, initial learning is the primary focus, and for experts, who are well-versed in the initial knowledge needed to perform the job or individual tasks, transfer of training related to task-specific knowledge, skills, and abilities is essential.

Actual assessment of fidelity is extremely difficult to obtain. Alessi gave an in-depth fidelity analysis of four different types of simulation to further investigate this effect: (1) situational simulation, (2) procedural simulation, (3) process simulation, and (4) physical simulation. Four dimensions of fidelity were identified and defined for this analysis: (1) the underlying model, (2) presentation, (3) user actions, and (4) system feedback.

It was found that for different trainees and different types of simulation, the requirement for fidelity varied greatly. As an instance of procedural simulation, flight simulators need to have high fidelity of the presentation, actions, and system feedback to result in significant increases in transfer of training efficiency.

With respect to the relationship between fidelity and transfer of training, other studies have also demonstrated that the law of diminishing returns holds true. This parallels a hypothesis put forth by Roscoe in 1971, where Roscoe hypothesized that there will be diminishing returns in transfer of training as the amount of simulator training increases. Thus, the first hours of simulation training have high amounts of positive transfer of training information, and latter hours will have lower amounts of positive transfer of training information (Roscoe, 1980).

Combining these two theories (Roscoe, 1971; Alessi, 1988), we can conclude that adding more fidelity, especially in the later stages of training, produces minimal gain in transfer of training. If this is the case, increasing fidelity may not always be necessary. Fidelity is expensive, and eliminating certain key elements of fidelity that do not necessarily increase transfer of training will reduce the cost of production. The question is: Which cues can we eliminate without reducing the amount of transfer?

This issue still remains vague and less understood. To answer this question, more research is necessary. This research would need to directly compare simulation training to traditional training done in the actual aircraft or real-world environment. The main obstacle that researchers face is cost. It is extremely expensive to conduct this type of research due to the high operational costs associated with real-world systems such as commercial aircraft, combat aircraft, surface ships, submarines, and other military and civilian systems. Additional issues revolve around the possible disruption of normal training. It may be difficult to find flight schools, student pilots, or military personnel and facilities that would participate, especially because training time is already scarce, necessary, time-consuming, and expensive.

SUMMARY

It is clear that simulation can provide great benefits regardless of the levels of fidelity (D'Asta et al., 2019; Labrague et al., 2019; Wenlock et al., 2020; Willie et al., 2016). The standard design approach for simulators is to incorporate the highest possible level of fidelity and hope for the best possible transfer of training outcome. This is the direct result of the belief (or assumption) that high levels of fidelity must equate to high levels of transfer of training despite the fact much evidence exists to indicate that this might not necessarily be true.

It is important to focus on the goal of simulation training, which is the transfer or translation of skills learned in one arena to another. The ever-present debate in the realm of simulation centers around how much fidelity is necessary to achieve a desired degree of training transfer. In other words, how "real" does the simulation need to be in order for the trainee to properly execute the skills learned in simulation to the real world? This question can only be answered with "it depends." As discussed previously, it depends on many factors, including the individual trainees, their levels of skill, and the instructor. It also depends on the particular skills to be learned and transferred.

Low-fidelity simulators maximize the initial learning rate of novice pilots and minimize cost, whereas costly, high-fidelity simulators predict the real-world in-flight performance of expert pilots (Kinkade & Wheaton, 1972; Fink & Shriver, 1978; Hays & Singer, 1989). This may be true because novice pilots may become overwhelmed by high fidelity, whereas experts will not. Initial pilot training focuses more on becoming familiar with the controls and the layout of the instruments, whereas experts will concentrate on more advanced operational aspects.

If economic constraints did not enter the picture, the question of the level of fidelity needed to obtain maximal transfer of training would not be an issue. Simulation designers would always include every possible attribute, increasing the level of fidelity as high as technology would allow. Unfortunately, financial resources are limited.

To get the most value from simulation, it is important to eliminate attributes that do not aid in the transfer of training. In other words, it has become imperative to get the most "bang for the buck." The goal is to give the trainee enough simulation fidelity to facilitate learning, without attaching costly and unnecessary simulation options and characteristics. There is no easy answer to this question at this time. Researchers in the areas of human factors, psychology, computer science, engineering, and many other fields involved in the simulation industry need to work closely together to conduct systematic, multidisciplinary research to achieve maximal training benefit while controlling and minimizing cost.

REFERENCES

Advani, S.K., and Mulder, J.A., 1995, Achieving high-fidelity motion cues in flight simulation, *AGARD FVP Symposium on Flight Simulation: Where are the Challenges*, Braunschweig, Germany.

Akalin, A., and Sahin, S., 2020, The impact of high-fidelity simulation on knowledge, critical thinking, and clinical decision-making for the management of pre-eclampsia, *International Journal of Gynecology and Obstetrics*, *150*(3), 354–360. https://doi.org/10.1002/ijgo.13243

Alconero-Camarero, A.R., Sarabia-Cobo, C.M., Catalán-Piris, M.J., González-Gómez, S., and González-López, J.R., 2021, Nursing students' satisfaction: A comparison between medium- and high-fidelity simulation training, *International Journal of Environmental Research and Public Health*, *18*(2), 804. https://doi.org/10.3390/ijerph18020804

Alessi, S.M., 1988, Fidelity in the design of instructional simulations, *Journal of Computer-Based Instruction*, *15*(2), 40–47.

Allen, J.A., 1986, Maintenance training simulator fidelity and individual difference in transfer of training, *Human Factors*, *28*(5), 497–509.

Andrews, D., Carroll, L., and Bell, H., 1995, The future of selective fidelity in training devices, Educational *Technology*, *35*, 32–36.

Bell, P.M., and Freeman, R., 1995, Qualitative and quantitative indices for simulation systems in distributed interactive simulation, *IEEE Proceedings of ISUMA-NAFIPS*, 745–748.

Bradley, D.R., and Abelson, S.B., 1995, Desktop flight simulators: Simulation fidelity and pilot performance, *Behavior Research Methods, Instruments, & Computers*, *27*(2), 152–159.

Bur, A.M., Gomez, E.D., Newman, J.G., Weinstein, G.S., O'Malley, B.W., Rassekh, C.H., and Kuchenbecker, K.J., 2017, Evaluation of high-fidelity simulation as a training tool in transoral robotic surgery, *The Laryngoscope*, *127*(12), 2790–2795. https://doi.org/10.1002/lary.26733

Chamberland, C., Hodgetts, H.M., Kramer, C., Breton, E., Chiniara, G., and Tremblay, S., 2018, The critical nature of debriefing in high-fidelity simulation-based training for improving team communication in emergency resuscitation, *Applied Cognitive Psychology*, *32*(6), 727–738. https://doi.org/10.1002/acp.3450

Connolly, T.J., Blackwell, B.B., and Lester, L.F., 1989, A simulator–based approach to training in aeronautical decision making, *Aviation, Space, and Environmental Medicine*, *60*, 50–52.

D'Asta, F., Homsi, J., Sforzi, I., Wilson, D., and de Luca, M., 2019, "SIMBurns": A high-fidelity simulation program in emergency burn management developed through international collaboration, *Burns*, *45*(1), 120–127.

Duncan, J.C., and Feterle, L.C., 2000, The use of personal computer-based aviation training devices to teach aircrew decision-making, teamwork, and resource management, *Proceedings of IEEE 2000 National Aerospace and Electronics Conference*, Dayton, OH, 421–426.

Ferguson, S.W., Clement, W.F., Hoh, R.H., and Cleveland, W.B., 1985, Assessment of simulation fidelity using measurements of piloting technique in flight: Part II, *41st Annual Forum of the American Helicopter Society*, Ft. Worth, TX, 1–23.

Field, E.J., Armor, J.B., and Rossitto, K.F., 2002, Comparison of in-flight and ground based simulations of large aircraft flying qualities, *AIAA 2002-4800, AIAA Atmospheric Flight Mechanics Conference and Exhibit*, Monterey, CA.

Fink, C., and Shriver, E., 1978, *Simulators for Maintenance Training: Some Issues, Problems and Areas for Future Research* (Tech. Rep. No. AFHRL-TR-78-27), Lowery Air Force Base, CO: Air Force Human Resources Laboratory. *Psychology*, *46*(4), 349–354.

Fortin, M., 1989, Cost/performance trade-offs in visual simulation, *Royal Aeronautical Society Conference on Flight Simulation: Assessing the Benefits and Economics*, London, 19.1–19.15.

Garrison, P., 1985, *Flying Without Wings: A Flight Simulation Manual*, Blue Ridge Summit, PA: TAB Books.

Gross, D.C., and Freeman, R., 1997, Measuring fidelity differentials in HLA simulations, *Fall 1997 Simulation Interoperability Workshop.*

Gross, D.C., Pace, D., Harmoon, S., and Tucker, W., 1999, Why fidelity? *Proceedings of the Spring 1999 Simulation Interoperability Workshop.*

Haddeland, K., Slettebø, Å., Svensson, E., Carstens, P., and Fossum, M., 2019, Validity of a questionnaire developed to measure the impact of a high-fidelity simulation intervention: A feasibility study, *Journal of Advanced Nursing*, *75*(11), 2673–2682. https://doi.org/10.1111/jan.14077

Hays, R.T., and Singer, M.J., 1989, *Simulation Fidelity in Training System Design*, New York: Springer.

Hays, R.T., Jacobs, J.W., Prince, C., and Salas, E., 1992, Flight simulator training effectiveness: A meta-analysis, *Military Psychology*, *4*(2), 63–74.

Hontvedt, M., and Øvergård, K.I., 2020, Simulations at work – A framework for configuring simulation fidelity with training objectives, *Computer Supported Cooperative Work*, *29*(1–2), 85–113. https://doi.org/10.1007/s10606-019-09367-8

Hughes, T., and Rolek, E., 2003, Fidelity and validity: Issues of human behavioral representation requirements development, *Proceedings of the 2003 Winter Simulation Conference*, New Orleans, LA.

Hutchins, E., and Klausen, T., 1996, Distributed cognition in an airline cockpit, in *Cognition and Communication at Work*, Engeström, Y., and Middleton, D., Eds., Cambridge University Press, 15–34.

Kaiser, M.K., and Schroeder, J.A., 2003, Flights of fancy: The art and science of flight simulation, in *Principles and Practice of Aviation Psychology*, Vidulich, M.A and Tsang, P.S., Eds., Mahwah, NJ: Lawrence Erlbaum Associates, 435–471.

Kinkade, R., and Wheaton, G., 1972, Training device design, in *Human Engineering Guide to Equipment Design*, Van Cott, H. and Kinkade, R., Eds., Washington, DC: Department of Defense, 668–699.

Labrague, L.J., McEnroe-Petitte, D.M., Bowling, A.M., Nwafor, C.E., and Tsaras, K., 2019, High-fidelity simulation and nursing students' anxiety and self-confidence: A systematic review, *Nursing Forum*, *54*(3), 358–368.

Lehmer, R.D., and Chung, W.W.Y., 1999, Image dynamic measurement system (IDMS-2) for flight simulation fidelity verification, American Institute of Aeronautics and Astronautics, 137–143.

Liu, D., and Vincenzi, D.A., 2004, Measuring simulation fidelity: A conceptual study, *Proceedings of the 2nd Human Performance, Situation Awareness and Automation Conference (HPSAA II)*, Daytona Beach, FL, 160–165.

Mania, K., Troscianko, T., Hawkes, R., and Chalmers, A., 2003, Fidelity metrics for virtual environment simulations based on spatial memory awareness states, *Presence, Teleoperators and Virtual Environments*, *12*(3), 296–310.

Nemire, K., Jacoby, R.H., and Ellis, S.R., 1994, Simulation fidelity of a virtual environment display, *Human Factors*, *36*(1), 79–93.

Noble, C., 2002, The relationship between fidelity and learning in aviation training and assessment, *Journal of Air Transport*, *7*(3), 34–54.

Parrish, R.V., McKissick, B.T., and Ashworth, B.R., 1983, *Comparison of Simulator Fidelity Model Predictions with In-simulator Evaluation Data* (Technical Paper 2106), Hampton VA: NASA Langley Research Center.

Ray, P.A., 1996, Quality flight simulation cueing—Why? *Proceedings of the AIAA Flight Simulation Technologies Conference*, San Diego, CA, 138–147.

Reed, E.T., 1996, The aerial gunner and scanner simulator "affordable virtual reality training for aircrews," *Training-Lowering the Cost, Maintaining Fidelity: Proceedings from the Royal Aeronautical Society*, London, 18.1–18.15.

Rehmann, A.J., Mitman, R.D., and Reynolds, M.C., 1995, *A Handbook of Flight Simulation Fidelity Requirements for Human Factors Research* (DOT/FAA/CT-TN95/46), Wright-Patterson AFB, OH: Crew System Ergonomics Information Analysis Center.

Rinalducci, E., 1996, Characteristics of visual fidelity in the virtual environment, Presence, *5*(3), 330–345.

Roscoe, S.N., 1971, Incremental transfer effectiveness, *Human Factors*, 13(6), 561–567. https://doi.org/10.1177/001872087101300607

Roscoe, S.N., 1980, *Aviation Psychology*, Ames, IA: Iowa State University Press.

Roza, M., 2000, Fidelity considerations for civil aviation distributed simulations, *Proceedings of the AIAA Modeling and Simulation Technologies Conference and Exhibit*, Denver, CO.

Roza, M., Voogd, J., and Jense, H., 2001, Defining, specifying and developing fidelity referents, *Proceedings of the 2001 European Simulation Interoperability Workshop*, London.

Schricker, B., Franceschini, R., and Johnson, T., 2001, Fidelity evaluation framework, *Proceedings of the IEEE 34th Annual Simulation Symposium*, Seattle, WA.

Scott, A., and Gartner, A., 2019, 9Low fidelity simulation in a high fidelity world, *Postgraduate Medical Journal*, *95*(1130), 687–688. https://doi.org/10.1136/postgradmedj-2019-FPM.9

Thomas, T.G., 2004, From virtual to visual and back? *AIAA Modeling and Simulation Technologies Conference and Exhibit*, Providence, RI: AIAA Paper 2004–5146.

Thorndike, E.L., 1903, *Educational Psychology*, New York: Lemke & Buechner.

Wenlock, R.D., Arnold, A., Patel, H., and Kirtchuk, D., 2020, Low-fidelity simulation of medical emergency and cardiac arrest responses in a suspected COVID-19 patient–an interim report, *Clinical Medicine*, *20*(4), e66.

Wickens, C.D., Lee, J.D., Liu, Y., and Becker, S.E.G., 2004, *An Introduction to Human Factors Engineering*, 2nd ed., Upper Saddle River, NJ: Pearson Prentice Hall.

Willie, C., Chen, F., Joyner, B.L., and Blasius, K., 2016, Using high-fidelity simulation for critical event training, *Medical Education*, *50*(11), 1161–1162.

Yu, Z.-G., 1997, Inquiry into concepts of flight simulation fidelity, in *First International Conference on Nonlinear Problems in Aviation and Aerospace Proceedings*, Sivasundaram, S., Ed., USA, 679–685.

Zhang, B., 1993, How to consider simulation fidelity and validity for an engineering simulator, *American Institute of Aeronautics and Astronautics*, 298–305.

4 Transfer of Training

Dahai Liu, Jacqueline McSorley, Elizabeth Blickensderfer, Dennis A. Vincenzi, and Nikolas D. Macchiarella

CONTENTS

INTRODUCTION

Commercial aviation and the military have long reaped the rewards provided by the use of flight simulation to train pilots, and more recently, the healthcare industry has seen tremendous growth in use of simulation training. Using simulators for training maximizes the use of operational systems for revenue-producing activities and minimizes the use of operational systems for non-revenue-producing activities such as training.

DOI: 10.1201/9781003401360-4

In addition, it eliminates the expense of aircraft fuel and maintenance costs associated with the training, and the loss of revenue, whereas the actual system is in use for training purposes (Moroney & Moroney, 1998). In the healthcare domain, simulations allow medical professionals to practice treatments and protocols without putting patients at risk; simulations in healthcare are often expensive, but the trade-off between economic cost and cost of patient lives is undeniably important (Ker et al., 2010).

Additionally, simulation can save training time. Training time-savings occur due to trainers positioning the trainee into the exact situation required to learn specific skills. For example, if a flight instructor was attempting to teach a student how to land in a crosswind, it would no longer be necessary to perform multiple takeoffs to practice crosswind landings. In this situation, the trainer can stop the simulation and place the student pilot on final approach repeatedly until they become proficient at the targeted skill. Likewise, in the healthcare domain, simulation can provide practice opportunities for students or practitioners to encounter a wide range of patient symptoms without the need of a live human actually presenting with a specific set of symptoms.

Also important, but more difficult to quantify, are increases in safety associated with using a simulated environment. In terms of flight training, mistakes made by a student pilot in flight simulation are regrettable, but not life-threatening. The instructor and student pilot can simply stop and reset the simulator to perform the flight task again. A clear example within in-flight training is unusual attitude recovery. In actual flight training, pilots are placed into unusual attitudes (e.g., extreme degrees of pitch, roll, and yaw) for training purposes (Federal Aviation Administration, 2006). All certified pilots have faced stressful and life-threatening situations during training and have recovered accordingly. However, when utilizing flight simulation, inexperienced pilots can encounter these dangerous situations without actually being in harm's way, increasing safety.

While it is clear that training via simulation offers numerous benefits, for simulation-based training to be successful, trainees must effectively apply learned knowledge, skills, and abilities gained from simulator training to their corresponding real-world task. In other words, transfer of training (ToT) must occur. This chapter describes the main concepts and theories pertaining to transfer of training with an emphasis on simulation-based training. We will first review the definition of transfer of training and discuss the factors involved. Next, we discuss research methods relating to transfer of training. Finally, we propose research issues regarding transfer of training for simulation-based training environments.

TRANSFER OF TRAINING: TERMS AND CONCEPTS

The ultimate goal of training is the trainee being able to apply or "transfer" what was learned in training to the actual real-world setting. Transfer of training refers to the extent which knowledge, skills, and abilities learned from training programs are generalized or applied to real-world situations and to the maintenance of these knowledge, skills, and abilities over time on the job (Blume et al., 2010).

Transfer of training can be classified into two types: positive and negative (Chapanis, 1996).

Positive Transfer

Positive transfer occurs when an individual correctly applies knowledge, skills, and abilities learned in one environment (e.g., in simulation) to a different setting (in the case of aviation, this would be real flight) (Burke, 1997). Positive transfer is the goal of any type of training. When referring to transfer of training, it denotes a positive transfer of training unless a different interpretation is indicated.

Negative Transfer

Negative transfer occurs when existing knowledge and skills (from previous experiences) impede proper performance in a different task or environment. For example, a skilled typist on a QWERTY keyboard would have difficulty using, or learning how to use, a non-QWERTY keyboard such as a Dvorak keyboard.

Negative transfer develops from at least two related reasons: (1) system design changes and (2) a mismatch between a training system and the actual task. First, system design changes (e.g., controls and software menus) can create one type of negative habit transfer interference. Specifically, the task performer has experience performing the task set up in one manner and has developed a certain degree of automaticity. If a design change occurs, it is likely that the task performer will revert to performing the task according to the previous system (i.e., "habit interference"). Avoiding habit interference should be a major design goal (Chapanis, 1996). Second, if the training system procedures do not match those in the transfer environment, negative transfer is likely to occur. Consider a pilot who is trained to pull back on the yoke to lower the nose of the aircraft in a simulator, while in the actual aircraft, pulling back on the yoke actually raises the nose. In the actual aircraft, if the pilot pulled back aiming to lower the nose, negative transfer would have occurred. The end result would be a dangerous situation for the pilot, any passengers, and the aircraft.

Thus, negative transfer occurs when the trainee reacts to the transfer stimulus with the correct response they learned during training; however, that response is incorrect in relation to the actual performance task—in this case, pulling back on the yoke thinking that this will lower the nose of the aircraft because that was what was trained, when in actuality it raises the nose.

Near Transfer

Another element of training transfer lies in the type of transfer: near, far, or both (Noe, 2006). Near transfer utilizes training strategies that are identical to the situations the trainee will encounter on the job. Equipment usage represents a skill that benefits from near transfer (Noe, 2006). The theory of identical elements, fundamental to near transfer, dictates that the training should mirror the task in factors like equipment and environment (Thorndike & Woodworth, 1901). This type of training promotes near transfer by structuring the training as close to the task as possible (van der Locht et al., 2013).

Far Transfer

Far transfer refers to the application of training to a more generalized environment (Noe, 2006). This occurs when the training environment and equipment may differ from what trainees will encounter in the real scenario. Far transfer is essential when the training cannot directly reflect on-the-job work, such as with interpersonal skills. The stimulus generalization approach supports the concept of far transfer by emphasizing broad principles to train rather than focusing on specific procedures (Noe, 2006).

Cognitive theory of transfer proposes the use of both near and far transfer. It emphasizes specific, meaningful material as well as schemas that encourage effective storage of the general content. This theory also states that providing trainees with application assignments can increase the chances of long-term recall (Noe, 2006). Appropriate conditions for the application of the cognitive theory include nearly all types of training and environments.

A MODEL OF FACTORS AFFECTING THE TRANSFER OF TRAINING

Over the past 30 years, the Baldwin and Ford (1988) model of transfer of training has been well-accepted in training research and practice. Due to its persistent impact, including that it has spawned expanded models (e.g., the Dynamic Model of Training Transfer (Blume et al., 2019)), it is the focus of the current chapter.

As shown in Figure 4.1, the model depicts transfer of training in terms of training input factors, training outcomes, and conditions of transfer. This section will

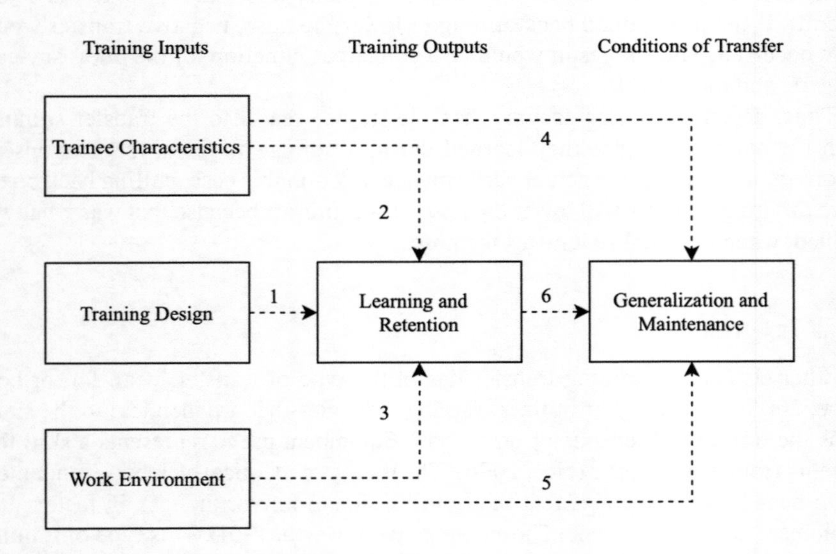

FIGURE 4.1 A model of training transfer. (Adapted from Baldwin, T. and Ford, J., *Personnel Psychol.*, 41, 63–105, 1988.)

review the Baldwin and Ford model with an emphasis on (1) research accomplished since publication of the model and (2) research performed in simulation and aviation studies.

Training Input Factors

Starting with the left side of Figure 4.1, the "Training Input" factors include training design, trainee characteristics, and work environments. The model depicts each of these three input factors as having a direct influence on learning in the training environment. Additionally, the model also connects trainee characteristics and work environment characteristics directly with the transfer performance. Thus, those factors are thought to exert a direct influence on performance in the transfer setting.

In terms of *trainee characteristics*, many research studies that suggest these characteristics affect training transfer efficiency (Smith-Jentsch, Salas, & Brannick, 2001). Numerous individual differences exist, including motivation, attitudes, and ability (e.g., cognitive and physical). In fact, Blume et al.'s (2010) meta-analysis found that cognitive ability was the number one predictor of training transfer. In an analysis specifically targeting flight simulation, Auffrey et al. (2001) argued that goal setting, planning, motivation, and attitudes were key factors in training effectiveness. In terms of motivation, a trainee must put forth effort to learn. A prerequisite for transfer is the trainee's motivation to successfully complete the training—to acquire the new skills and knowledge (Colquitt et al., 2000). In terms of ability, the trainee must also have the raw ability to improve his or her skills. For example, a pilot who is working on shortening his or her takeoff distance must analyze the situation and realize that he or she needs to adjust the flaps accordingly (depending upon the type of aircraft) and increase speed sufficiently to gain enough lift. The trainee must possess the cognitive ability to understand that a change in specific elements is necessary to accomplish the desired change in performance. These are just two examples of individual characteristics that relate to training effectiveness. More research is needed regarding the impact of trainee characteristics on training effectiveness.

Second, the *training design* or method plays a role in transfer of training. While an entire field of research on instructional systems design exists, of particular relevance to training transfer following simulation-based training are the principles of identical elements and the stimulus–response relationship. The first approach, "identical elements," dates back to the turn of the 20th century. Thorndike (1903) put forth a notion that is still held by simulator designers today. Thorndike argued that there would be transfer between the first task (simulation) and the second task (real world) if the first task contained specific component activities that were held by the second task. This approach is entirely dependent on the presence of shared identical elements (Thorndike, 1903). Although outside the realm of flight simulation, an illustration of this principle can be found in athletics. If an individual plays softball and then tries out for the baseball team, he or she will be better-off than the person who has previously played only golf. All three sports share common elements (e.g., ball, striking stick, and grass playing field); however, softball shares far more elements with baseball than golf does (e.g., number of players, bases, scoring, umpires, uniforms, fences, and dugouts). The more elements that are shared between the two environments, the better the transfer. Therefore, softball is a better form of simulation than golf for the training of baseball skills. This approach parallels the idea that simulators should duplicate the real-world situation to the greatest degree possible.

Another principle for training design can be seen in terms of "stimulus and response." In this respect, the idea is to examine the extent to which similarities exist between stimulus representation and the response demands of the training and those of the transfer task (Osgood, 1949). This perspective does not demand a duplication of elements. In contrast, the notion is that transfer of training can be obtained using training tasks and devices that do not duplicate the real world exactly, but that do maintain the correct stimulus–response relationship. Consider once again the example of athletics; golf transfers quite well to tennis, as the golf swing (low to high) and tennis hit (low to high) are similar. Even though the sports themselves are quite different (racquets versus golf clubs and holes), the stimulus response sequence for each of the sports is quite similar.

In simulation-based training, two major categories relating to the training design are the fidelity/authenticity of the simulation (Kaiser, 2003; Hays & Singer, 1989) and the use of a proven scenario-based training model (Salas et al., 2006). First, simulation fidelity should be considered from the perspective of how well the simulation includes appropriate stimulus–response relationships that create a high degree of cognitive fidelity with regard to performing the simulated task. For example, cognitive fidelity in flight simulation addresses the extent that the simulation engages the pilot in the types of cognitive activities encountered in real flight (Kaiser, 2003). Second, research has shown that higher training effectiveness occurs when simulation-based training follows a scenario-based model, which links training requirements, simulated events, performance measures, and feedback (Blickensderfer et al., 2012).

The third input factor in the model is *work–environment characteristics*. This refers to the overall organizational support for the learner, such as supervision, sponsorship, and subsequent reward for the training and skill development. It also refers to the skill that is being trained (i.e., task difficulty); some tasks are simpler and require little effort to transfer skills, whereas other tasks are difficult and require skills that are difficult to transfer and maintain (Blaiwes et al., 1973; Simon & Roscoe, 1984; Hays et al., 1992; Lathan et al., 2002). The simpler tasks or skills will be easier for a trainee to master and later transfer to the performance environment.

Other work has focused on the organizational characteristics involved in training transfer. In a review regarding flight simulation, Auffrey et al. (2001) reported that the characteristics of the work environment were important factors in training transfer. Additionally, Awoniyi et al. (2002) investigated the effect of various work environment factors on training transfer. These authors discussed the importance of the person–environment fit; the notion is that transfer will depend on the degree of fit between a worker and the particular work environment. In other words, two workers may attend training and acquire equivalent knowledge, skills, and abilities. Depending on the degree of person–environment fit in the organization, one worker may show a significantly higher degree of training transfer than the other. Five dimensions were studied: supervisory encouragement, resources, freedom, workload pressure, and creativity support. Results indicated that person–environment fit has a positive relationship with transfer of training and can be a moderate predictor for transfer of training.

Training Outputs

The middle of the model shown in Figure 4.1, "Training Outputs," focuses on the actual outputs of training or, in other words, the amount of learning that occurred during training and the amount retained after the training program was completed. The training outputs depend on the three inputs described above. In turn, the amount learned during training will directly influence ultimate training transfer.

Conditions of Transfer

Finally, the right side of the model, "Conditions of Transfer," refers to the post-training environment. At this stage, the learner is back in the actual work environment. Conditions of transfer include the real-world conditions surrounding the use, generalization, and maintenance of the knowledge, skills, and abilities learned in the training program; the degree to which the learner used the knowledge, skills, and abilities in a transfer setting; and the length of time the learner retained the knowledge, skills, and abilities. Blume et al.'s (2010) meta-analysis found moderate effect sizes for both transfer climate and support.

DYNAMIC MODELS OF TRAINING TRANSFER

Although most transfer of training research focuses on immediate results of the training instruction, more than one author has explained that transfer does not occur immediately, but instead occurs in a series of stages through which trainees pass (Foxon, 1993; Blume et al., 2019). According to Foxon (1993) (see Figure 4.2), the degree of transfer increases progressively, whereas the chance to transfer failure gets lower. Although this process model was proposed for organizational training programs, it can likely be generalized to any kind of training, including simulation training. Additionally, Blume and colleagues (2019) developed an organizational training model that illustrates a cycle of transfer that builds on itself each time (see Figure 4.3). This dynamic model shows the impact on both trainee behaviors and performance each time the transfer process repeats.

In summary, transfer of training is the combined result of input factors (characteristics of the trainee, training design, and work environment), the amount learned in training, and the conditions surrounding the transfer setting. Some of these factors are better understood than others. Simulation is one subfactor in this complex problem. Even with the best simulation available, if other variables do not exist in the appropriate manner, training will not be effective (it will not result in positive transfer). Overall, transfer of training remains a complex issue with numerous variables involved. Additional research is needed to further understand the differential

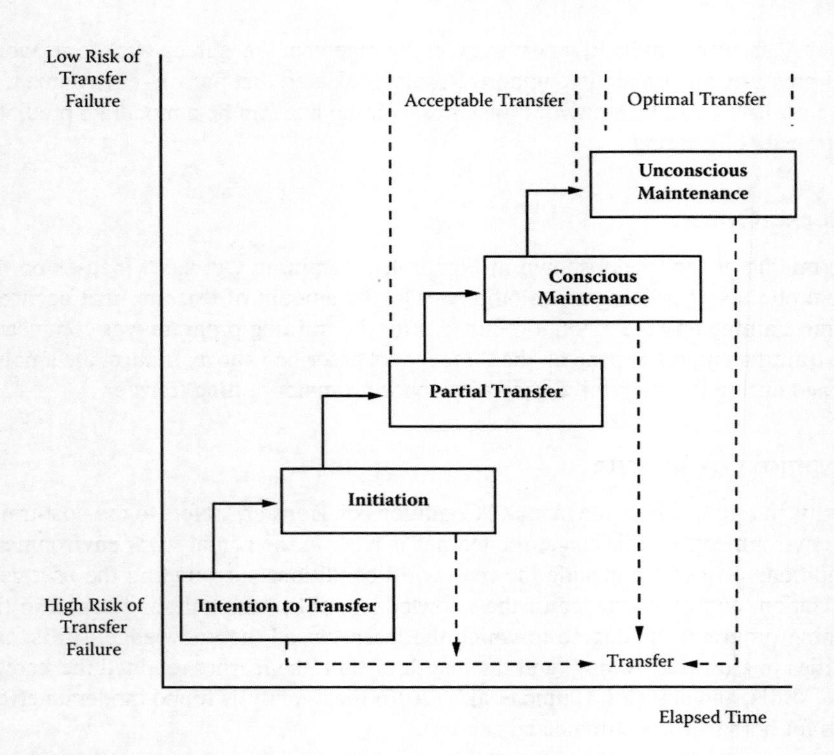

FIGURE 4.2 A model of the training transfer process. (Adapted from Foxon, M., *Aust. J. Educ. Technol.*, 9(2), 130–143, 1993.)

impacts and the interaction effects of the variables. We now turn to a discussion of the methodologies underlying transfer of training studies.

RESEARCH METHODS

How do researchers study transfer of training? Alternatively, how do designers validate their simulator as a training tool? It sounds straightforward enough: simply compare performance on the job before training and after training. Unfortunately, it is not that simple, and because of the complex nature of the problem, many unknowns still exist. Indeed, transfer of training research has been inconclusive due to the differences in concepts and definitions, differences in theoretical orientations, and methodological flaws. Perhaps the greatest problem is that transfer of training is difficult to measure (in terms of accurately determining whether transfer has occurred and how much has transferred). In other words, how transfer of training is conceptualized and when and how it is measured really do matter (Blume et al., 2010).

Two major issues are involved in assessing training transfer: performance measurement and research design. A full review of these topics is beyond the scope of this chapter, and several other chapters in this volume describe and discuss performance

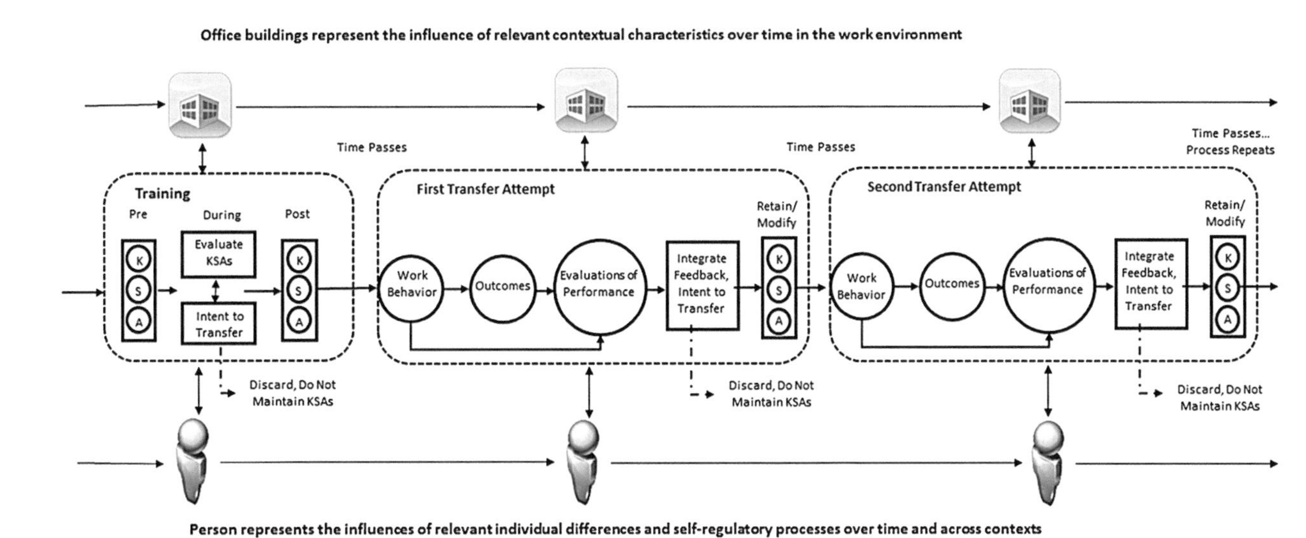

FIGURE 4.3 A model of the dynamic training transfer process. (Adapted from Blume, Ford, Surface, & Olenick, *Human Resource Management Review*, 29(2), 270–283., 2019.)

measurement. Since measuring training transfer is inherently linked to fundamental attributes of performance measurement for training effectiveness, we do offer some key terminology and issues.

TRANSFER OF TRAINING PERFORMANCE MEASUREMENT

Human performance (e.g., job performance) can be assessed in many ways. The most commonly used methods are objective performance measures and subjective judgments.

Objective Measures

Objective measures are accounts of various behaviors (e.g., number of errors) or the results of job behaviors (e.g., total passengers transported). Thus, objective measures are data that objectively reflect the trainee's performance level. An obvious problem with objective measurements is that they cannot provide insight into cognitive-related aspects of performance, such as workload, situational awareness, motivation, and attitudes. These variables are vital to transfer of training assessment because they are part of the transfer of training affected factors.

Subjective Measures

Another approach is to use subjective measures, which are ratings given by an expert: examples include surveys and questionnaires. Numerous examples of subjective performance measures can be found in the training literature.

SELECTING PERFORMANCE MEASURES

The selection of performance measures depends largely on the task. Indeed, it is rare for performance measures to be used in different tasks, and domains without at least some modification-training requirements are key (Hammerton, 1966; Bell et al., 2017). If the training purpose is to teach solely physical skills, objective measurement can be used as the primary tool. However, if cognitive training is involved, subjective performance assessment tools need to be applied to capture the cognitive skill learning and transfer. The notion of multiple measures also exists. For instance, Kraiger et al. (1993) advocated separate measures for cognitive, behavioral, and attitudinal factors, and the Kirkpatrick four levels of training evaluation model continue to be pervasive in the training and simulation literature (Kirkpatrick & Kirkpatrick, 2016). Combining the notions of multiple measures with the dynamic view of training transfer (e.g., Blume et al. (2019) suggest a need for multiple measures spread over multiple points in time.

Using Performance Measures to Indicate Transfer

Finally, as shown in Tables 4.1 and 4.2, Hammerton (1966), Roscoe and Williges (1980), and Taylor et al. (1999) present a number of different calculations that can be

TABLE 4.1
Trials/Time to Transfer or Trial to Criterion (TTC) Is the Number of Training Trials that a Subject Attempted for a Given Task before Reaching the Criterion Level of Proficiency on that Task, Given the Following Variables

Y_c	The control group's number of TTC for training conducted in aircraft
Y_x	The experimental group's number of TTC for training conducted in aircraft
X	The experimental group's number of TTC for training conducted in the simulator
F	The mean performance on the first simulator training trial
L	The mean performance on the last simulator training trial
T	The mean performance on the first post-transfer trial
C	The mean first trial performance of the real-situation training
S	The stable performance in real-situation training

TABLE 4.2
Transfer of Training Formulas

Name	Formula	Definition
Percent transfer	$\frac{Y_c - Y_x}{Y_c} \times 100$	Measures the saving of time or trials in aircraft training to criterion that can be achieved by use of a ground simulator
Transfer effective ration (TER)	$\frac{Y_c - Y_x}{X}$	Measures the efficiency of the simulation by calculating the saving of trials per time
First shot performance	$fsp = \cdot \frac{F - T}{F - L}$	Measures how much training will be retained on first transference to the real situation
Training retained (TR)	$tr = \cdot \frac{C - T}{C - S}$	Measures how much training is retained on the first post-transfer trial from simulator compared with that gained from real world

Source: Taylor, H.L. et al., 1997; Taylor, H.L., Lintern, G., Hulin, C.L., Talleur, D.A., Emanuel, T.W., Jr., & Phillips, , *Int. J. Aviat. Psychol.,* 9(4), 319–335, 1999; Hammerton, M., Proceedings of the IEE Conference, 113(11), 1881–1884, 1966.

done on the performance data. These include the following: percent transfer (the saving of time or trials in an aircraft by using a flight simulator); transfer effective ratio (measures the efficiency of the simulation); first shot performance (how much training will be retained on first transference to the real situation); and training retained (how much training is retained on the first post-transfer trial from the simulator compared with that gained from the real world).

In addition to performance measures, experimental design is another crucial issue to determine the degree of transfer of training that occurred. This will be discussed next.

EXPERIMENTAL DESIGN

Numerous training evaluation studies exist in the simulation-based training literature, some studies are dated back in 1940s (Valverde, 1973). Overall speaking, three types of studies include "Forward Transfer Study" (i.e., a predictive validation study), "Backward Transfer Study" (i.e., a concurrent validation study), and "Quasi-Experimental Study" (an approach similar to a construct validation study). 4.1

FORWARD TRANSFER STUDY

In the case of aviation, the classic experiment is to compare two matched groups of pilot trainees. One group (the control group) would be trained using only actual aircraft and the other (the experimental group) would receive their training via a simulator. The transfer environment is the actual aircraft. Pilot performance after training (e.g., TTC and flight technical performance) is captured, measured, and compared. This type of study is referred to as a Forward Transfer Study (Kaempf & Blackwell, 1990; Dohme, 1992; Darken & Banker, 1998), as it concurs with the transfer direction—first simulation, then the transfer environment. In some situations, this type of study can be expensive, time-consuming, or impossible to complete. For instance, in the occurrence of rare and dangerous events, such as engine failure, turbulence, or severe weather conditions, a forward transfer study is nearly impossible in a practical sense.

BACKWARD TRANSFER STUDY

To overcome the technical difficulties inherent in forward transfer studies, researchers use the Backward Transfer Study method. In a backward transfer study (i.e., concurrent validation study), current proficient aviators perform tasks on the job and performance measures are taken. Next, the same aviators perform the tasks in the simulation. Their performance in the simulation is compared with their performance on the job. The logic of the backward transfer study method lies in the following assumption: "If the aircraft proficient aviators cannot perform the flying tasks successfully in the simulator, the poor performance is attributed to deficiencies in the simulators" (Kaempf & Blackwell, 1990). Possible simulator deficiencies include the following: cues that may be different from the actual environment, controls that may be different from the actual environment, and the fact that flight simulators may require skills that are not required to fly aircraft (Kaempf & Blackwell, 1990).

Backward transfer studies use real pilots' performance in simulators only to predict forward transfer effectiveness for a particular simulator. If a low degree of backward transfer occurs, it implies that there are deficiencies in that simulator. Unfortunately, in the case of a high degree of backward transfer, it is not necessarily an indication of a high degree of forward transfer. It may simply mean that the pilots in the study were exemplary at getting the simulator to perform how they desired. Table 4.3 illustrates the formulas involved in backward transfer studies. Results from

Kaempf and Blackwell (1990) indicate that the inexpensive backward transfer studies may be employed to predict forward transfer tasks.

Quasi-Experimental Study

Lintern et al. (1997) and Stewart et al. (2002) completed a quasi-experimental study (i.e., similar to construct validation studies). In this method, transfer of training is compared between one of the configurations of simulation and another configuration of the same device. Quasi-experimental study is intended to investigate the basic knowledge about transfer of training principles and theories. Potential benefits include considerable savings in experimental cost and time, provided quasi-transfer methodology can be validated as the true transfer of training.

CURVE-FITTING METHOD

Another method used to assess transfer is the curve-fitting technique. One reason to use curve fitting is that the traditional transfer of training formulas does not take prior training amount or experience into account; the only data collected tends to be from immediately after one simulation exercise. Another shortcoming of traditional transfer of training formulas is that they only provide "crude" transfer by giving one global value. Additionally, the traditional estimates of transfer of training are not statistical tests (Damos, 1991).

In contrast, using the curve-fitting technique to assess transfer of training provides a more comprehensive measure of training transfer. In brief, the curve-fitting technique attempts to generate a learning curve for a particular task. Researchers find the best-fitting equation for the data. In this way, a more exact picture of the skill acquisition process is portrayed. Thus, the researcher is examining measures of learning over time rather than one individual measures.

TABLE 4.3
Index of Backward Transfer Formula (*B*)

$$B = \frac{\sum_{i=1}^{N} \left(\frac{A_i}{S_i} \right)}{N}$$

i = subject

N = total number of subjects

A = the mean of the subject's OPR scores for the last two trials in aircraft, and S = the subject's OPR score during second simulator check ride

B less than 1 indicates that performance in the simulator was substantially below that in the aircraft.

Source: Kaempf, G.L. and Blackwell, N.J., Transfer-of-training study of emergency touchdown maneuvers in the AH-1 flight and weapons simulator (Research Report 1561), 1990.

Curve fitting is claimed to be a major improvement for ToT estimates (Damos, 1991). Using this method, the curve equation provides estimates for rate of initial level of performance, rate of improvement, and the asymptotic level of performance for a particular group (e.g., type of training). Damos (1991) found that the curve-fitting method provided a much more detailed analysis of the data. This includes the following characteristics:

1. More insight into specific training effectiveness for differing training methods (i.e., calculating the inflection point and asymptote of the curves)
2. Transfer of training estimates without a control group
3. Statistical tests on curve parameters (to help assess differences between training interventions)

The typical steps involved in curve fitting include the following:

1. Visual inspection of data to "guess" an equation form (i.e., a general exponential equation (Damos, 1991) $dv = c \exp(gx) + h$, where dv is the dependent variable, c, g, and h are the parameters to be fit, and x is the trials block number)
2. Calculating the goodness-of-fit calculations by calculating the correlations between predicted and observed value
3. Performing a statistical test for each parameter

More curve-fitting techniques can be found in Spears (1985). Few curve-fitting studies have emerged over the years, possibly due to the time-consuming, expensive, and complex nature of this technique (Farmer et al., 2017).

SUMMARY

After reviewing the literature, a few points are quite clear. First, with 30 years of research behind it, the Baldwin and Ford model continues to provide guidance on the factors involved in transfer of training. Despite the presence of the overarching model, more research is needed before an exact understanding of the relative importance of the different variables will be achieved. Training transfer is linked acutely to measurement, and empirical studies with multiple measures of learning, transfer, and performance are needed to build our knowledge of training effectiveness in general (Bell et al., 2017). It is our hope that future empirical research will address the mechanism of the human learning process and the training transfer within different contexts and, in turn, will advance the understanding of the transfer of training.

REFERENCES

Auffrey, A.L., Mirabella, A., & Siebold, G.L., 2001, *Transfer of training revisited (Report ARI-RN-2001-10)*, Alexandria, VA: Army Research Institute for the Behavioral and Social Sciences.

Awoniyi, E.A., Griego, O.V., & Morgan, G.A., 2002, Person-environment fit and transfer of training, *International Journal of Training and Development*, 6(1), 25–35.

Baldwin, T., & Ford, J., 1988, Transfer of training: A review and direction for future research, *Personnel Psychology*, 41, 63–105.

Bell, B.S., Tannenbaum, S.I., Ford, J.K., Noe, R.A., & Kraiger, K., 2017, 100 years of training and development research: What we know and where we should go, *Journal of Applied Psychology*, 102(3), 305–323.

Blaiwes, A.S., Puig, J.A., & Regan, J.J., 1973, Transfer of training and the measurement of training effectiveness, *Human Factors*, 15(6), 523–533.

Blickensderfer, B., Strally, S., & Doherty, S., 2012, The effects of scenario-based training on pilots' use of a whole plane parachute, *International Journal of Aviation Psychology*, 22(2), 184–202.

Blume, B.D., Ford, J.K., Baldwin, T.T., & Huang, J.L., 2010, Transfer of training: A meta-analytic review, *Journal of Management*, 36(4), 1065–1105.

Blume, B.D., Ford, J.K., Surface, E.A., & Olenick, J., 2019, A dynamic model of training transfer, *Human Resource Management Review*, 29(2), 270–283.

Burke, L.A., 1997, Improving positive transfer: A test of relapse prevention training on transfer outcomes, *Human Resource Development Quarterly*, 8(2), 115–128.

Chapanis, A., 1996, *Human factors in systems engineering*, New York: John Wiley & Sons.

Colquitt, J.A., LePine, J.A., & Noe, R.A., 2000, Toward an integrated theory of training motivation: A meta-analytic path analysis of 20 years of research, *Journal of Applied Psychology*, 85(5), 678–707.

Damos, D.L., 1991, Examining transfer of training using curve fitting: A second look, *The International Journal of Aviation Psychology*, 1(1), 73–85.

Darken, R.P., & Banker, W.P., 1998, Navigating in natural environments: A virtual environment training transfer study, *Proceedings of VRAIS*, 12–19.

Dohme, J., 1992, Transfer of training and simulator qualification or myth and folklore in helicopter simulation (N93-30687), *NASA/FAA Helicopter Simulator Workshop Proceedings*, 115–121.

Federal Aviation Administration, 2006, Advisory circular. Retrieved from: https://www.faa.gov/regulations_policies/advisory_circulars/index.cfm/go/document.information / documentID/1030235

Farmer, E., van Rooij, J., Riemersma, J., & Jorna, P., 2017, *Handbook of simulator-based training*, Routledge.

Foxon, M., 1993, A process approach to the transfer of training. Part 1: The impact of motivation and supervisor support on transfer maintenance, *Australasian Journal of Educational Technology*, 9(2), 130–143.

Hammerton, M., 1966, Factors affecting the use of simulators for training, *Proceedings of the IEE Conference*, 113(11), 1881–1884.

Hays, R.T., Jacobs, J.W., Prince, C., & Salas, E., 1992, Flight simulator training effectiveness: A meta-analysis, *Military Psychology*, 4(2), 63–74.

Hays, R.T., & Singer, M.J., 1989, *Simulation fidelity in training system design*, New York: Springer.

Kaempf, G.L., & Blackwell, N.J., 1990, *Transfer-of-training study of emergency touch-down maneuvers in the AH-1 flight and weapons simulator (Research Report 1561)*, Alexandria, VA: U.S. Army Research Institute for the Behavioral and Social Sciences.

Kaiser, M.K., 2003, Flights of fancy: The art and science of flight simulation, in *Principles and practice of aviation psychology*, Tsang, P.S. & Vidulich, M.A., Eds., Mahwah, NJ: Lawrence Erlbaum Associates, pp. 435–471.

Ker, J., Hogg, G., Maran, N., & Walsh, K., 2010, Cost effective simulation, in *Cost effectiveness in medical education*, Radcliffe: Abingdon, pp. 61–71.

Kirkpatrick, J.D., & Kirkpatrick, W.K., 2016, *Kirkpatrick's four levels of training evaluation, 1st edition*, Association for Talent Development. ISBN-10: 1607280086

Kraiger, K., Ford, J.K., & Salas, E., 1993, Application of cognitive, skill-based, and affective theories of learning outcomes to new methods of training evaluation, *Journal of Applied Psychology*, 78, 311–328.

Lathan, C.E., Tracey, M.R., Sebrechts, M.M., Clawson, D.M., & Higgins, G.A., 2002, Using virtual environment as training simulators: Measuring transfer, in *Handbook of virtual environment*, Stanney, K.M., Ed., Mahwah, NJ: Lawrence Erlbaum Associates.

Lintern, G., Taylor, H.L., Koonce, J.M., Kaiser, R.H., & Morrison, G.A., 1997, Transfer and quasi-transfer effects of scene detail and visual augmentation in landing training, *The International Journal of Aviation Psychology*, 7(2), 149–169.

Moroney, W.F., & Moroney, B.W., 1998, Flight simulation, in *Handbook of aviation human factors*, Garland, D.J., Wise, J.A., & Hopkin, V.D., Eds., New York: Lawrence Erlbaum Associates.

Noe, R.A., 2006, *Employee training and development*, Irwin: McGraw-Hill.

Osgood, C.E., 1949, The similarity paradox in human learning: A resolution, *Psychological Review*, 56, 132–143.

Roscoe, S.N., & Williges, B.H., 1980, Measurement of transfer of training, in *Aviation psychology*, Roscoe, S.N., Ed., Ames, IA: Iowa State University Press, pp. 182–193.

Salas, E., Priest, H.A., Wilson, K.A., & Burke, C.S., 2006, Scenario-based training: Improving military mission performance and adaptability, in *Operational stress. Military life: The psychology of serving in peace and combat: Operational stress*, Adler, A.B., Castro, C.A., & Britt, T.W. Eds., Praeger Security International, pp. 32–53.

Simon, C.W., & Roscoe, S.N., 1984, Application of a multifactor approach to transfer of training research, *Human Factors*, 26(5), 591–612.

Smith-Jentsch, K.A., Salas, E., & Brannick, M.T., 2001, To transfer or not to transfer? Investigating the combined effects of trainee characteristics, team leader support, and team climate, *Journal of Applied Psychology*, 86(2), 279–292.

Spears, W., 1985, Measuring of learning and transfer using curve fitting, *Human Factors*, 27, 251–266.

Spector, P.E., 2003, *Industrial/organizational psychology: Research and practice*, 3rd ed., New York: John Wiley & Sons.

Stewart, J.E. II, Dohme, J.A., & Nullmeyer, R.T., 2002, U.S. Army initial entry rotary-wing transfer of training research, *The International Journal of Aviation Psychology*, 12(4), 359–375.

Taylor, H.L., Lintern, G., Hulin, C.L., Talleur, D.A., Emanuel, T.W. Jr., & Phillips, S.I., 1999, Transfer of training effectiveness of a personal computer aviation training device, *The International Journal of Aviation Psychology*, 9(4), 319–335.

Thorndike, E.L., 1903, *Educational psychology*, New York: Lemke & Buechner.

Thorndike, E.L., & Woodworth, R.S., 1901, The influence of improvement in one mental function upon the efficiency of other functions. II. The estimation of magnitudes, *Psychological Review*, 8(4), 384–395.

Valverde, H.H., 1973, A review of flight simulator transfer of training studies, *Human Factors*, 15(6), 510–523.

van der Locht, M., van Dam, K., & Chiaburu, D.S., 2013, Getting the most of management training: The role of identical elements for training transfer, *Personnel Review*, 42(4), 422–439. https://doi.org/10.1108/PR-05-2011-0072

5 Simulation-Based Training for Decision-Making

Providing a Guide to Develop Training Based on Decision-Making Theories

Richard J. Simonson, Kimberly N. Williams, Joseph R. Keebler, and Elizabeth H. Lazzara

CONTENTS

DOI: 10.1201/9781003401360-5

Decision-making is a core aspect of human work and performance. It includes a set of processes that individuals partake in every day of their lives to make choices and evaluate reality, both consciously and subconsciously. It is an activity that can both result in positive growth and beneficial outcomes, as well as lead to negative life-changing events and detrimental outcomes for the decision-maker and those influenced by their decisions. Additionally, decisions can compound themselves into habits that may further define individual and future judgments. Therefore, it needs to be trained to ensure that individuals are competent and knowledgeable about the success and failures associated with decision-making heuristics and biases.

One tool to mimic real-world scenarios in a safe, controlled, and non-consequential environment is simulation-based training (SBT; i.e., a methodology of providing structured experiences in a replicated environment to facilitate the acquisition of knowledge, skills, and attitudes; Salas et al., 2009). SBT can allow individuals an opportunity to practice a variety of decision-making tasks to gain insights into their own inherent biases, heuristics, and decision-making pathways. SBT can also be leveraged to train specific decision-making competencies, but the effectiveness of SBT relies just as much on the design of training and instructional strategy as it does on the efficacy of the equipment and tools used. Therefore, the purpose of this chapter is twofold: (1) to summarize decision-making theories and (2) to provide guidance for utilizing simulation-based training to enhance decision-making competencies. To accomplish these objectives, we lay out this chapter in the following order. First, we will provide a brief summary of the theoretical conceptualization of how people form decisions. Next, we will elucidate on how to design simulation-based training targeting decision-making. Finally, we conclude with the challenges and opportunities for research related to simulation-based training in decision-making.

THEORETICAL BACKGROUND OF DECISION-MAKING

Before we begin exploring how to best design SBT for decision-making, we provide a theoretical understanding of decision-making. Specifically, we describe the normative models of decision-making, expertise-based decision-making models, and common biases and heuristics associated with decision-making.

Normative Decision Models

Early models of decision-making were based on economic theories that took a normative or rational approach to decision theory. Two of these early theories include expected value and expected utility theories. The primary aspect of these normative theories posits that when humans are given a multitude of choices, we logically weigh those options against each other as a product of their relative values (or utilities) and the corresponding probabilities of occurrence (given inputs and possible outcomes), and ultimately we choose the path with the best total expected value. Normative theories would predict this type of rational decision-making for all decisions, regardless of whether the subject of the uncertainty was losses or gains. For example, imagine an individual is presented with two scenarios:

In the first scenario, the individual could choose either

A) a guaranteed gain of $200; or
B) have a 50% chance of gaining $500 and a 50% chance of gaining nothing at all.

In the second scenario, they could choose either

A) a guaranteed loss of $200; or
B) have a 50% chance of losing $500 and a 50% chance of losing nothing at all.

In each scenario's option B, the total expected value (Eq. 5.1) of the choice is $250 (gain in the first, loss in the second).

$$E(X) = \Sigma(X * P(X)) \tag{5.1}$$

Expected value theory would have predicted that in the first scenario an individual would choose option B because the expected value of this ($250) is higher than the option with the guaranteed gain ($200). In the second scenario, it would predict they would choose option A (guaranteed loss of $200) because the expected value of the guaranteed loss is lower than in option B (total expected value of the gamble is a loss of $250, which is a worse total outcome).

The example above demonstrates that normative theories do not align with how individuals typically make decisions. Humans are risk-averse, so framing based on losses and gains changes the way they make decisions. This is captured in the seminal work of Daniel Kahneman and Amos Tversky (1979), who modeled a more accurate picture of how and why humans actually make decisions with the groundbreaking Prospect Theory. This theory explains how humans systematically deviate from normative theories in their decision-making. One aspect of this is that individuals attribute different weights to equivalent gains versus losses because of their differential utility. In general, humans are risk-averse when presented with guaranteed versus uncertain gains and risk-seeking when faced with choices of guaranteed versus uncertain loss. So, in the example above, most individuals would choose option A in the first scenario (a risk-averse, "take the money and run" approach) and choose option B in the second scenario (risk-seeking, by risking a potentially large loss rather than be guaranteed to lose anything). Additional extensions of Prospect Theory have more to offer in understanding how humans deviate from normative theories and actually make decisions, primarily in their contribution to literature in heuristics and biases realm of decision-making literature, which will be discussed later in the chapter.

Decision Models: New Perspectives

Since Kahneman and Tversky (1979) highlighted the fallacies of normative models of decision-making, this field has evolved its research approaches in an attempt to accurately model human decision-making behaviors. As established, normative decision theories stipulate that individuals take the rational approach to all decisions, calculating the relative value of outcomes and probabilities associated with *n* number of options, regardless of the number and type of available options. Being so methodical requires substantial time and cognitive resources. Dedicating the amount of time and resources to make these types of calculations for all possible outcomes of all decisions is impractical.

Because it is not feasible to make rational comparisons for every option for every decision, individuals must have different strategies to make these decisions more efficiently. In the updated decision literature, there are two seemingly contradictory observations that give rise to two different realms of research: (1) that experts make remarkably fast and accurate decisions based on "intuitions" even when all of the evidence is unavailable and (2) individuals use shortcuts in their decision-making that lead them to make systematic errors in decisions and these errors change as a function of expertise. The next sections will discuss the intuition-based naturalistic decision-making (NDM) approach and the related recognition-primed decision (RPD) model, followed by evidence of systematic biases that occur in decision-making.

Naturalistic Decision-Making (NDM) Framework and Recognition-Primed Decision (RPD) Model

The NDM approach is an entire field of study that focuses on how individuals gain expertise and how this expertise (typically described as intuition, or pattern recognition) influences decision-making. NDM focuses more heavily on decision-making as it occurs in realistic scenarios, rather than laboratory environments. In fact, NDM first arose from observations of chess players and attempted to describe how expert chess players are able to make decisions quickly and accurately even in the inherently complex and uncertain context in which a chess game unfolds (Kahneman & Klein, 2009). More specifically, it attempts to elucidate what cues experts (i.e., chess masters) relied on to form recognizable patterns that aid their decision-making (Klein, 2015).

One of the primary models that has arisen from NDM is the RPD model. Klein et al. (1986) presented the RPD model to characterize how experts (i.e., firefighters) form rapid decisions under time pressure. RPD demonstrates how experience allows experts to refine their decision options to only one or two options that are likely to be effective; this refinement allows the expert to come to single decisions quickly and efficiently, rather than waste time identifying, comparing, and contrasting many possible options at once.

RPD postulates that this process occurs in four steps. The first step in this process is recognition—this is a phase in which the expert assesses the scene in front of them, and their assessment activates prior memories/experiences about similar scenarios they have encountered (either through training or direct experience; i.e., recognition priming) and draws forward a pattern associated with those scenarios. In the second step, experts then evaluate this pattern relative to the situation at hand (known as "progressive deepening"; Kahneman & Klein, 2009 as in deGroot, 1978) and if the pattern is accurate, decides on the action (or set of actions) that will be effective based on those associated with the pattern model they had formed in prior experience. In essence, experts are using heuristics (i.e., mental shortcuts) they have built based on their experiences with the domain by identifying patterns of *cues* that signal a course of action (*decision*) that will result in the desired *outcome*. In the third step, the expert can then challenge their own decision at various stages of the process to determine if it is still the best course of action based on the unfolding of events. If at any point the original pattern is no longer applicable to the scenario, the expert then enters the fourth step where they reevaluate and find an alternative pattern that fits the cues present and proceeds to mentally test and act on this situation accordingly. This entire process can repeat as many times as needed until the outcome is decided. This method of decision-making is highly beneficial due to its efficiency. Therefore, one major focus of decision training is to aid inexperienced individuals in learning cues that signal different patterns that dictate decision outcomes, so they can effectively mimic the decision skills of experts (e.g., Klein, 2015).

Biases in Decision-Making

Although this pattern recognition inherent in the RPD model may be able to lead to effective and efficient decision-making, it is also seen by some researchers as vulnerable to the influence of biases, which can ultimately lead to incorrect decision outcomes. The following sections will discuss various biases that have been known to affect decision-making abilities: confirmation, over- and under-confidence, framing, probability, and sunk costs.

Confirmation

Confirmation bias describes individuals' tendency to only seek information or cues that confirm already-held beliefs, ideas, or hypotheses (Fischer et al., 2008). A common representation of this bias is found in politics, where individuals tend to seek out evidence that only supports their party rather than discredits it. Cognitive bias occurs because it is difficult for individuals to interpret contradictory information (and experience cognitive dissonance as a result), so this bias simplifies the decision-making process by ignoring or down-weighting contradictory information. Although confirmation bias has historically been seen as negative, it is recently being viewed as potentially beneficial for making fast and accurate decisions in individuals who develop a high degree of metacognition (Rollwage & Fleming, 2021). Metacognition allows individuals to recognize when an initial conclusion may be incorrect and to seek out disconfirming evidence. Metacognition also allows individuals to gain the benefits of easier processing associated with confirmatory evidence for most decisions, while still permitting them to correct the effect of the bias when they realize their initial conclusion was incorrect.

Over- and Under-Confidence

Over- or under-confidence occurs when an individual fails to appropriately calibrate their perception of their own skill level to their actual abilities. These misperceptions can negatively influence decision-making. An example of this can be found in how individuals' financial knowledge and levels of confidence in the accuracy of this knowledge correlates with their investment behaviors. Pikulinaa et al. (2017) found that both types of inaccurate confidence relative to skill level affected individuals' decision-making on financial investments with very overconfident individuals making excessive investments and moderately under-confident individuals underinvesting. However, they also found that moderately overconfident individuals made the most accurate investments. These findings suggest that a slight degree of overconfidence may be beneficial when making decisions under conditions of uncertainty. Use of feedback, counter-thinking, and interventions that modify affect have been suggested as methods to calibrate confidence judgments; although, each of these methods has limited empirical support (Koellinger & Treffers, 2015).

Framing

Framing effects were first identified by Kahneman and Tversky (1979) in the context of Prospect Theory (discussed earlier in this chapter) and occur when the way potential outcomes are presented influences the decision. As a greatly simplified example, when physicians are presented with two different treatments and one is listed with a 98% survival rate while the other is listed as a 2% mortality rate, physicians may be more likely to choose the option listed with a 98% survival rate (see Fridmen et al., 2021 for a recent summary and extension of research exploring framing effects in the medical domain). This choice is because individuals tend to demonstrate risk-averse behavior when presented with a decision in a gain frame; meanwhile, individuals exhibit risk-seeking behavior when information is presented in a loss frame (e.g., Kahneman & Tversky, 1979). Empirical evidence for the framing effect is mixed. A meta-analysis supported its robustness (Steiger & Kühberger, 2018), but there is only partial support existing from statistical and mathematical modeling of the effect (e.g., Fan, 2017). Recent evidence, though, shows the impact of the framing effect on decisions may be reduced through explicit training in statistics/mathematics (Borracci et al., 2020).

Probability Perception (Gambler's Fallacy)

Humans sometimes demonstrate biases when predicting sequential outcomes based on chance probabilities. An example of this is often represented by a coin toss example from Kahneman and Tversky's 1972 paper; an individual is presented with the following sequences of heads/tails scenarios and asked to identify which is the most likely sequence given fair coin tosses: HHHHTH, HTHTTH, HHHTTT. In this paradigm, individuals will typically select HTHTTH, even though all of these scenarios are equally likely since each coin toss is independent of every other coin toss. This selection occurs because the "random" (more alternating) scenario is more representative of what individuals expect to see out of fair (i.e., random) coin tosses, though that perception does not match with how probability odds actually occur (e.g., Tversky & Kahneman, 1974).

An extension of probability perception is gambler's fallacy. Gambler's fallacy occurs when individuals see patterns in random previous events and mistakenly see these patterns as influencing the probabilities of future events. For example, if given a gambling scenario where an H is a "loss" and T is a "win," after a series of H results (HHHHHH), a chronic gambler is likely to incorrectly think that the next toss is more likely to be a T. The rationale for this choice is that the gambler believes that equivalent probability of occurrence of either H or T should equate to actual equivalent representation in reality; thus, they believe they are "due for a win" to even out the string of losing Hs. Recent research indicates that training that engages analytical thinking specific to these types of scenarios may reduce this effect and related gambling behaviors (Armstrong et al., 2020).

Sunk Costs

Once individuals invest something (e.g., time, effort, and money) into an outcome, they can fall victim to the sunk cost fallacy. A sunk cost is an investment that cannot be recovered, and sunk cost fallacy occurs when an individual (consciously or subconsciously) lets these sunk costs influence their future investment decisions. As an example, a small business owner decides to buy an expensive piece of machinery that will allow them to make items they think will sell in their shop. After several months, they fail to sell any of the products they have made with the machine thus far. The business owner knows how much money they spent on the machine and wants to put that money back into the products, so they invest more money to buy different materials to try to make the same products more appealing. This cycle may then repeat. Until the business owner can remove the previously incurred sunk costs (the machine, materials, etc.) from their decision to invest further, they are succumbing to the sunk cost fallacy. Sunk cost has been posited as a loss-oriented framing situation using Prospect Theory; that is, the frame of the sunk cost generates the risky investment behaviors associated with loss-framing (Thaler, 1980). This decision strategy is detrimental because it can lead to overinvestment in unprofitable projects and a high waste of resources (Roth et al. 2015 provide an overview of applied domains and potential varying effects). Activities that encourage introspection and suppression of future thinking toward projects have been seen as potentially successful in preventing sunk bias from influencing decisions (e.g., Strough et al., 2016).

Decision Theory Applied to Training

Thus far, we have presented several concepts of actual decision-making strategies with a focus on two approaches: (1) intuitive decision-making developed through expertise and (2) systematic biases that individuals may employ as decision-making strategies. The debate is ongoing as to which of these two approaches should be capitalized on most to train more effective decision-making skills (e.g., Anderson et al., 2019). Both approaches have merit; that is, experts can make good decisions based on intuition, *and* all individuals can experience systematic biases in decision-making. Proponents of each approach have come to agree that the primary determination lies in the predictability of the environment decisions taking place as well as how well the decision-maker has been able to learn the boundaries of that environment and the impact of their decisions (see Kahneman & Klein, 2009). In other words, which approach decision-makers rely on is based on how reliably cues present in the environment can predict outcomes of decisions and how well individuals have been exposed to these cues and outcomes to learn their associations correctly. Based on Kahneman and Klein's interpretations (described above), we might expect that in a more reliable environment, a higher degree of practice-based training focused on building experience may lead to effective intuition-based decision-making (see also Klein, 2015); meanwhile, in a low-reliability environment, the risk of bias negatively impacting decisions is high. Therefore, training should focus more explicitly on reducing the effects of bias. Regardless, training decision-making abilities provide trainees an opportunity to build upon their abilities to identify risks and biases as well as recognize and utilize heuristics to obtain an expert-based framework in various decision-making tasks. Thus, the following sections will describe how to create a decision training paradigm in a simulated environment.

STEPS IN DEVELOPING SIMULATIONS TO TRAIN DECISION-MAKING

Similar to other cognitive processes and skills, effective decision-making is not simply rooted in intuition, but rather effective decision-making requires a deliberate training curriculum. Due to the potential for disastrous outcomes associated with poor decision-making, decision-making is ripe for simulation-based training. Simulation-based training, when executed properly, provides the ability to replicate real-world tasks and environments while still within safe parameters void of actual catastrophes (Hochmitz & Yuviler-Gavish, 2011).

Thus, the following portion of this chapter will provide a guide to applying decision-making training in a simulation-based training context. We will specifically focus on the following: (1) determining what training is required, (2) identifying the learning objectives, (3) selecting the context of the simulation, (4) establishing the knowledge, skills, and attitudes (KSAs), (5) developing events to elicit KSAs, and (6) creating an assessment plan.

CONDUCT A NEEDS ASSESSMENT

The primary step in beginning any training endeavor is the implementation of a training needs assessment. Training needs assessments can assist in identifying the challenges that individuals face in completing various tasks as well as identify the competencies associated with said tasks. The application of a training needs assessment is typically associated with three objectives before initiating a training initiative. These include (1) identifying what competencies need to be trained, (2) establishing if the needs can be met by training, and (3) proposing a training solution (Brown, 2002). Although resource-intensive, a training curriculum that is not initiated by a needs assessment may lead to ineffective training and even negative organizational effects (Cekada, 2011). Additionally, as decision-making competencies are often dependent on the task and its associated cognitive processes, ignoring this needs assessment step can result in inappropriate training in decision-making that does not target the appropriate competencies.

Identify Learning Objectives

Once a needs assessment has been completed, one can begin to define the learning objectives. Learning objectives are the specifications of what the training is intending to accomplish. Consequently, learning objectives are the foundation of the curriculum and should be clear, concise, and measurable. Therefore, when crafting learning objectives, one needs to consider what specific metric will measure the corresponding learning objective as well as how to determine if the objective has been reached (e.g., at what point is the trainee considered successful?). If learning objectives do not adhere to those parameters, it becomes difficult to determine if the training is accomplishing its desired goals. Within the context of decision-making, a learning objective may relate to trainees significantly improving their knowledge of the biases that could disrupt their decision-making abilities in time-sensitive tasks.

Set the Simulation Context

When the learning objectives have been identified, the next step is to determine the simulation context. The simulation context provides a boundary to specify what should and should not be trained in the specified curriculum (Rosen et al., 2008). Establishing the simulation context assists in fostering cognitive fidelity. Cognitive fidelity refers to the extent that simulation requires the trainee to engage in the thought processes involved in the real-world task (Hochmitz & Yuviler-Gavish, 2011). Because decision-making is arguably a cognitive process, it is particularly relevant that the context fosters cognitive fidelity. Cognitive fidelity is essential for trained KSAs transferring to the real-world task (Hochmitz & Yuviler-Gavish, 2011); therefore, not all contexts are created equally. Rather, contexts should be carefully scrutinized such that the decision-making competencies targeted within the training can be elicited accordingly. For simulation-based training, setting contexts that can elicit specific competencies is crucial. Archer et al. (2006) provide an example of a decision-making storyline that utilizes decision-based nodes which dynamically change the story as the trainee moves through the scenario. Their specific context is that of soldiers planning a maneuver to a location but may end up in an ambush (failure), minefield (failure), or bridge (success).

Establish KSAs

Following the determination of the simulation context, one should determine the necessary, specific knowledge, skills, and attitudes (KSAs). Described as the key components that should be elicited during training and, more importantly, during task execution, KSAs serve as a mechanism to provide information on what the individual knows (knowledge), what the individual can do (skill), and what the individual feels (attitude) (Bloom, 1956). Targeting KSAs during training increases the potential that individuals will exhibit the KSAs in the real-world task (Bloom, 1956).

Due to the complexity of decision-making as described in the various cognitive models (i.e., RPD & NDM), training every KSA within the entire realm of decision-making is impractical from a pedagogical perspective. Key KSAs can be identified through various methods, such as an ethnographic study, cognitive task analyses of various work, or a review of commonly used decision-making competencies within the decision-making literature base (see Stanton et al., 2013). For example, a multitude of decision-making KSAs have been researched and utilized within the training and decision-making research domain and include (1) resistance to framing: the degree to which value assessments are affected by confounding factors, (2) recognizing social norms: the ability to gauge social norms, (3) under- and overconfidence: the level of understanding of one's own knowledge, (4) applying decision rules: the ability to apply elimination by aspects, satisficing, lexographic, and equal weights rules, (5) consistency in risk perception: the knowledge and ability to adhere to probability rules, (6) resistance to sunk costs: the degree to which one can ignore prior decisions regarding investments (emotional, financial, etc.), and (7) path independence: the ability to make consistent decisions regarding multi-stage events (c.f., Bruine de Bruine et al., 2007; Parker & Fishhoff, 2005; Parker et al., 2015; Finucane & Gullion, 2010).

Create Events to Elicit KSAs

Since the objective of SBT is to foster the ability to elicit specific KSAs by a trainee, one should rely on embedding trigger events. A trigger event serves as a cue to prompt the trainee to demonstrate the targeted KSAs (Rosen et al., 2008). Trigger events should correspond to the learning objectives of the training, where each learning objective is associated with at least one trigger event. Trigger events are arguably the building blocks of the simulation as they serve as a mechanism to consistently generate specific, observable key KSAs.

Each trigger event should be associated with at least one expected response in which the trainee exhibits the desired knowledge, skill, or attitude. Knowledge-based responses typically are used to elicit some type of mental store of information. These responses are usually followed by a verbal or written trigger requesting information from the trainee. Skill responses, on the other hand, necessitate a psychomotor reaction (i.e., physical movement) to a trigger that is typically observed via behavioral observation.

An example of a decision-making simulation scenario in the context of a soldier planning a route through a combat zone is presented below. Each trigger (Tr) is tied to the response (R) in which the trainee is given an opportunity to elicit a knowledge, skill, or attitude that can be observed and measured by the trainer.

Following the example from Archer et al. (2006), soldiers (trainees) are provided with a scenario in which they are asked to plan a route from some starting area to a recon bridge. During the training scenario they are asked to plan a route (Tr) and choose a route (Tr) which eventually spans out to a minefield (Tr), and ambush (Tr). However, every time they encounter one of these triggers, they are asked to plan a new route (R) based on the information garnered along their scenario.

Establish an Assessment Plan

Conducting the needs assessment, determining the learning objectives, selecting the simulation context, establishing KSAs, and finally creating the events and responses are all the steps one needs to develop a simulation-based training scenario. However, no training can be effective without assessment. Assessment facilitates the ability to determine if the training intervention was effective by providing evidence of demonstrable improvement in KSAs derived from the learning objectives. The first step in deciding on an assessment is to determine how to approach measuring training efficacy. The primary decision factor is based on three components.

One component is the type of competency being measured. Assessing knowledge, skills, and attitudes requires different assessment types. For example, knowledge and attitudinal assessments often rely on questionnaires; meanwhile, a psychomotor-based behavior (i.e., skill) cannot be measured and evaluated using such a questionnaire. Thus, determining when to utilize questionnaires, observational checklists, or even interview-based assessment should be carefully considered and correspond to the objectives and KSAs of the training.

To elaborate further, knowledge-based competencies rely on the recall or recognition of information typically assessed through questionnaires. Although the format may vary, knowledge-based assessments often consist of multiple-choice and free-response style questions. Regardless of the exact format, knowledge-based assessments offer the ability to quantitatively detect any changes in knowledge based on the training intervention.

Skill-based competencies require that some psychomotor task is observed to determine if a desired behavior was performed. There are a variety of formats of behavioral tools. One option is a checklist. Checklists are considered to be a dichotomous rating, wherein observers indicate the presence or absence of a behavior. Checklists are the least cognitively demanding on raters; however, dichotomous ratings are not very diagnostic by not providing any insights regarding the quality of the exhibited behavior. In other words, checklists indicate whether a behavior was present, but they do not indicate if the behavior was exhibited superbly or poorly. Another option is a frequency count. Frequency counts, on the other hand, require more attentiveness, are more cognitively demanding, and can be prone to more reliability issues compared to dichotomous ratings. Frequency counts, though, are particularly useful when the same behavior is expected to be demonstrated repeatedly. They can offer information regarding how often a trainee exhibited a behavior. A third option is behaviorally anchored rating scales (BARS). BARS typically have numeric ratings that are accompanied by observable behaviors. BARS can be challenging to develop as each numeric score requires a well-defined behavior determined a priori. Regardless of the format, each behavioral assessment option has its respective advantages and disadvantages. Consequently, which option is selected should be based on the specific competency being assessed as well as the available resources and practical constraints.

Finally, attitude-based competencies are often measured via the application of a survey. Surveys are the implementation and collection of perception-based questions that trainees can answer to provide subjective beliefs. To ensure that trainees have a meaningful and appropriate range of responses to subjective questions regarding their beliefs, Likert-type anchor responses are used. These Likert-type responses are common in surveys and a multitude of anchors have been devised (Vagias & Wade, 2006) to assist in capturing the correct feelings of the trainees. For example, Grol et al.'s (1990) attitudinal questionnaire assesses risk perception and Siebert and Kunz (2016) evaluate proactive attitude in decision-making.

Another component is whether there are psychometrically validated tools available. An important note on competencies and their associated measurable KSAs is the psychometric reliability and validity of the metrics utilized. Psychometrics consists of a set of processes to ensure that the metric is measuring constructs accurately and consistently (Shrout & Lane, 2012). Relying on metrics that lack reliability and validity may result in spurious assessments that do not provide any insight into the effectiveness of the training. One validated decision-making tool is the ADMC, which determines an adult's decision-making competency based on six factors (Bruine de Bruine et al., 2007).

A third component relates to the administration of the metric(s). One approach is referred to as a single measurement, and a second approach is multiple measurement. A single measurement approach relies on one method of measurement, either a knowledge, skill, or attitude, to assess a competency or set of competencies. This approach is great for trainings that have resource and time constraints, so the training relies on quick or simple assessments. As an example of a single measurement in decision-making, risk perception can be measured by a trainee exiting a bet. While this method does not require extensive time or resources, it is generally less informative in determining the efficacy of the training compared to the multiple method approach.

The multiple method approach, as the name suggests, uses multiple assessment types. This approach is far more effective at assessing proficiency as each learning objective and corresponding KSA is being measured in multiple instances. An extension of this approach, which is commonly believed to be a gold standard of assessment due to its ability to measure a variety of competency-based validities, is the multi-trait multi-method (MTMM) approach (Eid & Diener, 2006). MTMM uses a construct approach and attempts to correlate knowledge-, skill-, and attitude-based assessments to one another when targeting a specific set of competencies. In extending the above example to MTMM, we might first assess the knowledge of the trainees' risk perception, then test them in a behavioral context and assess their attitude toward withdrawing from or staying in the gamble. Once all these assessments are compiled, they would be constructed into a model and tested to determine if they are quantifiably associated with one another. Theoretically, if a trainee has a strong perception of risk, then all three measurements should be highly correlated to one another. As portrayed in this example, though, this method is substantially more time- and resource-intensive compared to the single measurement approach. Investing resources into assessment, though, is a worthwhile endeavor as assessment is foundational for effective training.

CONCLUSION

The purpose of this chapter was twofold. The first purpose was to provide a summary of decision-making theory. Our review highlighted milestones in decision-making research and described the cognitive processes behind decisions. More specifically, the Prospect Theory illustrated the concept that people do not make calculated decisions based on expected values in a strict probabilistic sense. Rather, people are risk-averse or risk-taking depending on the situation. Meanwhile, the NDM and RPD models provided a background into how people gain expertise in decision-making capabilities based on pattern recognition and experience. NDM and RPD also elucidate the idea that the process behind building this experience is susceptible to bias that influences decisions. Although these biases are inherent within cognitive processing, effective training can mitigate some of the potential problems.

The second purpose of this chapter was to offer some guidance on how to utilize simulation-based training to enhance decision-making. To maximize the benefits, though, of simulation-based training, curriculums should be rooted in the following: (1) determining what training is required, (2) identifying the learning objectives, (3) determining the context of the simulation, (4) establishing the knowledge, skills, and attitudes (KSAs), (5) developing events to elicit KSAs, and (6) creating an assessment plan. Adhering to the science of training, SBT can be one way to effectively train decision-making competencies and alleviate the biases that affect them all within the context of a safe environment.

REFERENCES

Anderson, N. E., Slark, J., & Gott, M. (2019). Unlocking intuition and expertise: Using interpretative phenomenological analysis to explore clinical decision making. *Journal of Research in Nursing*, *24*(1–2), 88–101. https://doi.org/10.1177/1744987118809528

Archer, R., Brockett, A. T., McDermott, P. L., Warwick, W., & Christ, R. E. (2006). *A simulation-based tool to train rapid decision-making skills for the digital battlefield*. Micro Analysis and Design Boulder Co.

Armstrong, T., Rockloff, M., Browne, M., & Blaszczynski, A. (2020). Training gamblers to re-think their gambling choices: How contextual analytical thinking may be useful in promoting safer gambling. *Journal of Behavioral Addictions JBA*, *9*(3), 766–784. Retrieved April 21, 2021, from https://akjournals.com/view/journals/2006/9/3/article-p766.xml

Bloom, B. S. (1956). *Taxonomy of educational objectives: The classification of educational goals. Cognitive domain*. Longman.

Borracci, R. A., Arribalzaga, E. B., & Thierer, J. (2020). Training in statistical analysis reduces the framing effect among medical students and residents in Argentina. *Journal of Educational Evaluation for Health Professions*, *17*(25). https://doi.org/10.3352/jeehp.2020.17.25

Brown, J. (2002). Training needs assessment: A must for developing an effective training program. *Public Personnel Management*, *31*(4), 569–578.

Bruine de Bruin, W., Parker, A. M., & Fischhoff, B. (2007). Individual differences in adult decision-making competence. *Journal of Personality and Social Psychology*, *92*(5), 938.

Cekada, T. L. (2011). Need training? Conducting an effective needs assessment. *Professional Safety*, *56*(12), 28.

Eid, M., & Diener, E. (Eds.). (2006). *Handbook of multimethod measurement in psychology*. American Psychological Association. https://doi.org/10.1037/11383-000

Fan, W. (2017). Education and decision-making: An experimental study on the framing effect in China. *Frontiers in Psychology*, *8*. https://doi.org/10.3389/fpsyg.2017.00744

Finucane, M. L., & Gullion, C. M. (2010). Developing a tool for measuring the decision-making competence of older adults. *Psychology and Aging*, *25*(2), 271.

Fischer, P., Jonas, E., Frey, D., & Kastenm¨uller, A. (2008). Selective exposure and decision framing: The impact of gain and loss framing on confirmatory information search after decisions. *Journal of Experimental Social Psychology*, *44*, 312–320.

Fridman, I., Fagerlin, A., Scherr, K. A., Scherer, L. D., Huffstetler, H., & Ubel, P. A. (2021). Gain–loss framing and patients' decisions: A linguistic examination of information framing in physician–patient conversations. *Journal of Behavioral Medicine*, *44*, 38–52. https://doi.org/10.1007/s10865-020-00171-0

Grol, R., Whitfield, M., Maeseneer, J. De, & Mokkink, H. (1990). Attitudes to risk-taking in medical decision-making among British, Dutch and Belgian General-Practitioners. British Journal of General Practice, 40(333), 134–136.

Hochmitz, I., & Yuviler-Gavish, N. (2011). Physical fidelity versus cognitive fidelity training in procedural skills acquisition. *Human Factors*, *53*(5), 489–501.

Kahneman, D., & Klein, G. (2009). Conditions for intuitive expertise: A failure to disagree. *American Psychologist*, *64*(6), 515–526. https://doi.org/10.1037/a0016755

Kahneman, D., & Klein, G. (2009). Progressive deepening. In deGroot, A. D. (1978). *Thought and choice in chess*. The Hague: Mouton. (Original work published 1946.)

Kahneman, D., & Tversky, A. (1979). Prospect theory: An analysis of decision under risk. *Econometrica*, *47*, 263–291.

Klein, G. (2015). A naturalistic decision making perspective on studying intuitive decision making. *Journal of Applied Research in Memory and Cognition*, *4*, 164–168.

Klein, G. A., Calderwood, R., & Clinton-Cirocco, A. (1986). Rapid decision making on the fire ground. *Proceedings of the Human Factors Society Annual Meeting*, *30*(6), 576–580. https://doi.org/10.1177/154193128603000616

Koellinger, P., & Treffers, T. (2015). Joy leads to overconfidence, and a simple countermeasure. *Plos One*, *10*(12). https://doi.org/10.1371/journal.pone.0143263

Parker, A. M., Bruine de Bruin, W., & Fischhoff, B. (2015). Negative decision outcomes are more common among people with lower decision-making competence: An item-level analysis of the decision outcome inventory (DOI). *Frontiers in Psychology*, *6*, 363.

Parker, A. M., & Fischhoff, B. (2005). Decision-making competence: External validation through an individual-differences approach. *Journal of Behavioral Decision Making*, *18*(1), 1–27.

Pikulina, E., Renneboog, L., & Tobler, P. N. (2017). Overconfidence and investment: An experimental approach. *Journal of Corporate Finance*, *43*, 175–192.

Rollwage, M., & Fleming, S. M. (2021). Confirmation bias is adaptive when coupled with efficient metacognition. *Philosophical Transactions of the Royal Society B*, *376*(1822), 20200131.

Rosen, M. A., Salas, E., Silvestri, S., Wu, T. S., & Lazzara, E. H. (2008). A measurement tool for simulation-based training in emergency medicine: The simulation module for assessment of resident targeted event responses (SMARTER) approach. *Simulation in Healthcare*, *3*(3), 170–179.

Roth, S., Robbert, T., & Straus, L. (2015). On the sunk-cost effect in economic decision-making: A meta-analytic review. *Business Research*, *8*, 99–138. https://doi.org/10.1007/s40685-014-0014-8

Salas, E., Rosen, M. A., Weaver, S. J., Held, J. D., & Weissmuller, J. J. (2009). Guidelines for performance measurement in simulation-based training. *Ergonomics in Design: The Quarterly of Human Factors Applications*, *17*(4), 12–18.

Shrout, P. E., & Lane, S. P. (2012). Psychometrics. In M. R. Mehl & T. S. Conner (Eds.), *Handbook of research methods for studying daily life* (pp. 302–320). The Guilford Press.

Siebert, J., & Kunz, R. (2016). Developing and validating the multidimensional proactive decision-making scale. *European Journal of Operational Research*, *249*(3), 864–877.

Stanton, N., Salmon, P. M., & Rafferty, L. A. (2013). *Human factors methods: A practical guide for engineering and design*. Ashgate Publishing, Ltd.

Steiger, A., & Kühberger, A. (2018). A meta-analytic re-appraisal of the framing effect. *Zeitschrift für Psychologie*, *226*(1), 45–55. https://doi.org/10.1027/2151-2604/a000321

Strough, J., Bruine de Bruin, W., Parker, A. M., Karns, T., Lemaster, P., Pichayayothin, N., Delaney, R., & Stoiko, R. (2016). What were they thinking? Reducing sunk-cost bias in a life-span sample. *Psychology and Aging*, *31*(7), 724–736. https://doi.org/10.1037/pag0000130

Thaler, R. (1980). Toward a positive theory of consumer choice. *Journal of Economic Behavior & Organization*, *1*(1), 39–60.

Tversky, A., & Kahneman, D. (1974). Judgment under uncertainty: Heuristics and biases. *Science*, *185*(4157), 1124–1131.

Vagias, W. M. (2006). *Likert-type scale response anchors*. Clemson International Institute for Tourism & Research Development, Department of Parks, Recreation and Tourism Management. Clemson University.

6 Almost Like the Real Thing – The Hidden Limits in Flight Simulation and Training

Shem Malmquist, Deborah Sater Carstens, and Nicklas Dahlstrom

CONTENTS

INTRODUCTION

Probably the most visible aspect of modern pilot training is the full-motion flight simulator. Far removed from the early Link trainers, with their motion attached only to the pilot controls and a sensation while "flying" them feeling similar to trying to balance on the head of a pin, today's simulators are a marvel of technology. With these impressive machines costing in the neighborhood of $10 million, it is not surprising that they are often showcased. These devices include a six-axis of motion via hydraulic or electric actuators, full "vision" for the pilots, as well as usually containing the actual flight controls and instruments manufactured for the real aircraft. Computer systems supply the simulator with an artificial but very detailed "world," and the instruments, systems, motion, and visual and sound systems react to that environment as they would if the environment were real. But how real are they really, and what problems might be present that are not described in the academic literature or the guidance material published by their manufacturers?

DOI: 10.1201/9781003401360-6

As with any modern endeavor, many researchers and flight training departments rely on the work of others. Previous work in the field serves as the foundation upon which new research is conducted and the foundation for how training programs are designed. The majority of literature describes the research and methods which can then form these foundations. This chapter is a bit different. Rather than describe research or methods, this chapter will delve into the aspects that the reader may not have considered. The intention is to enable the reader to cast a critical eye on the published literature. Does the research address internal and external validity? Are the instruments used to measure and analyze the training valid and reliable? Are the methods employed in the research appropriate to ensure internal validity? What limitations and delimitations might have been present? Were these reported or even known to the researchers?

Tuncman (2020) found that threats to the internal validity of research were not reported and possibly not addressed in 94% of the literature he reviewed (p. 120). Also, quasi-transfer of training studies, the lack of peer review for findings in accident reports, variations and limitations of simulator motion drive algorithms, and methods used in research can create a perfect storm to yield research or a training program that, on the surface, appears robust, but may not be providing the outcome that was intended.

As is described in Chapter 4, the goal of training is to transfer what is learned in training to the real world. This depends on the ability of the training environment to be generalized to the real world and the general population. In this chapter, we review the limitations that researchers and those working in training development should consider in the hopes that knowledge can improve both the quality of the research and the training.

INTRODUCTION TO SIMULATOR MOTION

Although flight simulators are used for various purposes, from initial aircraft handling qualities to development of pilot procedures, flight deck design, airspace management, and accident investigation, this chapter will address their use relevant to pilot training. There has been debate in research circles about whether or not simulator motion is beneficial in a training context (Burki-Cohen et al., 1998; Burki-Cohen & Go, 2005; Winter et al., 2012). The goal for a simulator in a training environment is to provide *cognitive fidelity*, which is the extent that the simulator can require the pilot to engage in the "same sort of cognitive activities … as the actual modern fight deck" (Kaiser & Schroeder, 2003, p. 440). To assure that the simulator fidelity is beneficial for pilot training, it needs to be valid, meeting the goal of accurately modeling the real aircraft to the extent necessary to accomplish the goals for transfer of training reliably. To identify the higher-level fidelity necessary for training, the simulators are evaluated with either rating methods or *quasi-transfer* methods. "In a quasi-transfer experiment, transfer of training is tested in the simulator with motion as a stand-in for the airplane" (Burki-Cohen & Go, 2005, p. 6). What are the limitations of these approaches?

Much of the research for flight simulator motion has involved *quasi-transfer* exercises in testing the validity of simulator motion. In such a study, transfer of training is tested using the same simulator throughout the research and not on an aircraft. This is because testing in an aircraft presents many challenges, some obvious ones such as cost and safety for certain maneuvers and the control of extraneous variables. The ability, in research, to control aspects such as daylight or darkness and other factors makes simulation ideal for ensuring reliability and replicability of studies. Unfortunately, the same aspects represent threats to internal validity that need to be addressed. The increased control over the experiment comes at a cost to external validity (Ary et al., 2010).

A simulation is, in the end, a simulation. It is necessarily a model of reality. As George E. P. Box stated, "Since all models are wrong, the scientist must be alert to what is importantly wrong" (Box, 1976, p. 792). Understanding what is "importantly wrong" with simulation is crucial for any organization conducting training for pilots using these high-technology devices. Flight simulation fidelity requirements for aircraft certification require higher standards, although these are not codified. The simulators used for training must meet regulatory standards. In the United States, these requirements are delineated in Title 14 CFR 60 of the Federal Aviation Administration (FAA) regulations. These regulations require that the key simulator equipment is able to duplicate the feel and appearance of the actual aircraft (FAA, 1992; Rehmann et al., 1995). To accomplish this, advanced modern simulators often use actual aircraft components in the cockpit, these include the same instruments and controls as the real airplane. The simulators also model the visual information and add motion to increase the fidelity of the experience for pilots (Kaiser & Schoeder, 2003). The main point of this is to emulate the control and feedback that pilots experience in the real world. As described in system engineering, pilots (controllers in the system) use their control to change the state of the aircraft and feedback to know what the current state is and what corrections they might need to make (Checkland, 1971).

Even if the instruments and controls are the exact components used in a real airplane, the simulation is still a model. As stated by Weisberg (2012), "A model is similar to its target…when it shares certain highly valued features … and when the target doesn't have many significant features the model lacks" (pp. 144–145). The instruments and components are being "fed" information from an artificial world that is simpler than the real world (Kaiser & Schroeder, 2003). This affects the simulation in several important ways, which include both visual and motion.

The visual systems are not as varied and detailed as in the real world; using several "cheats" to somewhat mitigate these differences, such as using "actual components of the environment, most notably the cockpit interior" (p. 446). Adding in the indications and "force feedback to the controls complete the emulation" (p. 446). The second "cheat" as described by Kaiser and Schroeder is "to not even try" (p. 447). Simulators use computer-generated imagery, which, although it looks good, is nowhere close to reality. Still, it is "good enough" to create impressions of aircraft motion in many cases. Nonvisual senses, coming from the inner ear, skin, and joint receptors, supplement the visual sense, and are particularly useful for detecting aircraft aerodynamic, propulsive, and gear-specific forces. These specific forces are immediately perceived with motion and they provide the pilot with immediate feedback to make a much quicker control response (Kaiser & Schroeder, 2003).

In simple terms, to design how the simulator should move, the aircraft model is programmed with data provided by the aircraft manufacturer, from which one can infer how the real aircraft would move, or more specifically, how the cockpit of the aircraft would move, in the real world. All modern simulators are programmed to replicate flight by moving the simulator in various ways. The design of that motion is a motion-drive algorithm. Motion cueing algorithms transform that data into simulator motion (Telban et al., 2000; Telban & Cardullo, 2005).

Modern advanced flight simulators use hexapod actuators to create the motion necessary to create the sensations of flight for pilots in all six possible directions (pitch, roll, yaw, heave, surge, and sway). The amount of motion that can be provided is limited by the physical length of the hexapod actuators. Combinations of the actuator motions are used to mimic flight sensations. Designers employ several strategies to maximize the value of the motion given these limitations. Within the human vestibular system, the otolith organs sense the three translational specific forces (i.e., heave, surge, sway). As the otoliths cannot differentiate between platform acceleration and tilt, the designer of the motion can take advantage of this to "trick" the human brain into believing that acceleration is occurring when it is not (Telban & Cardullo, 2005). One technique is to move the platform just "a percentage of the full motion" (Kaiser & Schroeder, 2003, p. 456). An additional technique is to move the platform at high frequencies (p. 456). High-frequency accelerations, unlike low-frequency accelerations, do not require a considerable simulator platform motion. The use of the simulator roll and pitch can be used to replicate low-frequency accelerations that would require considerable platform motion.

As an example, consider a takeoff, which has significant surge acceleration. After applying full thrust in the simulator, the simulator cab moves forward to mimic the aircraft motion. However, the simulator will quickly reach the physical limits of its actuators as the real aircraft continues to accelerate down the runway. The "trick" now is to slowly pitch up the simulator cab. With this pitch-up motion, a pilot's back pushes on the seat, just as would happen on a takeoff. This pitch-up motion has to be done at a slow rate, one below the pitch rate threshold of the inner ear; otherwise, the pilot will believe the aircraft is pitching during the takeoff, which would be a false cue. This combination of fore-aft simulator motion and cab tilt is often a compromise.

One consequence of these compromises is that it can adversely impact a pilot's ability to recover from some very dynamic conditions, such as Dutch roll in the lateral-direction axes, or even respond to certain upsets or failure modes (Kaiser & Schroeder, 2003). In fact, the assumption that "incorrect motion cues are worse than no motion cues," especially for large-amplitude motion, has influenced the use of motion in centerline thrust military simulators such that most do not have moving platforms (Allerton, 2009).

As the design of the motion cueing algorithms requires trade-offs, choices need to be made on what to optimize. In some cases, this has resulted in motion that adversely affected training. Wu and Cardullo (1997) reported that, in response to this, sometimes motion is even turned off to eliminate the problem. The optimal motion algorithm is actually dependent on what facet of flight is trying to be replicated (Zaal et al., 2015). Although a full discussion of the advantages and disadvantages of each type of motion algorithm is outside the scope of this chapter, the point here is that the differences should not be ignored.

Motion algorithms are chosen based on a combination of compromises, which include assumptions about which aspects of flight are the most important, physical limitations of the flight simulator platform, the costs, the availability of the data from the manufacturer, and more (Nahon & Reid, 1990). Can using a less-than-optimal motion cueing algorithm affect the transfer of training? Can the choice of the model affect the outcome? Schroeder (2012) identified several challenges to increasing simulator fidelity to simulate aircraft upsets and the difficulties pilots can face while attempting to recover. Proper simulation can make a difference. In one evaluation where airline pilots were placed into a flight simulator that contained much more accurate (but still necessarily limited due to physical constraints of simulation) aircraft stall characteristics, "less than one-quarter" of those pilots performed the recovery correctly (Schroeder et al., 2014, p. 15). It was postulated that the cues in the stall with updated algorithms (which were more closely aligned with the real aircraft) may have created a surprise effect that influenced the pilot's performance (2014).

As previously described, quasi-transfer of training tests motion against itself. If the motion model used is itself not valid, that can result in skewed results that are then published in the research. Those that are designing training programs need to be aware of these issues. The choice of which model is used may be a limitation or a delimitation, or possibly both. Researchers should report these aspects, and those who are using the research should look for these, and if they are not reported, consider the possible impact of these factors. Consider that the assumptions made about pilot behavior and the choice of which model to use to balance the different needs can directly affect flight safety. Accidents often result from flawed assumptions (Leveson, 2015), so the implications of these problems have a much deeper aspect than just academic integrity.

HOW DO HUMANS PERCEIVE MOTION DRIVE?

Motion is perceived by pilots through their vestibular, visual, and proprioceptive systems (Hosman & Advani, 2016). "Humans sense motion through the vestibular system located in the inner ear, which consists of semi-circular channels sensing angular motion and otoliths sensing translational motion" (Volkaner et al., 2016, p. 2). In addition to sensing motion, these are also responsible for humans' ability to balance. The visual system can perceive images, depth, slope, and changes in visual feedback, which all contribute to motion perception in simulators. This includes optical workflow useful in predicting velocity, distance, and relative spatial range. As the proprioceptive system is responsible for body awareness – being able to sense where the body is positioned in space – the motion model is critical.

There are many advantages and disadvantages to flight simulators. Myers et al. (2018) describe the advantages as providing a safe way to practice routine emergency procedures, reduced training costs, reducing the carbon footprint, and providing a research platform. There are also disadvantages. Although training costs are reduced, flight simulators are still a costly purchase and can be overregulated. Simulators can also cause pilots to become less motivated because their stress levels are lowered because they know that they are not in a real aircraft. It can also induce adaptation and compensatory skills. Another disadvantage is that poor motion cueing can be a cause of motion sickness such as "sweating, fatigue, dizziness, and vomiting" (Myers et al., 2018, p. 4).

Research suggests that pilot tracking behavior and performance was improved when motion feedback was available (Hosman & Advani, 2016). However, Jones (2018) suggests that aspects of motion systems in flight simulation training devices may not be necessarily suited for pilot training because "the actual response of the motion platform to pilot control is not objectively considered" (p. 488-489). This is because the motion tuning is conducted only through a simulator assessment pilot and subject matter experts (SMEs) without manufacturer constraints. Therefore, the outcome of the motion tuning process is highly reliant on the evaluation pilot and simulator software engineer who assess the final configuration with changes to the motion parameter settings occurring from open-loop and closed-loop testing with SMEs and a motion filter expert (Hosman & Advani, 2016; Jones, 2018). This outcome is often less than ideal because of the process being dependent on humans with the potential for variability and errors. This has brought about research in motion criteria to determine motion fidelity.

Myers et al. (2018) describe the transfer of training as the process that a pilot goes through in acquiring knowledge, abilities, or skills from the simulator training. This describes how the training in the simulated environment transfers into the real-world environment. There was more emphasis placed on the importance of calibrating motion cueing systems when conducting transfer of training research that was generalizable versus just for the specific simulator that was part of a specific study (Hosman & Advani, 2016). Volkaner et al. (2016) examined human perception and how to improve it in flight simulators. Myers et al. (2018), referencing Vaden and Hall (2005), suggest that pilot performance is improved using a motion simulator if their performance depends on motion in flight, e.g. pilots desired motion simulators for the task of controlling an unstable aircraft. A lack of motion made it difficult for pilots to develop flight control strategies successfully. Another aspect of improving the transfer of training is for instructors to explain to their flight students the differences between the simulator and aircraft so that pilots are mindful of this. Knowing where simulators are weak in terms of lacking realism is a necessary step for future pilot performance.

Fidelity is another challenge with regard to simulators. "Fidelity, the degree to which the simulator looks like the real aircraft and the similarity to which it acts like the real aircraft, is closely linked to training transfer" (Myers et al., 2018). Fidelity can be further broken down into three components: physical fidelity, cognitive fidelity, and functional fidelity. Physical fidelity comes down to the look and feel of the simulator in comparison to a real aircraft and associated flight deck. As a simulator cannot fully simulate real-world motion, physical fidelity is critical to ensure that the motion feel is realistic. Cognitive fidelity is how well the simulator can make the pilot have a similar cognitive state as when in a real aircraft. The cognitive state includes how realistic situational awareness, decision-making, anxiety, and stress are compared to flying a real aircraft. The last part of fidelity is functional fidelity. This is merely how well the simulator acts in comparison to an actual aircraft's equipment. The next section focuses on fidelity.

LET THEM EAT HUMBLE PIE—OR NOT?

A common self-imposed limitation in current training using high-fidelity simulators is the avoidance of consequences of trainee performance, i.e., letting the simulator "crash" or come close to this following trainee actions or inactions. This is rarely documented or stated policy but rather accepted practice in most pilot training environments across the industry. Overall, this can be seen as reflective of a more positive training culture than previous generations of pilots got to experience, i.e., more focused on developing pilot competencies than checking them. Still, in many other domains failure can be seen as an integral and productive way of learning, and most people have experienced this in learning writing, sports, music, etc. Specific examples of studies on this can be found for learning mathematics (Kapur, 2014) and other STEM topics (Simpson & Maltese, 2017), business studies (P.R. & Gupta, 2019), and art and design (Sawyer, 2018). A meta-study of failure and learning from failure as an instructional strategy (Darabi, Arrington & Sayilir, 2018) found 187 publications, with 62 of relevance, but only 12 based on an experimental design. The limited number of experimental studies demonstrated a lack of research on the role of failure for learning, but these studies overall showed a moderately positive result for learning from failure. There is even a specific learning design named "Productive Failure" (Kapur, 2015).

Concerns about how failure in the simulator may affect the self-confidence of a pilot are relevant but should be put in context. When this concern is expressed, it often relates to how pilots, following a severe failure in the simulator, react in a challenging in-flight situation. The potential risk is then that a lack of self-confidence may result in hesitation or lack of decisive action to resolve the situation. Self-confidence is certainly a desired trait in pilot selection (ECA, 2013), and in challenging situations, it can be supportive of taking action and a successful outcome. However, it should be remembered that pilot overconfidence has contributed to accidents, arguably more often than lacking self-confidence, most recently linked to the accident of PIA 8303 (BBC, 2020). In the context of recent aviation history, Crew Resource Management (CRM) training was in part introduced and promoted to counter the self-confidence of captains who would not let first officers question their actions. The argument to favor confidence over consequence by not letting pilots fail severely in the simulator implies that pilots would have fragile egos. In contrast, the industry experience implies that overconfidence is the bigger problem.

Considerations about letting pilots fail severely in the simulator must consider the context of pilot training. As mentioned previously, failure in simulator training can lead to not only re-training, but if repeated, also to actions regarding performance and ultimately loss of employment and licence. It is no surprise that pilots and their trainers, who normally are colleagues in an airline, want to avoid severe failure. For the airline, the administration and re-training related to failure requires resources and adds cost. There is nothing inherently positive about the experience of a serious safety event or an accident in the simulator. Designing scenarios that reliably produced such outcomes would be counterproductive to effective learning and successful training outcomes. On the other hand, there is most likely nothing inherently positive about avoiding this experience if prompted by poor pilot performance. Provided that the framework and setup for the training are supportive of learning from errors and failure, seeing the consequences of shortcomings in performance can be a motivating learning experience. While avoiding severe failure makes sense in the current context of pilot training in the industry, it is possible to imagine a different context, where this limitation was removed and just seen as a consequence that can be used for learning – like in many other domains.

SAME, SAME BUT DIFFERENT

With the accident of Air France 447 (BEA, 2012) "startle effect" came into focus as a contributing factor in accidents. It was seen as contributing to "The excessive nature of the PF's inputs" (p. 175), and "Poor management of the startle effect" was stated as having "generated a highly charged emotional factor for the two copilots" (p. 203). Also highlighted was "its potential to disrupt the ability of the crew to work as a team" (p. 188), and in conclusion, the accident report stated that "The startle effect played a major role in the destabilisation of the flight path and in the two pilots understanding the situation" (p. 211). In response, BEA (p.157-158) refers in section "1.18.4.6 Work currently underway on simulator fidelity and training" to a working group and one of its conclusions:

> The design of the training must be such that it generates surprise and startle effect to teach the pilots how to react to these phenomena and how to work in stressful situations, in order to prepare the trainees for the actual operating environment.

Since around the time of AF 447, startle and the related concepts of surprise have been identified as contributing factors in accidents, e.g., Colgan Air 3407 (NTSB, 2010), Air Asia 8501 (KNKT, 2014), Lion Air 610, and Ethiopian 302 (Committee on Transportation and Infrastructure, 2019), as well as implied at the time or in hindsight for many more. Not that startle or surprise was not known before this time, "automation surprise" had been a similar contributing factor to accidents previous to this, especially related to the increase in flight deck automation from the 1980s and forward (Dekker, 2009). It should be noted that startle and surprise are not the same, although they can be overlapping. As outlined by Landman (2019), startle is primarily related to the intensity of input, whereas surprise is related to a mismatch between what is expected to happen and what actually happens in a situation. Startle and surprise are, however, at times used without this distinction when describing pilots' reactions to unusual or unexpected events (Figure 6.1).

The paradox of our times is that as aviation continues to improve its safety record, startle and surprise will remain a challenge and probably one of increasing importance to address. Referring to Figure 6.1 (Dahlstrom, 2019), experience has always been seen as leading to expertise for pilots. Accumulating flight hours was the way to gain experience, and given the exposure to different events to manage, this would lead to expertise. This assumption has been well aligned with the reality of the pilot profession for generations of pilots. However, if exposure to unusual and risk events decreases, it may take longer for experience to lead to expertise. Also, we can expect that events that happened relatively often for previous generations of pilots may these days be a source of startle and surprise, e.g., considering increasingly reliable technology and automation. The paradox is then that the safer we get, the more we may have to train pilots to be able to manage unexpected and unusual situations to avoid negative consequences of startle and surprise.

The problem here is that a high-fidelity simulator does not perfectly replicate real flight regarding some aspects of flying, as is part of the focus of this chapter, but the high-fidelity simulator may not produce the same effect in regard to startle and surprise. Dahlstrom and Nahlinder (2007) illustrated this by using heart rate to measure workload (see Figure 6.2). The simulator, in this case, was not high fidelity, but the difference in heart rate shows that in an aircraft peak mental workload occurs

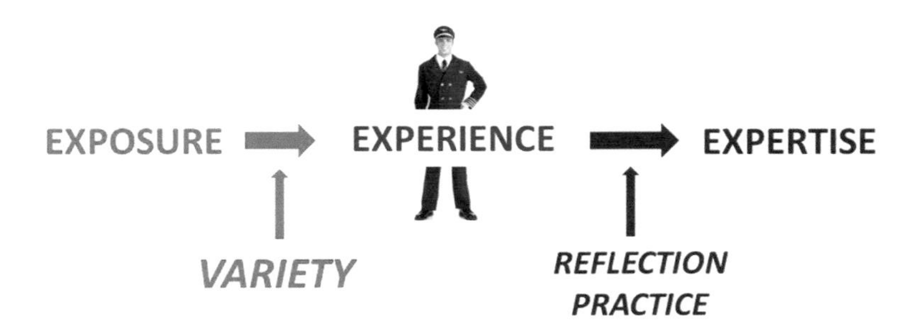

FIGURE 6.1 Argument about pilot expertise, slide from presentation at conference (Dahlstrom, 2019).

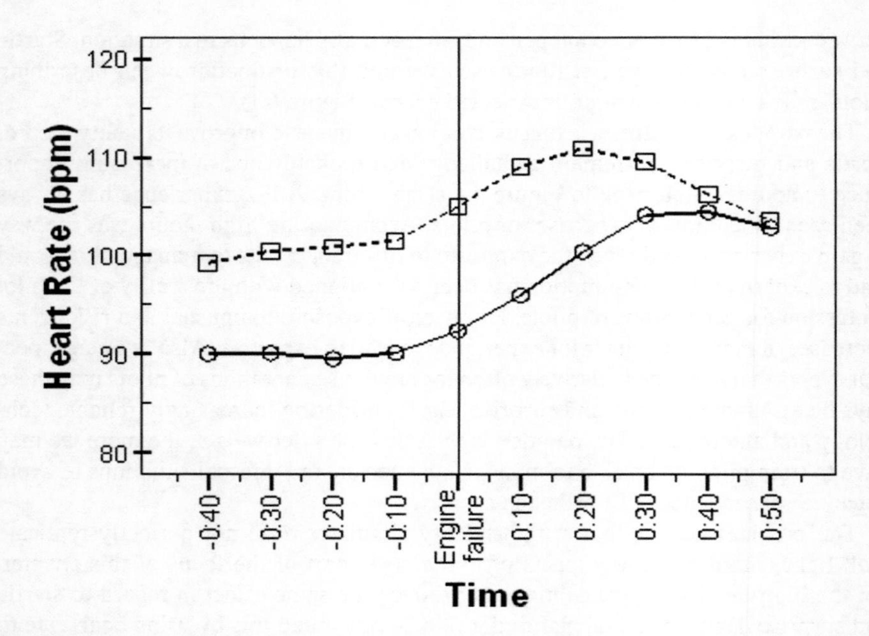

FIGURE 6.2 Average heart rate for eight participants during the flight segment rejected takeoff. Solid lines are aircraft flight, dashed lines are simulator flight (Dahlstrom & Nahlinder, 2007).

when managing a rejected takeoff, rather than when it was called, as was the case in the simulator. Dahlstrom and Nahlinder (2007) found more differences of a similar kind, and while these differences can be interpreted and challenged, the fact that there is a difference in the reactions of pilots between the aircraft and the simulator remains. This is not surprising to any experienced pilot, and should not be to anyone else, as the reality of being in an aircraft represents risks that are not present in the simulator (although the threat of being checked in the simulator adds some risks which are not present in the aircraft).

The difficulty of producing startle or surprise effects in a simulator does not only have to do with the different contexts or the challenge to reproduce in-flight conditions in the simulator. It also has to do with how the simulator is used for pilot training. Anyone who has worked with pilot training knows that as soon as new training has been implemented, the first trainees will tell their other pilot colleagues about the exercises used in training. Training departments have tried to get around this by keeping training exercises secret, decreasing explicit and detailed descriptions of the training, providing multiple alternative scenarios for trainers to choose from, etc. Even so, given that a pilot's career depends on successful simulator training and checks, few will arrive at the training without some idea about what will play out in the session in the simulator. This means that it remains an industry challenge to realistically reproduce startle and surprise effects in training to prepare pilots for what can happen in real flight. This is and remains one of the current limits of simulator training.

There has been a trend in recent years that airlines, some using the framework of evidence-based training (EBT), are moving to training that does not penalize failure and allows for learning from errors. Still, the professional pilot culture and industry culture are highly protective of performance standards and far from accepting with regard to errors and failures in training. This is as it should be but may continue to make it difficult to use high-fidelity simulators for effective training of pilots on how to manage effects of startle and surprise. Pilots need to be confident that they are supported to learn from what goes wrong in startle and surprise scenarios, rendering preparation for them counterproductive. The limitation of simulators in this respect can only be overcome by using them differently for training, in a manner that allows for the effects of startle and surprise to play out as fully as possible.

FIDDLING WITH FIDELITY WHILE MISSING THE STORY

One hidden limit with current high-fidelity simulation stems from its success in convincingly providing a replication of an aircraft's real-life environment and performance. It is indeed very close to the "real thing," and for most involved in-flight simulation, this can only be a "good thing." However, this replication of reality via high-fidelity environments and performance can lead to a focus on ever more detailed realistic features (Jackson, 1993) and limit in-depth considerations of training content and methodology. Dahlstrom, Dekker, van Winsen and Nyce (2009, p. 309) expressed this problem in this way:

> The emphasis on photorealism in visual and task contexts may retard or limit the development of skill sets critical for creating safety in domains where not all combinations of technical and operational failure can be foreseen or formalized (and for which failure strategies then cannot be proceduralised and simulated). The assumption that photorealism can capture all possible naturalistic cues in addition to the skills necessary to act competently in these domains may be overly optimistic. Competencies the aviation community recognises as important and significant (e.g. communication, coordination, problem solving, management of unanticipated and escalating situations) are thought to emerge directly from context-fixed simulator training. It is assumed that photorealism can achieve these ends.

Caird (1996, p. 127) stated that "For decades, the naïve but persistent theory of fidelity has guided the fit of simulation systems to training." This theory seems to still be influential in the aviation industry. What it is missing is that the most important aspect of a training scenario is the fidelity of "the story." When one pilot in a group speaks up with words such as "This is what happened to me the other day," something happens in the room. A low-level simulation has started, with all other pilots in the room transported into the situation, thinking about what they would have done if they were there. The same can happen when a pilot reads a report from an event or when it is presented in training. It is the fidelity of the story, in this case, its realism and relevance, that creates the simulation in the minds of the pilots, even without any tool to support this.

With the current focus on evidence-based training, operational events are often turned into training scenarios. There is indeed a focus on making the best possible use of scenarios in training departments across the industry. Still, the impact on training from carefully choosing and constructing scenarios seems to be underestimated, especially regarding the time and effort it takes to fully develop a credible and effective training scenario. This may be due to regulatory requirements (focusing more on the level of a simulator than content), limited training resources (time, manpower, etc.), limited knowledge about the use of scenarios (e.g., how to build an effective scenario from a human factors perspective), and in general limited knowledge about learning and training beyond traditional methods of pilot training. Also, it may be due to reliance on the idea that the high fidelity of the full flight simulator will be enough to make a scenario effective.

The untapped potential of cleverly constructed scenarios has been demonstrated by the effectiveness of simple training tools and lower fidelity levels of simulation throughout aviation history. More recently, this has been shown in research and in the industry with the use of Mid-Fidelity Simulation (MFS) scenarios (Dahlstrom, 2020). These are scenarios in the style of "microworlds," presented on a screen and with simplified representations of environments, whether they are domain-specific (representing aircraft controls and displays) or represent other environments (an industrial plant, a ship, a spacecraft, etc.). One example of this is the use of a single-parameter simulation in a simulation named "Cold Store," which put pilots in a new problem-solving situation and allowed the collection of data that was then used to identify patterns of problem-solving style (Rosa et al., 2021). Another example comes from simply putting a moving aircraft symbol on a map, inserting a loss of cabin pressure, and then allowing for search for information related to a diversion decision simulates most of the cognitive process of such a decision (Dahlstrom, 2020). These applications are low on environmental fidelity but high on the fidelity of the story, making them cost-effective training tools. They show that deeper thinking about scenarios still offers untapped potential for improving training, but this requires going beyond reliance on simulator fidelity as a proxy for quality of training.

The critical importance of advanced high-fidelity simulation for pilot training and thus for aviation safety cannot be overstated. Still, today the reality is that while current high-fidelity simulation remains the most fundamental and powerful training tool the industry has available, the continued quest for incremental increases in fidelity does not seem to represent the most productive focus to further improve training. It should be possible to agree that high-fidelity simulation is well-designed, tried, and tested as a tool. At this point, we should start balancing how much time we are spending on improving the tool with how much time we spend on asking ourselves "how" we can use it better to achieve training outcomes that make operations safer and more efficient. The success of high-fidelity flight simulation should be a foundation for effective pilot training, not a limitation for further development of it.

A PERFECT TOOL OR A TOOL THAT CAN BE PERFECTED?

The limitations outlined in this chapter do not take away from the critical contribution of all levels of simulation to the improvement of aviation safety, especially the advanced high-fidelity simulation available today. However, even for an almost perfect tool, it is useful, and at times necessary to be aware of its limitations so it can be used optimally. There is potential for further perfection of simulation by using data from simulators and other technology that can augment data collection to gain an in-depth understanding of pilot performance.

With Simulator Operational Quality Assurance (SOQA), data from simulator training sessions can be collected, analyzed, and used to provide feedback to the individual pilot as a trainee and the whole organization. This has been shown with the promotion of the tool CAE Rise (CAE, 2021) and a similar application for collecting and presenting simulator data. According to EASA (2019, p.45), SOQA is "still in its infancy, but it offers interesting prospects." This is probably an understatement, as having extensive data from simulator sessions would offer the opportunity not only to see what pilots did in training but also by inferences make it possible to analyze why they took certain actions.

Eye-tracking equipment has been used in research for a long time to understand pilot performance, but new, high-precision, and nonintrusive solutions (no need to wear any equipment) can, with simple means, be integrated into simulators. Recent research (Niehorster et al., 2020) and proof of concept trials at airlines (Behrend, 2020, Cameron & Nolan, 2018; Behrendt, 2020; Saleh & Myers, 2020) have demonstrated the potential this holds for pilot training. However, eye-tracking has still not yet been integrated into a day-to-day operational training environment and holds the potential to improve understanding and improvement of pilot performance.

Finally, speech analysis could be used to provide an additional source of data (Bresee, 2019). Taken together, data from the simulator, eye-tracking, and speech analysis could provide new insights on pilot performance as well as support trainer focus on competencies rather than on SOP compliance (which would be recorded via the combined data from interaction with the simulator, gaze data from eye-tracking and recording of callouts). This could bring simulator training to the next level regarding the in-depth understanding of pilot performance and behavior and assist trainers' efforts to support and develop their trainees. Although there are hidden limitations for simulators, the potential for further improvement of them as training tools is in plain view.

SUMMARY

This chapter focused on the use of flight simulators for pilot training and specifically addressed the motion argument in the context of training. The chapter consisted of five sections. These sections discussed an introduction of simulator motion, human perception in motion drive, high-fidelity simulator limitations in current training, contributing factors in accidents caused by limitations, and how simulators are useful but can also be improved. There are many advantages and disadvantages to flight simulators. In order to assure that the simulator fidelity is beneficial for pilot training, it needs to be valid, meeting the goal of being able to accurately model the real aircraft to the extent necessary to accomplish the goals for transfer of training, which is how the training in the simulated environment transfers into the real-world environment. The authors hope to have provided the reader with insight into current limitations and essentially existing gaps within high-fidelity training environments. It is our hope that these simulators can continue to be improved to enhance pilot performance with regard to aviation safety, especially when making safety-critical decisions. Optimizing pilots' transfer of training with regard to high-fidelity training environments is crucial and continues to be an essential field of research.

ACKNOWLEDGMENT

The authors wish to acknowledge Jeffery Schroeder for his contribution in reviewing this chapter and providing input.

REFERENCES

Allerton, D. (Ed.) (2009). *Principles of Flight Simulation.* Chichester, UK: Wiley.

Ary, D., Jacobs, L. C., Sorensen, C., & Razavieh, A. (2010). *Introduction to Research in Education*, 8th edition. Canada: Wadsworth Cengage Learning.

BBC. (2020). Pakistan Plane Crash was "Human Error" – Initial Report. Retrieved 2021-04-17: https://www.bbc.com/news/world-asia-53162627

BEA – Bureau Bureau d'Enquêtes et d'Analyses pour la sécurité de l'aviation civile. (2012). Final Report on the Accident on 1st June 2009 to the Airbus A330-203Registered F-GZCP Operated by Air France Flight AF 447 Rio de Janeiro. Retrieved 2021-04-11: https://www.bea.aero/docspa/2009/f-cp090601.en/pdf/f-cp090601.en.pdf

Behrend, J. (2020). The Influence of Role Assignment on Pilot Decision-Making: An Eye-Tracking Study. *Presentation at the 73rd International Air Safety Summit*, Arranged by Flight Safety Foundation, 19–22 October 2020, Online. Alexandria, VA: Flight Safety Foundation.

Box, G.E. (1976). Science and statistics. *Journal of the American Statistical Association*, 71(356), 791–799.

Bresee, J. (2019). Beyond Freeze and Reset: Objectively Assessing Crew Interaction Without Interrupting Task Performance. Presentation at Asia Pacific Aviation Training Symposium (APATS), 3–4 September 2019, Singapore. Retrieved 2021-04-17: https://apats-event.com/wp-content/uploads/2019/09/Jerry-Bresee.pdf

Bürki-Cohen, J., Soja, N. N., & Longridge, T. (1998). Simulator Platform Motion-The Need Revisited. *The International Journal of Aviation Psychology*, 8(3), 293–317.

Bürki-Cohen, J., & Go, T. (2005). The Effect of Simulator Motion Cues on Initial Training of Airline Pilots. AIAA Modeling and Simulation Technologies Conference and Exhibit, 15–18 August 2005, San Francisco, CA. Retrieved 2021-03-01: https://doi.org/10.2514/6.2005-6109

CAE. (2021). CAE Rise™ Training System. Information of website of CAE. Retrieved 2021-04-17: https://www.cae.com/civil-aviation/aviation-software/cae-rise/

Caird, J. K. (1996). Persistent Issues in the Application of Virtual Environment Systems to Training. Proceedings of HICS'96: Third annual symposium on human interaction with complex systems. Los Alamitos, CA: IEEE Computer Society Press, 124–132.

Cameron, M., & Nolan, P. (2018). Eyes Are Never Quiet – Towards Eye Tracking as a Practical Training Tool. Presentation at the 71st International Air Safety Summit, Arranged by

Flight Safety Foundation. Retrieved 2021-04-17: https://flightsafety.org/wp-content/uploads/2018/11/Cameron-day-2-IASS-20181112-Eye-Tracking-Presentation.pdf

Checkland, P. (1971). *Systems Thinking, Systems Practice*. New York: Wiley.

Committee on Transportation and Infrastructure. (2019). Hearing Before the Subcommittee on Aviation of the Committee on Transportation and Infrastructure, House of Representatives, One Hundred Sixteenth Session. Retrieved 2021-04-16: https://www.govinfo.gov/content/pkg/CHRG-116hhrg37277/pdf/CHRG-116hhrg37277.pdf

Dahlstrom, N. (2019). Human Factors and CRM – Successes, Shortcomings and Solutions. Conference Presentation at Flight Operational Forum, 1–3 April 2019, Oslo, Norway. Retrieved 2021-04-12: http://www.fof.aero/presentations/F/o/F/2/0/1/9/2019_FoF_0206_dahlstrom.pdf

Dahlstrom, N. (2020). The Future Training EcoSystem: A New Normal for Flight Crew Training. Presentation at the "International Air Safety Summit", Arranged by Flight Safety Foundation. Presentation available from author.

Dahlstrom, N., & Nahlinder, S. (2007). Mental Workload in Simulator and Aircraft During Basic Civil Aviation Training. *International Journal of Aviation Psychology*, 19(4), 309–325.

Dahlstrom, N., Dekker, S., van Winsen, R., & Nyce, J. (2009). Fidelity and Validity of Simulator Training. *Theoretical Issues in Ergonomics Science*, 10(4), 305–314.

Darabi, A., Arrington, T. L., & Sayilir, E. (2108). Learning from Failure: A Meta-Analysis of the Empirical Studies. *Educational Technology, Research and Development*, 66, 1101–1118.

Dekker, S. W. A. (2009). Flight Crew Human Factors Investigation Conducted for the Dutch Safety Baard into the Accident of TK1951, Boeing 737–800 near Amsterdam Schiphol Airport, February 25, 2009. Retrieved 2021-04-13: https://www.onderzoeksraad.nl/nl/media/inline/2020/1/21/human_factors_report_s_dekker.pdf

De Winter, J. C., Dodou, D., & Mulder, M. (2012). Training Effectiveness of Whole Body Flight Simulator Motion: A Comprehensive Meta-Analysis. *The International Journal of Aviation Psychology*, 22(2), 164–183.

EASA. (2019). "Breaking the Silos" – Fully Integrating Flight Data Monitoring into the Safety Management System. Report from European Operators Flight Data Monitoring Forum, Woking Group C. Retrieved 2021-04-17: https://www.easa.europa.eu/sites/default/files/dfu/BreakingTheSilos%20Issue.pdf

ECA – European Cockpit Association. (2013). Pilot Training Compass – "Back to the Future". Retrieved 2021-04-17: https://www.eurocockpit.be/sites/default/files/eca_pilot_training_compass_back_to_the_future_13_0228.pdf

Federal Aviation Administration (FAA). (1992). Advisory Circular Number 120-45A. https://www.faa.gov/documentLibrary/media/Advisory_Circular/AC_120-45A.pdf

Hosman, R., & Advani, S. (2016). Design and Evaluation of the Objective Motion Cueing Test and Criterion. *The Aeronautical Journal*, 120(1227), 873–891. http://dx.doi.org.portal.lib.fit.edu/10.1017/aer.2016.35

Jackson, P. (1993). Applications of Virtual Reality in Training Simulation. In K. Warwick, J. Gray, & D. Roberts (Eds.), *Virtual Reality in Engineering* (pp. 121–136). London: The Institution of Electrical Engineers.

Jones, M. (2018). Enhancing Motion Cueing Using an Optimisation Technique. *The Aeronautical Journal*, 122(1249), 487–518. http://dx.doi.org.portal.lib.fit.edu/10.1017/aer.2017.141

Kaiser, M. K., & Schroeder, J. A. (2003). Flights of Fancy: The Art and Science of Flight Simulation. In M. A. Vidulich & P. S. Tsang (Eds.), *Principles and Practice of Aviation Psychology* (pp. 435–471). Mahwah, NJ: Lawrence Erlbaum Associates.

Kapur, M. (2014). Productive Failure in Learning Math. *Cognitive Science – A Multidisciplinary Journal*, 38(5), 1008–1022.

Kapur, M. (2015). Learning from Productive Failure. *Learning: Research and Practice*, 1(1), 51–65.

KNKT (Komite Nasional Keselamatan Trasportasi) – National Transportation Safety Committee of Indonesia. (2014). Final Report KNKT.14.12.29.04 PT. Indonesia Air Asia Airbus A320-216; PK-AXC Karimata Strait Coordinate 3°37’19”S–109°42’41”E Republic of Indonesia 28 December 2014. Jakarta, Indonesia: KNKT. Retrieved 2021-01-06: https://www.bea.aero/uploads/tx_elydbrapports/Final_Report_PK-AXC-reduite.pdf

Landman, A. (2019). Managing Startle and Surprise in the Cockpit. Ph.D. thesis at Delft University. Delft, Netherlands: Delft University. Retrieved 2021-04-14: https://research.tudelft.nl/files/55707836/dissertation_startle.pdf

Leveson, N. (2015). A Systems Approach to Risk Management Through Leading Safety Indicators. *Reliability Engineering & System Safety*, 136, 17–34.

Myers III, P. L., Starr, A. W., & Mullins, K. (2018). Flight Simulator Fidelity, Training Transfer, and the Role of Instructors in Optimizing Learning. *International Journal of Aviation, Aeronautics and Aerospace*, 5(1), 6. http://dx.doi.org.portal.lib.fit.edu/10.15394/ijaaa.2018.1203

National Transportation Safety Board (NTSB). (2010). Loss of Control on Approach, Colgan Air, Inc., Operating as Continental Connection Flight 3407, Bombardier DHC-8-400, N200WQ, Clarence Center, New York, 12 February 2009. NTSB/AAR-10/01. Washington, DC: NTSB. Retrieved 2021-04-16: https://www.ntsb.gov/investigations/AccidentReports/Reports/AAR1001.pdf

Nahon, M. A., & Reid, L. D. (1990). Simulator motion-drive algorithms-A designer’s perspective. *Journal of Guidance, Control, and Dynamics*, 13(2), 356–362.

Niehorster, D. C., Hildebrandt, M. Smoker, A. Jarodzka, H., & Dahlstrom, N. (2020). Towards eye tracking as a support tool for pilot training and assessment. Eye-Tracking in Aviation. Proceedings of the 1st International Workshop (ETAVI 2020), 17–28. Retrieved 2021-04-17: https://www.research-collection.ethz.ch/bitstream/handle/20.500.11850/407625/ETAVI_paper_2_407625.pdf?sequence=10&isAllowed=y

PR, S., & Gupta, D. (2019). Introducing Failure as a Deliberate Instructional Strategy to Enhance Learning and Academic Outcomes. In 2019 IEEE Tenth International Conference on Technology for Education (T4E), Goa, India, 67–70.

Rehmann, A. J., Mitman, R. D., & Reynolds, M. C. (1995). *A Handbook of Flight Simulation Fidelity Requirements for Human Factors Research*. Crew System Ergonomics Information Analysis Center Wright-Patterson AFB OH.

Rosa, E., Dahlstrom, N., Knez, I., Ljung, R., Cameron, M., & Willander, J. (2021). Dynamic Decision-Making of Airline Pilots in Low-Fidelity Simulation. *Theoretical Issues in Ergonomics Science*, 22(1), 83–102.

Saleh, P., & Myers, R. (2020). How Eye Tracking Supported Tools Enhance “Monitoring” Training and Improve Flight Safety. Presentation at the 73rd International Air Safety Summit, arranged by Flight Safety Foundation, 19–22 October 2020, Online. Alexandria, VA: Flight Safety Foundation.

Sawyer, R. K. (2018). Teaching and Learning How to Create in Schools of Art and Design. *Journal of the Learning Sciences*, 27(1), 137–181.

Schroeder, J. (2012, August). Research and Technology in Support of Upset Prevention and Recovery Training. AIAA Modeling and Simulation Technologies Conference, 4567.

Schroeder, J. A., Bürki-Cohen, J. S., Shikany, D., Gingras, D. R., & Desrochers, P. P. (2014). An Evaluation of Several Stall Models for Commercial Transport Training. AIAA Modeling and Simulation Technologies Conference, 1002.

Simpson, A., & Maltese, A. (2017). "Failure Is a Major Component of Learning Anything": The Role of Failure in the Development of STEM Professionals. *Journal of Science Education & Technology*, 26, 223–237.

Telban, R. J., & Cardullo, F. M. (2005). Motion Cueing Algorithm Development: Human-Centered Linear and Nonlinear Approaches.

Telban, R. J., Wu, W., Cardullo, F. M., & Houck, J. A. (2000). Motion Cueing Algorithm Development: Initial Investigation and Redesign of the Algorithms.

Tuncman, I. (2020). Assessing the Methodological Quality of Aviation Research: A Content Analysis of Articles Published in Subscription and Open Access Aviation Journals 2014–2016 [Doctoral dissertation, Florida Institute of Technology]. Scholarship Repository. https://repository.lib.fit.edu/handle/11141/3217

Vaden, E. A., & Hall, S. (2005). The Effect of Simulator Platform Motion on Pilot Training Transfer: A Meta-Analysis. *The International Journal of Aviation Psychology*, *15*(4), 375–393. https://doi.org/10.1207/s15327108ijap1504_5

Volkaner, B., Sozen, S. N., & Omurlu, V. E. (2016). Realization of a Desktop Flight Simulation System for Motion-Cueing Studies. *International Journal of Advanced Robotic Systems*, *13*(3) http://dx.doi.org.portal.lib.fit.edu/10.5772/63239

Weisberg, M. (2012). *Simulation and Similarity: Using Models to Understand the World*. Oxford: Oxford University Press.

Wu, W., Cardullo, F., Wu, W., & Cardullo, F. (1997). Is there an optimum motion cueing algorithm? *Modeling and Simulation Technologies Conference*, 3506.

Zaal, P. M., Schroeder, J. A., & Chung, W. W. (2015). Transfer of training on the vertical motion simulator. *Journal of Aircraft*, 52(6), 1971–1984.

7 Cybersickness in Immersive Training Environments

Kay M. Stanney, Claire L. Hughes, Peyton Bailey, Ernesto Ruiz, and Cali Fidopiastis

CONTENTS

INTRODUCTION

The availability of low-cost, high-quality wearable and mobile displays has led to a surge in application areas for immersive eXtended Reality (XR) technologies such as virtual reality (VR), augmented reality (AR), and mixed reality (MR) outside of traditional gaming genres, especially in training and education (Wu, Yu, & Gu, 2020). Immersive training systems surround a trainee within a multisensory environment, providing contextualized and situated learning opportunities with tangible objects to be manipulated and venues to be traversed. These training experiences can take on different forms. For example, VR aims to fully occlude the real world with a computer-generated world that fully immerses, isolates, and engages users in other than real-world reality. AR/MR, on the other hand, blend real and virtual worlds such that they coexist in the same space, allowing users to maintain awareness of the real-world while being presented with augmentations from beyond reality (Stanney, Nye, Haddad, Padron, Hale, & Cohn, 2021). All of these forms of extended reality, to a greater or lesser extent, allow for replication of relevant environmental (e.g., battlefield context) and task (e.g., applying a tourniquet) cues that may be necessary for training transfer. This blending of contextualized information and virtual cues has been shown to lead to faster knowledge acquisition and retention, increased

DOI: 10.1201/9781003401360-7

engagement and presence, and more immersive learning (Giamellaro, 2017; Lee, 2012). VR is perhaps best at immersion, as it allows an individual to experience a myriad of computer-generated worlds from a first-person perspective. AR is perhaps best at conveying knowledge, as it allows information to be overlaid onto a real-world location to instruct, advise, and inform. MR is perhaps best suited for internalizing physical behaviors, as it affords users the ability to embody psychomotor behaviors by requiring trainees to enact requisite actions (e.g., apply a physical tourniquet) on physical entities (e.g., a real manikin) while constructing knowledge and skills in a real-world context. Combining AR and MR allows for the coupling of embodied psychomotor behavior with information overlays that can guide knowledge acquisition tasks, which may speed skill acquisition and increase retention (Kahol, Vankipuram, & Smith, 2009; Levine et al., 2018). If AR/MR are further coupled with VR, true extended reality is achieved, which blends all three forms of technologically enabled reality together. Yet, even with modern XR headsets, minimizing cybersickness is still a potential barrier to the adoptability and utility of these state-of-the-art immersive technologies and their respective authoring tools (Caserman et al., 2021; Stanney, Fidopiastis, & Foster, 2020).

CYBERSICKNESS BACKGROUND

There are many theories related to cybersickness, which is a form of motion sickness related to immersive environments (Keshavarz, Hecht, & Lawson, 2014; Stanney, Lawson et al., 2020). Sensory conflict theory, which is the most widely cited, suggests that cybersickness is due to mismatches between what the human sensory systems expect and what sensory cues in immersive environments provide (Reason, 1970, 1978; Reason & Brand, 1975; Nooij et al., 2017). Related is the evolutionary (or poison) theory, which suggests that the human brain has evolved such that any conflicts in real versus apparent motion stimuli are processed in the same manner as a toxin-related malfunction in the central nervous system (CNS) and thus they initiate an emetic response as a defense mechanism (Money, 1970; Treisman, 1977). The ecological theory suggests that postural instability is a necessary precursor to motion sickness (Riccio & Stoffregen, 1991). A relatively recent theory in the area is the multisensory re-weighting theory, which suggests that sensory dynamics can be re-shaped in XR to support adaptation to sensory perturbations, with those individuals better able to down-weight non-veridical XR cues being less susceptible to cybersickness (Weech, Calderon, & Barnett-Cowan, 2020). Taken together, these theories suggest that altered sensory cues, whether they be in the XR system (e.g., visual–vestibular mismatch, vergence–accommodation conflict) or in the human (e.g., postural instability, sensory re-weighting) drive cybersickness. Other individual differences beyond postural stability and sensory re-weighting capacity have been associated with cybersickness susceptibility, with demographic factors and technological variables each accounting for about half the variance in cybersickness (Rebenitsch & Owen, 2021). Previous experience with virtual motion, field of view (FOV), interpupillary distance (IPD), field dependence, female hormonal cycle, state/trait anxiety, migraine susceptibility, ethnicity, aerobic fitness, and body mass index have all been associated with cybersickness (Stanney, Fidopiastis, & Foster, 2020). One of the most common factors linked to cybersickness susceptibility is sex, but the results have been mixed: some have reported that females have increased susceptibility to cybersickness (Allen et al., 2016; Moroz et al., 2019; Shafer et al., 2019); others have suggested increased female susceptibility varies with experimental conditions (Munafo et al., 2017), while others have found no sex-linked risk for cybersickness (Al Zayer et al., 2019; Bracq et al., 2019; Clifton & Palmisano, 2020; Curry et al., 2020; Melo et al., 2018; Wilson & Kinsela, 2017). Recent research suggests that it may be the design of the head-worn displays (HWDs), specifically an inability to fit the anthropometrics of certain end users, that drives cybersickness and not sex (Stanney, Fidopiastis, & Foster, 2020).

Cybersickness resulting from the exposure to immersive environments comprises a host of associated problems, including vomiting (about 1% in VR; none is expected in AR/MR because of the availability of real-world rest frames which resolve visual–vestibular mismatches), nausea, disorientation, and oculomotor problems, as well as sleepiness (i.e., sopite syndrome) and visual flashbacks (Stanney & Hughes, 2021; Stanney, Kennedy, & Hale, 2014). At least 5% of those exposed to VR will not be able to tolerate prolonged exposure (Lawson, 2014); but no such limitation is expected in AR/MR due to a lack of nausea-related symptoms. Thus, while VR is oftentimes self-limiting (i.e., users have to withdraw from exposure due to nausea and disorientation), AR is expected to be tolerable for very long duration exposure (Stanney & Hughes, 2021).

Cybersickness effects can be quite pervasive. For example, ~80–95% of those exposed to VR and ~50–80% of those exposed to AR report some level of symptomatology postexposure, which may be as minor as a headache and visual fatigue to as severe as vomiting or intense vertigo (Stanney & Hughes, 2021; Stanney, Salvendy, et al., 1998). What's more troubling is that the problems do not cease immediately upon cessation of exposure, posing lingering adverse physiological aftereffects that can affect the safety of users hours after exposure (Smyth et al., 2018; Stanney, Fidopiastis & Foster, 2020; Stanney & Kennedy, 1998). Table 7.1 provides a synopsis of the extent and severity of adverse effects associated with exposure to immersive training environments.

Based on the studies summarized in Table 7.1, the most conservative prediction would be that a substantial proportion (>50%) of users, in general, would experience some level of adverse effects both during and after XR exposure, with prolonged aftereffects associated with safety concerns.

Why does cybersickness matter? Immersive technologies such as AR, MR, and VR are entering the mainstream in industry and the military. The global XR market is estimated at ~$31 billion in 2021, and anticipated to grow to ~$300 billion by 2024 (Alsop, 2021). In terms of adoption, ~35% of businesses report having adopted XR technology at a small scale, with ~13% having adopted at a broader scale (Vailshery, 2021). Adoption has, however, been considerably slower than other interactive technologies (Doolani et al., 2020; see Figure 7.1). One has to wonder if people have been slow to embrace XR technology due, in part, to cybersickness.

TABLE 7.1
Problems Associated with Exposure to Immersive Training Systems

- 80–95% of individuals interacting with a VR system report some level of side effects, with 5–50% experiencing symptoms severe enough to end participation, approximately 50% of those dropouts occurring in the first 20 min and nearly 75% by 30 min (Caserman et al., 2021; Cobb et al., 1999; DiZio & Lackner, 1997; Howarth & Finch, 1999; Regan & Price, 1994; Singer et al., 1998; Stanney, Kennedy, & Kingdon, 2002; Stanney, Lanham, Kennedy, & Breaux, 1999; Stanney, Kingdon, Graeber, & Kennedy, 2002; Wilson et al., 1995; Wilson et al., 1997).
- Unlike VR, AR/MR are typically not associated with dropouts, likely due to low levels of nausea experienced (Stanney & Hughes, 2021).
- VR presents with higher levels of disorientation and nausea, with lesser oculomotor disturbances. With prolonged (>45 min) VR exposure, oculomotor problems may become more pronounced, while nausea and disorientation level off (Stanney, Kingdon, Graeber, Kennedy, 2002).
- ~50–80% of AR/MR users experience adverse symptoms, mostly in terms of oculomotor disturbances, such as visual discomfort and fatigue, eyestrain, double vision, headaches, adaptation of vestibulo-ocular reflex (VOR), with lesser disorientation and very little nausea (Stanney & Hughes, 2021).
- Fully immersive VR exposure can cause people to vomit (about 1%), and approximately three-quarters of those exposed tend to experience some level of nausea, disorientation, and oculomotor problems (Cobb, Nichols, Ramsey, & Wilson, 1999; DiZio & Lackner, 1997; Howarth & Finch, 1999; Regan & Price, 1994; Singer, Ehrlich, & Allen, 1998; Lawson, Graeber, Mead, & Muth, 2002; Stanney, Kingdon, Graeber, & Kennedy, 2002; Stanney, Lawson et al., 2020; Stanney, Salvendy et al., 1998; Wilson, Nichols, & Haldane, 1997; Wilson, Nichols, & Ramsey, 1995).
- AR/MR are not typically associated with vomiting (Stanney & Hughes, 2021).
- Females exposed to immersive systems have previously been expected to be more susceptible to motion sickness than males and to experience higher levels of oculomotor and disorientation symptoms as compared to males (Graeber, 2001; Stanney, Kingdon Graeber, & Kennedy, 2002). However, recent studies have demonstrated that this may be more related to the fit of the technology than sex-related (Stanney, Fidopiastis, & Foster, 2020). Specifically, when a user's inter-pupillary distance (IPD) cannot be properly aligned in the XR headset, this may lead to higher levels of cybersickness. Current VR headsets, in particular, accommodate the IPD of males better than females.
- Individuals susceptible to motion sickness can be expected to experience more than twice the level of adverse effects to XR exposure as compared to non-susceptible individuals (Stanney, Kingdon, Graeber, & Kennedy, 2002).
- Individuals exposed to XR systems can be expected to experience lowered arousal (e.g., drowsiness, fatigue; a.k.a. sopite syndrome) upon post-exposure (Lawson et al., 2002; Stanney, Kingdon, Graeber, & Kennedy, 2002).
- Flashbacks (i.e., visual illusion of movement or false sensations of movement, when *not* in the XR environment) can be expected to occur (Lawson et al., 2002; Stanney, Kingdon, Graeber, & Kennedy, 2002).
- Prolonged aftereffects may occur after XR exposure, with symptoms potentially lasting more than 24 h (Baltzley, Kennedy, Berbaum, Lilienthal, & Gower, 1989; Stanney & Hughes, 2021; Stanney & Kennedy, 1998; Stanney, Kingdon, & Kennedy, 2002; Stanney, Kingdon, Graeber, & Kennedy, 2002).
- Increased postural instability while in a seated position seems to precede cybersickness (Mehri, 2009; Riccio & Stoffregen, 1991).

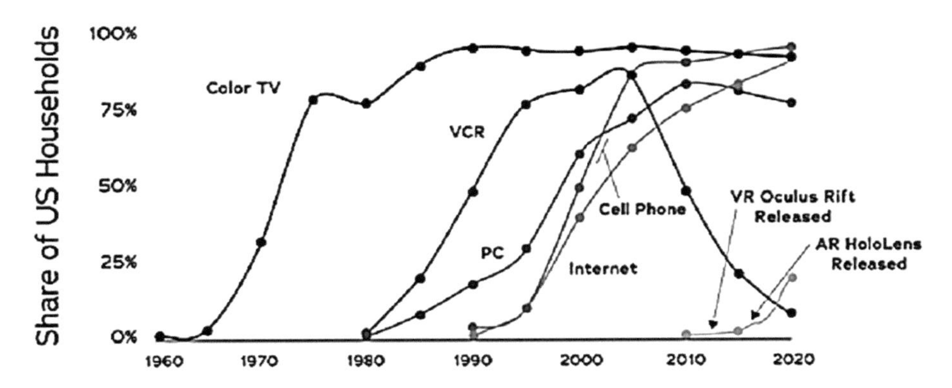

FIGURE 7.1 Interactive technology adoption rates. (Adapted from Llamas, 2019.)

With adoption going mainstream, ignoring the adverse effects of cybersickness could become problematic, with the potential for

- creating unequal opportunities for immersive accessibility among the moderate to highly motion sickness susceptible population, with those who can handle immersive exposure advancing due to better, more contextualized, embodied training, while those who are susceptible to cybersickness are left to train with increasingly outdated technology (Allen et al., 2016; Stanney, Kingdon, & Kennedy, 2002; Stanney, Fidopiastis, & Foster, 2020);
- decreased trainee acceptance and use of immersive training systems (Biocca, 1992; Sagnier et al., 2020);
- decreased human performance (An et al., 2018; Kolasinski, 1995; Lawson et al., 2002; Stanney, Kingdon, & Kennedy, 2002); and
- acquisition of improper behaviors (Kennedy, Hettinger, & Lilienthal, 1990).

XR designers and developers must mitigate the effects of cybersickness in order to reduce the possibility of these powerful immersive technologies never achieving their full potential.

INDIVIDUAL SUSCEPTIBILITY AND STIMULUS INTENSITY

There is a substantial individual component to cybersickness. As Lackner (2014, p. 2495) noted:

> the range of sensitivity in the general population varies about 10–1, and the adaptation constant also ranges from 10 to 1. By contrast, the decay time constant varies by 100–1. The import of these values is that susceptibility to motion sickness in the general population varies by about 10,000–1, a vast range.

The fundamental question that needs to be addressed is as follows: Can an understanding of human physiological responses to immersive training technology be developed and incorporated into design guidelines and usage protocols rendering immersive systems safe and effective to use for all? Research conducted over the past several decades has made tremendous gains in this regard (Chinn & Smith, 1953; Crampton, 1990; DiZio & Lackner, 2002; DiZio & Lackner, 1997; Heutink et al., 2019; Howarth & Finch, 1999; Hughes et al., 2020; Kennedy & Fowlkes, 1992; Keshavarz et al., 2014; Lawson, 2014; McCauley & Sharkey, 1992; McNally & Stuart, 1942; Pot-Kolder et al., 2018; Reason, 1970, 1978; Reason & Brand, 1975; Sjoberg, 1929; Stanney, Fidopiastis, & Foster, 2020; Stanney, Salvendy et al., 1998; Tyler & Bard, 1949; Weech et al., 2019; Welch & Mohler, 2014; Wendt, 1968). These researchers set out to achieve a number of challenging objectives, including developing tools to measure the adverse effects of immersive training system exposure (Kennedy & Stanney, 1996; Stanney, Kennedy, Drexler, & Harm, 1999), examining the psychometrics of cybersickness (Hildebrandt et al., 2018; Hughes et al., 2020; Kennedy, Stanney, & Dunlap, 2000; Kingdon, Stanney, & Kennedy, 2001; Stanney & Kennedy, 1997a, 1997b; Stanney, Lanham et al., 1999; Stanney, Kingdon, Nahmens, & Kennedy, 2003), developing usage protocols (Stanney, Kennedy, & Hale, 2014) and screening tools (Kennedy, Lane, Stanney, Lanham, & Kingdon, 2001), investigating system-related issues that influence cybersickness (Lucas et al., 2020; Park & Lee, 2020; Stanney, Fidopiastis, & Foster, 2020; Stanney & Hash, 1998; Stanney, Kingdon, Graeber, & Kennedy, 2002; Stanney, Salvendy et al., 1998; Widdowson et al., 2021; Wilson, 2016), examining the efficacy of readaptation mechanisms for recalibrating those exposed to immersive systems (Champney et al., 2007; Smither, Mouloua, & Kennedy, 2003), as well as examining the influences of cybersickness on human performance (Stanney, Kingdon, & Kennedy, 2001; Kennedy, French, Ordy, & Clark, 2003), among other related pursuits. These studies have led to an understanding that the response to immersive exposure varies directly with the capacity of the individual exposed (e.g., susceptibility, experience), dose (i.e., stimulus intensity), exposure duration (Hughes et al., 2020; Kennedy, Stanney, & Dunlap, 2000), and individual fit of the technology (Stanney, Fidopiastis, & Foster, 2020). These findings suggest that effective usage protocols that address the screening of individuals, strength of the immersive stimulus, usage instructions, and design of the HWD can minimize problems associated with immersive technologies.

From the individual susceptibility perspective, age, prior experience, individual factors (e.g., unstable binocular vision; individual variations in inter-pupillary distance (IPD); susceptibility to photic seizures and migraines), drug/alcohol consumption, health status, and ability to adapt to novel sensory environments are all thought to contribute to the extent of symptoms experienced (Kennedy, Dunlap, & Fowlkes, 1990; Kolasinski, 1995; McFarland, 1953; Mirabile, 1990; Reason & Brand, 1975; Stanney, Kennedy, & Kingdon, 2002; Stanney, Salvendy et al., 1998) (see Table 7.2).

While considerable research into the exact causes is requisite and has been ongoing for decades in its various forms (seasickness, motion sickness, simulator sickness, space sickness, cybersickness) (Chinn & Smith, 1953; Crampton, 1990; DiZio & Lackner, 2002; DiZio & Lackner, 1997; Heutink et al., 2019; Howarth &

TABLE 7.2
Factors Affecting Individual Capacity to Resist Adverse Effects of Immersive Exposure

- Age: Expect little motion sickness for those under age 2; expect greatest susceptibility to motion sickness between the ages of 2 and 12; expect motion sickness to decline after 12, with those over 25 being about half as susceptible as they were at 18 years of age.
- Anthropometrics: Consider setting immersive stimulus intensity in proportion to body weight/stature; ensure that hardware accommodates a large range of IPDs.
- Individual susceptibility: Expect individuals to differ greatly in motion sickness susceptibility and use the Motion History Questionnaire (MHQ) (Kennedy & Graybiel, 1965; Kennedy, Lane, Grizzard, Stanney, Kingdon, & Lanham, 2001) or another instrument (cf. Golding, Rafiq, & Keshavarz, 2021) to gauge the susceptibility of the target trainee population.
- Motion sickness history: Individuals who have experienced an emetic response associated with carnival rides can be expected to experience more than twice the level of adverse effects to immersive exposure as compared to those who do not experience such emesis (Stanney, Kingdon, Graeber, & Kennedy, 2002).
- HWD design: Expect IPD fit to be a primary driver of cybersickness, especially in VR (Stanney, Fidopiastis, & Foster, 2020). Some VR headsets on the market today are expected to not properly fit up to ~40% of females and ~18% of males, which is expected to drive high levels and persistent cybersickness, particularly for females who tend to have a larger IPD mismatch.
- Sensory plasticity: Cybersickness is expected to be less severe for those individuals who can rapidly re-weight (Oman, 1990) conflicting multisensory cues in XR environments, such as visual–vestibular mismatches (Gallagher & Ferrè, 2018).
- Physiological state: Individuals with higher pre-exposure drowsiness will be more likely to experience drowsiness upon postimmersive exposure, and those exposed to VR for 60 min or longer can be expected to experience more than twice the level of drowsiness as compared to those exposed for a shorter duration (Lawson et al., 2002; Stanney, Kingdon, Graeber, & Kennedy, 2002). As drowsiness increases, one can expect a greater severity of flashbacks (Lawson et al., 2002; Stanney, Kingdon, Graeber, & Kennedy, 2002).
- Body mass index: BMI does not tend to be related to cybersickness symptoms; however, those with higher BMIs may be less prone to experience an emetic response (Stanney, Kingdon, Graeber, & Kennedy, 2002).
- Drug/alcohol consumption: Limit immersive exposure to those individuals who are free from drug or alcohol consumption.
- Rest: Encourage individuals to be well rested before commencing immersive exposure
- Ailments: Discourage those with cold, flu, or other ailments (e.g., headache, diplopia, blurred vision, sore eyes, or eyestrain) from participating in immersive exposure; encourage those susceptible to photic seizures and migraines, as well as individuals with pre-existing binocular anomalies to avoid exposure.
- Clinical trainee groups: Obtain informed sensitivity to the vulnerabilities of these trainee groups (e.g., unique psychological, cognitive, and functional characteristics). Encourage those displaying comorbid features of various psychotic, bipolar, paranoid, substance abuse, claustrophobic, or other disorders where reality testing and identity problems are evident to avoid exposure.

TABLE 7.3
System and Usage Factors Influencing Immersive Stimulus Intensity

- Exposure duration: Adverse effects associated with immersive exposure in VR are positively correlated with exposure duration (Kennedy, Stanney, & Dunlap, 2000; Saredakis et al., 2020). Lanham (2000) has shown that sickness increases linearly at a rate of 23% per 15 min. Dropouts in VR occur in as little as 15 min of exposure (Cobb et al., 1999; DiZio & Lackner, 1997; Duzmanska et al., 2018; Howarth & Finch, 1999; Hughes et al., 2020; Regan & Price, 1994; Singer et al., 1998; Stanney, Kingdon, & Kennedy, 2002; Stanney, Lanham et al., 1999; Stanney, Kingdon, Graeber, & Kennedy, 2002; Wilson et al., 1995; Wilson et al., 1997). However, AR exposure does not appear to follow this same pattern. Recent research suggests that those who undergo protracted, long duration (2 h+) AR exposure may actually experience habituation, feeling more comfortable the longer they don the AR headset (Stanney & Hughes, 2021).
- Intersession intervals: VR studies have indicated that intersession intervals of two to five days are effective in mitigating adverse effects, while intervals less than or greater than two to five days are ineffective in reducing symptomatology (Kennedy, Lane, Berbaum, & Lilienthal, 1993; Watson, 1998). On the other hand, repeated, intermittent brief AR exposures appear to drive sensitization, where the sensory conditions in the AR headset continue to elicit cybersickness (Stanney & Hughes, 2021).
- Movement control: As the amount of trainee movement control in terms of degrees of freedom (DoF) and head tracking within fully immersive environments increases so too does the level of nausea experienced (Clifton & Palmisano, 2020; Grassini & Laumann, 2020; So & Lo, 1999; Stanney & Hash, 1998; Stanney, Kingdon, Graeber, & Kennedy, 2002). Complete trainee movement control (six DoF) can be expected to lead to 2.5 times more dropouts than streamlined control (three DoF). Further, movements in the rotational (i.e., roll, pitch, and yaw) axes may be more provocative than those in the translational (i.e., x, y) axes. Further, the type of movement initiation, whether initiated by a controller or walking, matters, with walking potentially being advantageous when trying to minimize cybersickness (Saredakis et al., 2020). Artificial continuous locomotion techniques (e.g., joystick-based movement) are typically associated with higher levels of cybersickness than discrete locomotion techniques (e.g., teleportation, rotation snapping, translation snapping; Caserman et al., 2021; Farmani & Teather, 2020).
- Visual scene complexity: The rate of visual flow (i.e., visual scene complexity) may influence the incidence, and more so the severity of motion sickness experienced by an individual (Kennedy & Fowlkes, 1992; McCauley & Sharkey, 1992). Complex visual scenes may be more nauseogenic than simple scenes with complex scenes possibly resulting in 1.5 times more emetic responses; however, scene complexity does not appear to affect dropout rates (Dichgans & Brandt, 1978; Kennedy, Berbaum, Dunlap, & Hettinger, 1996; Stanney, Kingdon, Graeber, & Kennedy, 2002). Immersive gaming content and 360 videos may be the most provocative (Saredakis et al., 2020). Such affects may be exacerbated by a large field-of-view (FOV) (Kennedy & Fowlkes, 1992), high spatial frequency content (Dichgans & Brandt, 1978), and visual simulation of action motion (i.e., vection) (Kennedy, Berbaum et al., 1996).
  - When large FOVs are used, determine if it drives high levels of vection (i.e., perceived self-motion). If high levels of vection are found and they lead to high levels of sickness, then reduce the spatial frequency content of visual scenes.

(*Continued*)

TABLE 7.3 (CONTINUED)
System and Usage Factors Influencing Immersive Stimulus Intensity

- Sensory mismatch: Visual–vestibular mismatches are probably one of the most provocative drivers of cybersickness, being associated with both substantially higher severity (perhaps double the severity) and dropout levels (Caserman et al., 2021). Further, differences in virtual versus physical head pose may be particularly provocative (Palmisano, Allison, & Kim, 2020).
  - Consider use of concordant motion (e.g., a motion base to reduce visual–vestibular conflicts; Kuiper et al., 2019), limiting or slowing forward speed and acceleration to reduce visual scene motion (Dennison & D'Zmura, 2017), rest frames (e.g., adding inertially stable visual motion cues, such as a fixed-horizon; Cao, Grandi, & Kopper, 2021), teleportation (Caserman et al., 2021), rotation and translation snapping (Farmani & Teather, 2020), depth-of-field or peripheral blur (Carnegie & Rhee, 2015), and dynamic field of view (e.g., modifying field of view based on speed and angular velocity; Fernandes & Feiner, 2016) techniques to minimize sensory conflicts.
- Visual display: Various visual display factors thought to influence how provocative an immersive environment includes system consistency (Uliano, Kennedy, & Lambert, 1986); lag (So & Griffin, 1995); update rate (So & Griffin, 1995); mismatched inter-pupillary distance (IPDs) (Mon-Williams, Rushton, & Wann, 1995, Stanney, Fidopiastis, & Foster, 2020); vergence–accommodation conflict (Zhan et al., 2020), restricted field-of-view (Fernandes & Feiner, 2016); unimodal and intersensorial distortions (both temporal and spatial) (Welch, 1978); and lighting conditions, with darker displays reported to be more conducive to cybersickness (Hoesch et al., 2018; Kobayashi et al., 2018). To minimize the effects of visual display on cybersickness:
  - Ensure any system lags/latencies are stable; variable lags/latencies can be debilitating (Palmisano, Allison, & Kim, 2020; Stauffert, Niebling, & Latoschik, 2020).
  - Minimize display/phase lags (i.e., end-to-end tracking latency between head motion and resulting update of the display; Kämäräinen et al., 2017).
  - Optimize frame rates, with a minimum frame rate of 60 Hz (frames per second) and upwards of 90 Hz recommended to minimize cybersickness (Freiwald, Katzakis, & Steinicke, 2018).
  - Provide adjustable IPD, with a range of ~ 50–77 mm recommended to capture >99% of people (Stanney, Fidopiastis, & Foster, 2020).
  - Provide multimodal feedback that minimizes sensory conflicts (i.e., provide visual, auditory, and haptic/kinesthetic feedback appropriate for situation being simulated; Kemeny, Chardonnet, & Colombet, 2020).
  - When appropriate, make the display as light as possible (minimize the use of dark scenarios; Hoesch et al., 2018; Kobayashi et al., 2018).

Finch, 1999; Hughes et al., 2020; Kennedy & Fowlkes, 1992; Keshavarz et al., 2014; Lawson, 2014; McCauley & Sharkey, 1992; McNally & Stuart, 1942; Reason, 1970, 1978; Reason & Brand, 1975; Sjoberg, 1929; Stanney, Fidopiastis, & Foster, 2020; Stanney, Salvendy et al., 1998; Tyler & Bard, 1949; Weech et al., 2019; Welch & Mohler, 2014; Wendt, 1968), there is still a lack of understanding about the factors that drive immersive environment stimulus intensity such that this knowledge can be used to identify usage protocols that minimize adverse effects. In fact, current usage of immersive technology generally treats trainees as if they are immune to cybersickness or possess low motion sickness susceptibility and are capable of rapid acclimation to novel sensory environments. This is not always the case, as evidenced by the intensity and extent of side effects listed in Table 7.1 and individual susceptibility factors summarized in Table 7.2. Beyond individual predisposition, one can see from Table 7.3 that there are a number of system and usage factors influencing stimulus intensity. By developing an understanding of these factors, means of acclimating trainees to an immersive experience could potentially be identified.

QUANTIFYING IMMERSIVE STIMULUS INTENSITY

The summary in Table 7.3 suggests that the intensity of an immersive stimulus could be reduced for VR by shortening exposure duration, maintaining an intersession interval of two to five days, reducing degrees of freedom (DoF) of trainee movement control – particularly avoiding rotational movements – and simplifying visual scenes. While the latter are expected to be the same for AR, usage protocol recommendations for minimizing stimulus intensity in AR are divergent, with repeated long duration (>2 h) exposures potentially being the most advantageous. Table 7.3 further suggests that the exact conditioning strategy that is most effective may depend on individual susceptibility. If these tactics are coupled with conditioning approaches, and improvements to the design of HWDs to include a larger segment of the population with respect to adjustable IPD ranges, enlarged FOV, and visual display advancements, reductions in adverse effects and associated dropout rates should result.

There are several published case reports on effective interventions to alleviate moderate-to-severe cybersickness. Rine and colleagues (1999) reported on a successful ten-week intervention combining visuo-vestibular habituation and balance

TABLE 7.4
Steps to Quantifying Immersive Stimulus Intensity

1. Get an initial estimate: Talk with target trainees (not developers) of the system and determine the level of adverse effects they experience
2. Observe: Watch trainees during and after exposure and note comments and behaviors
3. Try the system yourself: Particularly if you are susceptible to cybersickness or have an IPD that does not match the HWD, obtain a firsthand assessment of the adverse effects
4. Measure dropout rate: If most people can stay in for an hour without symptoms, then the system is likely benign; if most people drop out within 10 min, then the system is probably in need of redesign
5. Monitor: Use simple rating scales to assess sickness (Kennedy, Lane, Berbaum, & Lilienthal, 1993) and visual, proprioceptive, and postural measures to assess aftereffects (Kennedy, Stanney, Compton, Drexler, & Jones, 1999)
6. Compare: Determine how the system under evaluation compares to other immersive systems
7. Report: Summarize the severity of the problem, specify required interventions (e.g., warnings, instructions), and set expectations for use (e.g., target exposure duration, intersession intervals)
8. Expect dropouts: With a high-intensity immersive stimulus, dropout rates can be high

TABLE 7.5
Immersive Training Systems Usage Protocol

1. Reviewing information in Table 7.2, identify individual capacity of target trainees to resist adverse effects of immersive training exposure
2. Considering the factors in Table 7.3, design immersive training stimuli to minimize adverse effects
3. Following the guidelines in Table 7.4, quantify the stimuli intensity of immersive system
4. Warnings: Provide warnings for those with severe susceptibility to motion sickness, photic seizures, and migraines, as well as those with preexisting binocular anomalies, cold, flu, or other ailments (see Table 7.2)
5. Educate trainees as to how the occurrence or intensity of adverse effects may be mitigated by limiting drug/alcohol consumption and ensuring ample rest (see Table 7.2)
6. Educate trainees on the potential risks of immersive training exposure. Inform trainees of the insidious effects they may experience during exposure, including nausea, malaise, disorientation, headache, dizziness, vertigo, eyestrain, drowsiness, fatigue, pallor, sweating, increased salivation, and vomiting
7. Educate trainees as to the potential adverse aftereffects of immersive training exposure (see Table 7.1). Inform trainees that they may experience lowered arousal (e.g., drowsiness, fatigue), disturbed visual functioning, visual flashbacks, as well as unstable locomotor and postural control for prolonged periods following exposure. Relating these experiences to excessive alcohol consumption may prove instructional in understanding safety concerns with performing complex tasks (e.g., operating heavy machinery, or driving an automobile) shortly after exposure
8. Inform trainees that if they start to feel ill, they should terminate their immersive training exposure because extended exposure is known to exacerbate adverse effects (Kennedy, Stanney, & Dunlap, 2000)
9. Prepare trainees for their transition to the immersive training by informing them that there will be an adjustment period
10. Adjust environmental conditions: Provide adequate air flow and comfortable thermal conditions (Kloskowski, Medeiros, & Schöning, 2019). Sweating often precedes an emetic response, thus proper air flow can enhance trainee comfort. In addition, extraneous noise should be eliminated, as it can exacerbate ill-effects
11. Adjust equipment to minimize fatigue: Fatigue can exacerbate the adverse effects of immersive training exposure. To minimize fatigue, ensure all equipment is comfortable and properly adjusted for fit, including adjustment for IPD and performing any available visual calibrations to mitigate visual fatigue
12. Gauge initial exposure duration: For strong VR training stimuli, limit initial exposures to a short duration (e.g., 10 min or less) and allow an intersession recovery period of two to five days (Table 7.3). For AR, consider protracted long-duration exposure to foster habituation
13. Avoid provocative movements: For strong immersive training stimuli, warn trainees to avoid movements requiring high rates of linear or rotational acceleration and extraordinary maneuvers (e.g., flying backward) during initial interaction (McCauley & Sharkey, 1992)
14. Monitor trainees: Throughout immersive training exposure, an attendant should be available at all times to monitor trainees' behavior and ensure their well-being

(Continued)

TABLE 7.5 (CONTINUED)
Immersive Training Systems Usage Protocol

15. Look for red flags: Indicators of impending trouble include excessive sweating, verbal frustration, lack of movement within the environment for a significant amount of time, and less overall movement (e.g., restricting head movement). Trainees demonstrating any of these behaviors should be observed closely, as they may experience an emetic response. Extra care should be taken with these individuals, postexposure. *Note:* It is beneficial to have a plastic bag or garbage can located near trainees in the event of an abrupt emetic response
16. Termination: Set criteria for terminating exposure. Exposure should be terminated immediately if trainees verbally complain of symptoms and acknowledge they are no longer able to continue. Also, to avoid an emetic response, if telltale signs are observed (i.e., sweating, increased salivation), exposure should be terminated. Some individuals may be unsteady upon postexposure. These individuals may need assistance when initially standing up after exposure
17. Debriefing: After exposure, the well-being of trainees should be assessed. Measurements of their hand–eye coordination and postural stability should be taken. Similar to field sobriety tests, these can include measures of balance (e.g., standing on one foot, walking an imaginary line, leaning backward with eyes closed), coordination (e.g., alternate hand clapping and finger-to-nose touch while the eyes are closed), and eye nystagmus (e.g., follow a light pen with the eyes without moving the head). Do not allow individuals who fail these tests to conduct high-risk activities until they have recovered (e.g., have someone drive them home)
18. Releasing: Set criteria for releasing trainees. Specify the amount of time after exposure that trainees must remain on premises before driving or participating in other such high-risk activities. In our lab, a 2-to-1 ratio is used; postexposure trainees must remain in the laboratory twice the amount of exposure time to allow recovery
19. Follow-up: Call trainees the next day, have them call, or complete online surveys to report any prolonged adverse effects

training, which successfully rid a patient of strong visually induced motion sickness. Another effective strategy for reducing motion sickness is autogenic feedback training, in which subjects are taught to control physiological responses to environmental stressors (Cowings & Toscano, 2000). Thus, while technological improvements and sound design principles that ensure that the hardware developed reflects human physiological variation are essential, components for expanding the use of immersive environments while minimizing unwanted side effects should carefully consider usage protocols that foster habituation (Stanney & Hughes, 2021).

Focusing on the factors in Table 7.3, system developers can identify the primary factors that are inducing adverse effects in their system and adjust accordingly. Further, the steps in Table 7.4 can be used to establish the stimulus strength of a particular immersive system.

USAGE PROTOCOL

Integrating the issues reviewed above, Table 7.5 provides a systematic usage protocol that can be used by system developers and system administrators to minimize risks to trainees exposed to immersive training systems.

CONCLUSIONS

To facilitate adoption and minimize the risks associated with exposure to immersive training systems, developers and system administrators should identify the capacity of end users to resist the adverse effects of exposure, quantify and minimize stimulus intensity, and follow a systematic usage protocol. This protocol should focus on warning, educating, and preparing trainees, setting appropriate environmental and equipment conditions, gauging initial exposure duration based on system type, monitoring trainees while looking for red flags, and setting criteria for terminating exposure, debriefing, and terms for release. Adopting such a protocol can minimize the risk factors associated with immersive training system exposure, thereby enhancing the safety of trainees, while limiting the liability of system developers and administrators.

ACKNOWLEDGMENTS

This chapter is dedicated to the late Robert S. Kennedy, a world-leading expert in motion sickness and many other things. His tireless work ethic, unrivaled intellect, and selfless mentorship were a gift to all those whose lives he touched. This material is based upon work supported in part by the Office of Naval Research (ONR) under grant No.N000149810642, the National Science Foundation (NSF) under grants No. DMI9561266 and IRI-9624968, the National Aeronautics and Space Administration (NASA) under grants No. NAS9-19482 and NAS9-19453, and the U.S. Army Medical Research & Development Command (USAMRDC) under the guidance of the Joint Program Committee – JPC-1 at Ft. Detrick, MD under Contract number: MTEC-W81XWH1990005. The views, opinions, and/or findings contained in this research/presentation/publication are those of the authors/company and do not necessarily reflect the views of the ONR, NSF, NASA, or USAMRDC and should not be construed as an official DoD/NSF/NASA position, policy or decision unless so designated by other documentation. No official endorsement should be made. Reference herein to any specific commercial products, process, or service by trade name, trademark, manufacturer, or otherwise does not necessarily constitute or imply its endorsement, recommendation, or favoring by the U.S. government.

REFERENCES

Allen, B., Hanley, T., Rokers, B., & Green, C. S. (2016). Visual 3D motion acuity predicts discomfort in 3D stereoscopic environments. *Entertainment Computing, 13*, 1–9.

Alsop, T. (2021, Mar 22). Extended reality (XR): AR, VR, and MR – Statistics & facts. *Statista*. https://www.statista.com/topics/6072/extended-reality-xr/#dossierSummary

Al Zayer, M., Adhanom, I. B., MacNeilage, P., & Folmer, E. (2019, May). The effect of field-of-view restriction on sex bias in VR sickness and spatial navigation performance. *Proceedings of the 2019 CHI Conference on Human Factors in Computing Systems* (pp. 354:1–354:12). Glasgow, Scotland UK: ACM.

An, B., Matteo, F., Epstein, M., & Brown, D. E. (2018). Comparing the performance of an immersive virtual reality and traditional desktop cultural game. In *CHIRA* (pp. 54–61). In *Proceedings of the 2nd International Conference on Computer-Human Interaction*

Research and Applications – Volume 1: CHIRA (pp. 54–61). https://doi.org/10.5220/0006922800540061

Baltzley, D. R., Kennedy, R. S., Berbaum, K. S., Lilienthal, M. G., & Gower, D. W. (1989). The time course of postflight simulator sickness symptoms. *Aviation, Space, and Environmental Medicine*, *60*(11), 1043–1048.

Biocca, F. (1992). Will simulation sickness slow down the diffusion of virtual environment technology? *Presence: Teleoperators and Virtual Environments*, *1*(3), 334–343.

Bracq, M. S., Michinov, E., Arnaldi, B., Caillaud, B., Gibaud, B., Gouranton, V., & Jannin, P. (2019). Learning procedural skills with a virtual reality simulator: An acceptability study. *Nurse Education Today*, *79*, 153–160.

Cao, Z., Grandi, J., & Kopper, R. (2021). Granulated rest frames outperform field of view restrictors on visual search performance. *Frontiers in Virtual Reality*, https://doi.org/10.3389/frvir.2021.604889

Carnegie, K. C., & Rhee, T. (2015). Reducing visual discomfort with HMDs using dynamic depth of field. *IEEE Computer Graphics and Applications*, *35*(5), 34–41.

Caserman, P., Garcia-Agundez, A., Gámez Zerban, A. et al. (2021). Cybersickness in current-generation virtual reality head-mounted displays: Systematic review and outlook. *Virtual Reality*. https://doi.org/10.1007/s10055-021-00513-6

Champney, R., Stanney, K. M., Hash, P., Malone, L., Kennedy, R. S., & Compton, D. (2007). Recovery from virtual environment exposure: Expected time-course of symptoms and potential readaptation mechanisms. *Human Factors*, *49*(3), 491–506.

Chinn, H. I., & Smith, P. K. (1953). Motion sickness. *Pharmacological Review*, *7*, 33–82.

Clifton, J., & Palmisano, S. (2020). Effects of steering locomotion and teleporting on cybersickness and presence in HMD-based virtual reality. *Virtual Reality*, *24*(3), 453–468.

Cobb, S. V. G., Nichols, S., Ramsey, A. D., & Wilson, J. R. (1999). Virtual Reality-Induced Symptoms and Effects (VRISE). *Presence: Teleoperators and Virtual Environments*, *8*(2), 169–186.

Cowings, P. S., & Toscano, W. B. (2000). Autogenic-feedback training exercise is superior to promethazine for control of motion sickness symptoms. *Journal of Clinical Pharmacology*, *40*, 1154–1165.

Crampton, G. H. (Ed.). (1990). *Motion and space sickness*. Boca Raton, FL: CRC Press.

Curry, C., Li, R., Peterson, N., & Stoffregen, T. A. (2020). Cybersickness in virtual reality head-mounted displays: Examining the influence of sex differences and vehicle control. *International Journal of Human–Computer Interaction*, *36*(12), 1161–1167. https:/doi.org/10.1080/10447318.2020.1726108

Dennison, M. S., & D'Zmura, M. (2017). Cybersickness without the wobble: Experimental results speak against postural instability theory. *Applied Ergonomics*, *58*, 215–223.

Dichgans, J., & Brandt, T. (1978). Visual-vestibular interaction: Effects on self-motion perception and postural control. In R. Held, H. W. Leibowitz, & H. L. Teuber (Eds.), *Handbook of sensory physiology, Vol. VIII: Perception* (pp. 756–804). Heidelberg: Springer-Verlag.

DiZio, P., & Lackner, J. R. (1997). Circumventing side effects of immersive virtual environments. In M. Smith, G. Salvendy, & R. Koubek (Eds.), *Design of computing systems: Social and ergonomic considerations* (pp. 893–896). Amsterdam, Netherlands: Elsevier Science Publishers, San Francisco, CA, August 24–29.

DiZio, P., & Lackner, J. R. (2002). Proprioceptive adaptation and aftereffects. In K. M. Stanney (Ed.), *Handbook of virtual environments: Design, implementation, and applications* (pp. 791–806). Mahwah, NJ: Lawrence Erlbaum Associates.

Doolani, S., Wessels, C., Kanal, V., Sevastopoulos, C., Jaiswal, A., Nambiappan, H., & Makedon, F. (2020). A review of extended reality (XR) technologies for manufacturing training. *Technologies*, *8*, 77. https://doi.org/10.3390/technologies8040077

Duzmanska, N., Strojny, P., & Strojny A. (2018). Can simulator sickness be avoided? A review on temporal aspects of simulator sickness. *Frontiers in Psychology, 9*, 2132. https://doi.org/10.3389/fpsyg.2018.02132

Farmani, Y., & Teather, R. J. (2020). Evaluating discrete viewpoint control to reduce cybersickness in virtual reality. *Virtual Reality, 24*, 645–664. https://doi.org/10.1007/s10055-020-00425-x

Fernandes, A. S., & Feiner, S. K. (2016). Combating VR sickness through subtle dynamic field-of-view modification. In 2016 *IEEE Symposium on 3D User Interfaces (3DUI)*, Greenville, SC. https://doi.org/10.1109/3DUI.2016.7460053

Freiwald, J. P., Katzakis, N., & Steinicke, F. (2018). Camera time warp: Compensating latency in video see-through head-mounted-displays for reduced cybersickness effects. *Proceedings of VRST '18*, November 29–December 1, 2018, Tokyo, Japan.

Gallagher, M., & Ferrè, E. R. (2018). Cybersickness: A multisensory integration perspective. *Multisensory Research, 31*(7), 645–674. https://doi.org/10.1163/22134808-20181293

Giamellaro, M. (2017). Dewey's Yardstick: Contextualization as a crosscutting measure of experience in education and learning. *Sage Open, 7*(1). https://doi.org/10.1177/2158244017700463

Golding, J. F, Rafiq, A., & Keshavarz, B. (2021). Predicting individual susceptibility to visually induced motion sickness by questionnaire. *Frontiers in Virtual Reality, 2*, 576871. https://doi.org/10.3389/frvir.2021.576871

Graeber, D. A. (2001). *Use of incremental adaptation and habituation regimens for mitigating optokinetic side effects*. Unpublished doctoral dissertation, University of Central Florida.

Grassini, S., & Laumann, K. (2020). Are modern head-mounted displays sexist? A systematic review on gender differences in HMD-mediated virtual reality. *Frontiers in Psychology, 11*, 1604. https://doi.org/10.3389/fpsyg.2020.01604

Heutink, J., Broekman, M., Brookhuis, K. A., Melis-Dankers, B. J., & Cordes, C. (2019). The effects of habituation and adding a rest-frame on experienced simulator sickness in an advanced mobility scooter driving simulator. *Ergonomics, 62*(1), 65–75.

Hildebrandt, J., Schmitz, P., Valdez, A. C., Kobbelt, L., & Ziefle, M. (2018, July). Get well soon! human factors' influence on cybersickness after redirected walking exposure in virtual reality. *International Conference on Virtual, Augmented and Mixed Reality* (pp. 82–101). Cham: Springer.

Hoesch, A., Poeschl, S., Weidner, F., Walter, R., & Doering, N. (2018). The relationship between visual attention and simulator sickness: A driving simulation study. 2018 *IEEE Conference on Virtual Reality and 3D User Interfaces (VR)* (pp. 1–2). https://doi.org/10.1109/VR.2018.8446240.

Howarth, P. A., & Finch, M. (1999). The nauseogenicity of two methods of navigating within a virtual environment. *Applied Ergonomics, 30*, 39–45.

Hughes, C. L., Bailey, P. S., Ruiz, E., Fidopiastis, C. M., Taranta, N. R., & Stanney, K. M. (2020). The psychometrics of cybersickness in augmented reality. *Frontiers in Virtual Reality: Virtual Reality & Human Behavior, 1*, 602954. https://doi.org/10.3389/frvir.2020.602954

Kahol, K., Vankipuram, M., & Smith, M. L. (2009). Cognitive simulators for medical education and training. *Journal of Biomedical Informatics, 42*(4), 593–604. https://doi.org/10.1016/j.jbi.2009.02.008

Kämäräinen, T., Siekkinen, M., Ylä-Jääski, A., Zhang, W., & Hui, P. (2017). Dissecting the end-to-end latency of interactive mobile video applications. *Proceedings of the 18th International Workshop on Mobile Computing Systems and Applications – HotMobile '17* (pp. (pp. 61–66). Sonoma, CA: ACM Press. https://doi.org10.1145/3032970.3032985

Kemeny, A., Chardonnet, J.-R., & Colombet, F. (2020). *Getting rid of cybersickness in virtual reality, augmented reality, and simulators*. Switzerland: Springer International. https:/doi.org/ 10.1007/978-3-030-59342-1

Kennedy, R. S., Berbaum, K. S., Dunlap, W. P., & Hettinger, L. J. (1996). Developing automated methods to quantify the visual stimulus for cybersickness. *Proceedings of the Human Factors and Ergonomics Society 40th Annual Meeting* (pp. 1126–1130). Santa Monica, CA: Human Factors & Ergonomics Society.

Kennedy, R. S., Dunlap, W. P., & Fowlkes, J. E. (1990). Prediction of motion sickness susceptibility: A taxonomy and evaluation of relative predictor potential. In G. H. Crampton (Ed.), *Motion and space sickness* (pp. 179–215). Boca Raton, FL: CRC Press.

Kennedy, R. S., & Fowlkes, J. E. (1992). Simulator sickness is polygenic and polysymptomatic: Implications for research. *International Journal of Aviation Psychology*, *2*(1), 23–38.

Kennedy, R. S., French, J., & Ordy, J. M., & Clark, J. (2003). *Visually induced motion sickness, cognitive performance, saliva melatonin, and cortisol*. Paper accepted for presentation at the Society for Neuroscience 33rd Annual Meeting, November 8–12, New Orleans, LA.

Kennedy, R. S., & Graybiel, A. (1965). *The Dial test: A standardized procedure for the experimental production of canal sickness symptomatology in a rotating environment* (Rep. No. 113, NSAM 930). Pensacola, FL: Naval School of Aerospace Medicine.

Kennedy, R. S., Hettinger, L. J., & Lilienthal, M. G. (1990). Simulator sickness. In G. H. Crampton (Ed.), *Motion and Space Sickness* (pp. 247–262). Boca Raton, FL: CRC Press.

Kennedy, R. S., Lane, N. E., Berbaum, K. S., & Lilienthal, M. G. (1993). Simulator sickness questionnaire: An enhanced method for quantifying simulator sickness. *International Journal of Aviation Psychology*, *3*(3), 203–220.

Kennedy, R. S., Lane, N. E., Grizzard, M. C., Stanney, K. M., Kingdon, K., & Lanham, S. (2001). Use of a motion history questionnaire to predict simulator sickness. *Proceedings of the Sixth Driving Simulation Conference- DSC2001* (pp. 79–89). France: INRETS/Renault.

Kennedy, R. S., Lane, N. E., Stanney, K. M., Lanham, D. S., & Kingdon, K. (2001). Use of a motion experience questionnaire to predict simulator sickness. *Usability evaluation and interface design: Cognitive engineering, intelligent agents and virtual reality* (pp. 1061–1065). Mahwah, NJ: Lawrence Erlbaum Associates.

Kennedy, R. S., & Stanney, K. M. (1996). Postural instability induced by virtual reality exposure: Development of a certification protocol. *International Journal of Human-Computer Interaction*, *8*(1), 25–47.

Kennedy, R. S., Stanney, K. M., Compton, D. E., Drexler, J. M., & Jones, M. B. (1999). *Virtual environment adaptation assessment test battery* (Phase II Final Report, Contract No. NAS9-97022). Houston, TX: NASA Lyndon B. Johnson Space Center.

Kennedy, R. S., Stanney, K. M., & Dunlap, W. P. (2000). Duration and exposure to virtual environments: Sickness curves during and across sessions. *Presence: Teleoperators and Virtual Environments*, *9*(5), 463–472.

Keshavarz, B., Hecht, H., & Lawson, B. D. (2014). Visually-induced motion sickness: Causes, characteristics, and countermeasures. In K. S. Hale & K. M. Stanney (Eds.), *Handbook of virtual environments: Design, implementation, and applications* (2nd edition, pp. 647–698). New York, NY: CRC Press.

Kingdon, K., Stanney, K. M., & Kennedy, R. S. (2001). Extreme responses to virtual environment exposure. *The 45th Annual Human Factors and Ergonomics Society Meeting* (pp. 1906–1910). Minneapolis/St. Paul MN, October 8–12, 2001.

Kloskowski, H., Medeiros, D., & Schöning, J. (2019). OORT: An air-flow based cooling system for long-term virtual reality sessions. *Proceedings of VRST '19*, November 12–15, 2019, Parramatta, NSW, Australia.

Kobayashi, N., Yamashita, H., Matsuura, A., & Ishikawa, M. (2018). Effects of illuminance environment on visual induced motion sickness. *2018 IEEE 7th Global Conference on Consumer Electronics (GCCE)* (pp. 429–430). https://doi.org/10.1109/GCCE.2018.8574778

Kolasinski, E. M. (1995). *Simulator sickness in virtual environments* (ARI Technical Report 1027). Alexandria, VA: U.S. Army Research Institute for the Behavioral and Social Sciences.

Kuiper, O. X., Bos, J. E., Diels, C., & Cammaerts, K. (2019). Moving base driving simulators' potential for carsickness research. *Applied Ergonomics*, *81*, 102889.

Lackner, J. R. (2014). Motion sickness: More than nausea and vomiting. *Experimental Brain Research*, *232*(8), 2493–2510. https://doi.org/10.1007/s00221-014-4008-8

Lanham, S. (2000). *The effects of motion on performance, presence, and sickness in a virtual environment*. Master's Thesis, University of Central Florida.

Lawson, B. D. (2014). Motion sickness symptomatology and origins. In K. S. Hale & K. M. Stanney (Eds.), *Handbook of virtual environments: Design, implementation, and applications* (2nd edition, pp. 531–600). New York, NY: CRC Press.

Lawson, B. D., Graeber, D. A., Mead, A. M., & Muth, E. R. (2002). Signs and symptoms of human syndromes associated with synthetic experiences. In K. M. Stanney (Ed.), *Handbook of virtual environments: Design, implementation, and applications* (pp. 791–806). Mahwah: NJ: Lawrence Erlbaum Associates.

Lee, K. (2012). Augmented reality in education and training. *TechTrends*, *56*(2), 13–21.

Levine, S., Goldin-Meadow, S., Carlson, M., & Hemani-Lopez, N. (2018). Mental transformation skill in young children: The role of concrete and abstract motor training. *Cognitive Science*, *42*, 1207–1228. https://doi.org/10.1111/cogs.12603

Llamas, S. (2019). XR by the numbers: What the data tells us. *Gaming & Entertainment Track at AWE USA 2019*. Santa Clara, CA. https://www.slideshare.net/AugmentedWorldExpo/stephanie-llamas-superdata-xr-by-the-numbers-what-the-data-tells-us

Lucas, G., Kemeny, A., Paillot, D., & Colombet, F. (2020). A simulation sickness study on a driving simulator equipped with a vibration platform. *Transportation Research Part F: Traffic Psychology and Behaviour*, *68*, 15–22.

McCauley, M. E., & Sharkey, T. J. (1992). Cybersickness: Perception of self-motion in virtual environments. *Presence: Teleoperators and Virtual Environments*, *1*(3), 311–318.

McFarland, R. A. (1953). *Human factors in air transportation: Occupational health & safety*. New York: McGraw-Hill.

McNally, W. J., & Stuart, E. A. (1942). Physiology of the labyrinth reviewed in relation to seasickness and other forms of motion sickness. *War Medicine*, *2*, 683–771.

Melo, M., Vasconcelos-Raposo, J., & Bessa, M. (2018). Presence and cybersickness in immersive content: Effects of content type, exposure time and sex. *Computers & Graphics*, *71*, 159–165.

Merhi, O. A. (2009). Motion sickness, virtual reality and postural stability. Retrieved from the University of Minnesota Digital Conservancy, https://hdl.handle.net/11299/58646.

Mirabile, C. S. (1990). Motion sickness susceptibility and behavior. In G. H. Crampton (Ed.), *Motion and space sickness* (pp. 391–410). Boca Raton, FL: CRC Press.

Money, K. E. (1970). Motion sickness. *Psychological Reviews*, *50*(1), 1–39.

Mon-Williams, M., Rushton, S., & Wann, J. P. (1995). Binocular vision in stereoscopic virtual-reality systems. *Society for Information Display International Symposium Digest of Technical Papers*, *25*, 361–363.

Moroz, M., Garzorz, I., Folmer, E., & MacNeilage, P. (2019). Sensitivity to visual speed modulation in head-mounted displays depends on fixation. *Displays*, *58*, 12–19.

Munafo, J., Diedrick, M., & Stoffregen, T. A. (2017). The virtual reality head-mounted display Oculus Rift induces motion sickness and is sexist in its effects. *Experimental Brain Research*, *235*, 889–901. https://doi.org/10.1007/s00221-016-4846-7

Nooij, S. A., Pretto, P., Oberfeld, D., Hecht, H., & Bülthoff, H. H. (2017). Vection is the main contributor to motion sickness induced by visual yaw rotation: Implications for conflict and eye movement theories. *PloS One*, *12*(4), e0175305.

Oman, C. M. (1990). Motion sickness: A synthesis and evaluation of the sensory conflict theory. *Canadian Journal of Physiology and Pharmacology*, *68*(2), 294–303. https://doi.org/10.1139/y90-044

Palmisano, S., Allison, R. S., & Kim, J. (2020). Cybersickness in head-mounted displays is caused by differences in the user's virtual and physical head pose. *Frontiers in Virtual Reality*, *1*, 587698. https://doi.org/10.3389/frvir.2020.587698

Park, S., & Lee, G. (2020). Full-immersion virtual reality: Adverse effects related to static balance. *Neuroscience Letters*, *733*, 134974.

Pot-Kolder, R., Veling, W., Counotte, J., & Van Der Gaag, M. (2018). Anxiety partially mediates cybersickness symptoms in immersive virtual reality environments. *Cyberpsychology, Behavior, and Social Networking*, *21*(3), 187–193.

Reason, J. T. (1970). Motion sickness: A special case of sensory rearrangement. *Advancement in Science*, *26*, 386–393.

Reason, J. T. (1978). Motion sickness adaptation: A neural mismatch model. *Journal of the Royal Society of Medicine*, *71*, 819–829.

Reason, J. T., & Brand, J. J. (1975). *Motion sickness*. New York: Academic Press.

Rebenitsch, L., & Owen, C. (2021). Estimating cybersickness from virtual reality applications. *Virtual Reality*, *25*, 165–174. https://doi.org/10.1007/s10055-020-00446-6

Regan, E. C., & Price, K. R. (1994). The frequency of occurrence and severity of side-effects of immersion virtual reality. *Aviation, Space, and Environmental Medicine*, *65*, 527–530.

Riccio, G. E., & Stoffregen, T. A. (1991). An ecological theory of motion sickness and postural instability. *Ecological Psychology*, *3*(3), 195–240.

Rine, R. M., Schubert, M. C., & Balkany, T. J. (1999). Visual-vestibular habituation and balance training for motion sickness. *Physical Therapy*, *79*(10), 949–57.

Sagnier, C., Loup-Escande, E., Lourdeaux, D., Thouvenin, I., & Valléry, G. (2020). User acceptance of virtual reality: An extended technology acceptance model. *International Journal of Human–Computer Interaction*, *36*(11), 993–1007. https://doi.org/10.1080/10447318.2019.1708612

Saredakis, D., Szpak, A., Birckhead, B., Keage, H., Rizzo, A., & Loetscher, T. (2020). Factors associated with virtual reality sickness in head-mounted displays: A systematic review and meta-analysis. *Frontiers in Human Neuroscience*, *14*, 96. https://doi.org/10.3389/fnhum.2020.00096

Shafer, D. M., Carbonara, C. P., & Korpi, M. F. (2019). Factors affecting enjoyment of virtual reality games: A comparison involving consumer-grade virtual reality technology. *Games for Health Journal*, *8*(1), 15–23.

Singer, M. J., Ehrlich, J. A., & Allen, R. C. (1998). Virtual environment sickness. Adaptation to and recover from a search task. *Proceedings of the 42nd Annual Human Factors and Ergonomics Society Meeting* (pp. 1506–1510). Chicago, IL, October 5–9.

Sjoberg, A. A. (1929). Experimental studies of the eliciting mechanism of sea sickness. *Acta oto-laryngolica*, *13*, 343–347.

Smither, J. A., Mouloua, M., & Kennedy, R. S. (2003). *Reducing symptomatology of visually-induced motion sickness through perceptual training*. Manuscript submitted for publication.

Smyth, J., Jennings, P., Mouzakitis, A., & Birrell, S. (2018, November). Too sick to drive: How motion sickness severity impacts human performance. 2018 *21st International Conference on Intelligent Transportation Systems (ITSC)* (pp. 1787–1793). IEEE.

So, R. H., & Griffin, M. J. (1995). Effects of lags on human operator transfer functions with head-coupled systems. *Aviation, Space, and Environmental Medicine*, *66*, 550–556.

So, R .H. Y., & Lo, W. T. (1999). Cybersickness: An experimental study to isolate the effects of rotational scene oscillations. *Proceedings of the IEEE Virtual Reality Conference* (pp. 237–241). Los Alamitos, CA: IEEE Computer Society.

Stanney, K., Fidopiastis, C., & Foster, L. (2020). Virtual reality is sexist: But it does not have to be. *Frontiers in Robotics & AI – Virtual Environments*, *7*, 4. https://doi.org/10.3389/frobt.2020.00004

Stanney, K. M., & Hash, P. (1998). Locus of user-initiated control in virtual environments: Influences on cybersickness. *Presence: Teleoperators and Virtual Environments*, *7*(5), 447–459.

Stanney, K. M., & Hughes, C. (2021). *Final report: Assessment of psychological and physiological effects of augmented reality: Development of the Dual-Adaptation Protocol for Augmented Reality (DAPAR)*. Orlando, FL: Design Interactive.

Stanney, K. M., & Kennedy, R. S. (1997a). Cybersickness is not simulator sickness. *Proceedings of the 41st Annual Human Factors and Ergonomics Society Meeting* (pp. 1138–1142). Albuquerque, NM, September 22–26.

Stanney, K. M., & Kennedy, R. S. (1997b). The psychometrics of cybersickness. *Communications of the ACM*, *40*(8), 67–68.

Stanney, K. M., & Kennedy, R. S. (1998). Aftereffects from virtual environment exposure: How long do they last? *Proceedings of the 42nd Annual Human Factors and Ergonomics Society Meeting* (pp. 1476–1480). Chicago, IL, October 5–9.

Stanney, K. M., Kennedy, R. S., Drexler, J. M., & Harm, D. L. (1999). Motion sickness and proprioceptive aftereffects following virtual environment exposure. *Applied Ergonomics*, *30*, 27–38.

Stanney, K. M., Kennedy, R. S., & Hale, K. (2014). Virtual environments usage protocols. In K. S. Hale & K. M. Stanney (Ed.), *Handbook of virtual environments: Design, implementation, and applications* (2nd edition, pp. 797–809). Boca Raton, FL: CRC Press.

Stanney, K. M., Kingdon, K., Graeber, D., & Kennedy, R. S. (2002). Human performance in immersive virtual environments: Effects of duration, user control, and scene complexity. *Human Performance*, *15*(4), 339–366.

Stanney, K. M., Kingdon, K., & Kennedy, R. S. (2002). Dropouts and aftereffects: Examining general accessibility to virtual environment technology. *The 46th Annual Human Factors and Ergonomics Society Meeting* (pp. 2114–2118). Baltimore, MD, September 29–October 4, 2002.

Stanney, K. M., Kingdon, K., & Kennedy, R. S. (2001). Human performance in virtual environments: Examining user control techniques. In M. J. Smith, G. Salvendy, D. Harris, & R. J. Koubek (Eds.), *Usability evaluation and interface design: Cognitive engineering, intelligent agents and virtual reality* (Vol. 1 of the Proceedings of HCI International 2001) (pp. 1051–1055). Mahwah, NJ: Lawrence Erlbaum.

Stanney, K. M., Kingdon, K., Nahmens, I., & Kennedy, R. S. (2003). What to expect from immersive virtual environment exposure: Influences of gender, body mass index, and past experience. *Human Factors*, *45*(3), 504–522.

Stanney, K. M., Lanham, S., Kennedy, R. S., & Breaux, R. B. (1999). Virtual environment exposure drop-out thresholds. *The 43rd Annual Human Factors and Ergonomics Society Meeting* (pp. 1223–1227). Houston, TX, September 27-October 1, 1999

Stanney, K. M., Lawson, B. D., Rokers, B., Dennison, M., Fidopiastis, C., Stoffregen, T., Weech, S., & Fulvio, J. M. (2020). Identifying causes of and solutions for cybersickness in immersive technology: Reformulation of a research and development agenda. *International Journal of Human-Computer Interaction*, *36*(19), 1783–1803.

Stanney, K. M., Nye, H., Haddad, S., Padron, C. K., Hale, K. S., & Cohn, J. V. (2021, in press). eXtended reality (XR) environments. In G. Salvendy & W. Karwowski (Eds.), *Handbook of human factors and ergonomics* (5th edition) (pp. 782–815). New York: John Wiley.

Stanney, K. M., Salvendy, G., Deisigner, J., DiZio, P., Ellis, S., Ellison, E., Fogleman, G., Gallimore, J., Hettinger, L., Kennedy, R., Lackner, J., Lawson, B., Maida, J., Mead, A., Mon-Williams, M., Newman, D., Piantanida, T., Reeves, L., Riedel, O., Singer, M., Stoffregen, T., Wann, J., Welch, R., Wilson, J., & Witmer, B. (1998). Aftereffects and sense of presence in virtual environments: Formulation of a research and development agenda. Report sponsored by the Life Sciences Division at NASA Headquarters. *International Journal of Human-Computer Interaction*, *10*(2), 135–187.

Stauffert, J.-P., Niebling, F., & Latoschik, M. E. (2020) Latency and cybersickness: Impact, causes, and measures. A review. *Frontiers in Virtual Reality*, *1*, 582204. https://doi.org/10.3389/frvir.2020.582204

Treisman, M. (1977). Motion sickness: An evolutionary hypothesis. *Science*, *197*(4302), 493–495. https://doi.org/10.1126/science.301659

Tyler, D. B., & Bard, P. (1949). Motion sickness. *Physiological Review*, *29*, 311–369.

Uliano, K. C., Kennedy, R. S., & Lambert, E. Y. (1986). Asynchronous visual delays and the development of simulator sickness. *Proceedings of the Human Factors Society 30th Annual Meeting* (pp. 422–426). Dayton, OH: Human Factors Society.

Vailshery, L. S. (2021, Jan 22). Share of business executives adopting augmented or virtual reality technology worldwide as of December 2018, by stage. *Statista*. https://www.statista.com/statistics/1097137/ar-vr-adoption-levels-among-global-business-executives/

Watson, G. S. (1998). The effectiveness of a simulator screening session to facilitate simulator sickness adaptation for high-intensity driving scenarios. *Proceedings of the 1998 IMAGE Conference*. Chandler, AZ: The IMAGE Society.

Weech, S., Calderon, C. M., & Barnett-Cowan, M. (2020). Sensory down-weighting in visual-postural coupling is linked with lower cybersickness. *Frontiers in Virtual Reality*, *1*, 10. https://doi.org/10.3389/frvir.2020.00010

Weech, S., Kenny, S., & Barnett-Cowan, M. (2019). Presence and cybersickness in virtual reality are negatively related: A review. *Frontiers in Psychology*, *10*, 158. https://doi.org/10.3389/fpsyg.2019.00158

Welch, R. B. (1978). *Perceptual modification: Adapting to altered sensory environments.* New York: Academic Press.

Welch, R. B., & Mohler, B. J. (2014). Adapting to virtual environments. In K. S. Hale & K. M. Stanney (Eds.), *Handbook of virtual environments: Design, implementation, and applications* (2nd edition, pp. 627–646). New York, NY: CRC Press.

Wendt, G. R. (1968). *Experiences with research on motion sickness* (NASA Special Publication No. SP-187). Pensacola, FL: Fourth Symposium on the Role of Vestibular Organs in Space Exploration.

Widdowson, C., Becerra, I., Merrill, C., Wang, R. F., & LaValle, S. (2021). Assessing postural instability and cybersickness through linear and angular displacement. *Human Factors*, *63*(2), 296–311. https://doi.org/10.1177/0018720819881254

Wilson, M. L. (2016). *The effect of varying latency in a head-mounted display on task performance and motion sickness*. Clemson University. Retrieved from http://tigerprints.clemson.edu/all_dissertations/1688/

Wilson, M. L., & Kinsela, A. J. (2017, September). Absence of gender differences in actual induced HMD motion sickness vs. pretrial susceptibility ratings. *Proceedings of the Human Factors and Ergonomics Society Annual Meeting* (Vol. 61, No. 1, pp. 1313–1316). Los Angeles, CA: SAGE Publications.

Wilson, J. R., Nichols, S., & Haldane, C. (1997). Presence and side effects: Complementary or contradictory? In M. Smith, G. Salvendy, & R. Koubek (Eds.), *Design of computing systems: Social and ergonomic considerations* (pp. 889–892). Amsterdam, Netherlands: Elsevier Science Publishers, San Francisco, CA, August 24–29.

Wilson, J. R., Nichols, S. C., & Ramsey, A. D. (1995). Virtual reality health and safety: Facts, speculation and myths. *VR News*, *4*, 20–24.

Wu, B., Yu, X., & Gu, X. (2020). Effectiveness of immersive virtual reality using head-mounted displays on learning performance: A meta-analysis. *British Journal of Educational Technology*, *51*, 1991–2005. https://doi.org/10.1111/bjet.13023

Zhan, T., Xiong, J., Zou, J., & Wu, S. T. (2020). Multifocal displays: review and prospect. *PhotoniX*, *1*, 1–31.

8 Distributed Debriefing for Simulation-Based Training

Cullen D. Jackson, Di Qi, Anna Johansson, Emily E. Wiese, William J. Salter, Emily M. Stelzer, and Suvranu DeJared Freeman

CONTENTS

DOI: 10.1201/9781003401360-8

INTRODUCTION

When we read the word "debrief," we often imagine a group of soldiers around a map discussing the battle just fought, pilots slicing their hands through the air describing a dogfight, or a medical team discussing a difficult case. In each of these instances, the situation the team is discussing could have been a training exercise or a real-world operation. For the purposes of this chapter, we are focused on a version of the latter—debriefing for simulation-based training, and specifically on considerations for conducting debriefs when the team is working together in a distributed manner.

A "distributed" team is one that is geographically dispersed—teammates could be in different rooms of the same building or thousands of miles apart. Regardless of the manner in which they are separated, team members in these situations usually are connected only by the technology available (i.e., phones, chat rooms, simulation-based trainers) as part of the training event. It also may be the case that small teams are situated together, and they are working with other distributed teams as a multilevel system (or team-of-teams) (Kozlowski & Klein 2000); in this case, teams would be interacting with other teams (or individuals within distributed teams) at a distance.

A "debrief" (aka debriefing, after-action review [AAR]) is a facilitated discussion of training performance in which the basic goal is to enhance subsequent trainee performance, generally conducted soon after the training event (Sawyer et al. 2016), and it is considered the most fundamental aspect of simulation-based training for effective learning (Issenberg et al. 2005). During a *distributed* training event, the learners (or teams of learners) are separated from each other, and the instructors may be together or distributed among the learners. In the case of the former, the instructors may have similar observation viewpoints of the learners depending on whether they are observing the learners as a whole, in small groups, or one-to-one. In the case in which the instructors also are distributed, they definitely will have different perspectives on the event, and since they are not collocated, they (and the learners) will need to use some collaborative technology to interact during the debriefing; this could range from simple teleconferences to sophisticated computer-based approaches.

"Simulation-based training" is a methodology that allows learners to immerse themselves in realistic situations for the purpose of gaining experience in a safe (physically, psychologically) manner compared to real operations or live exercises (Lateef 2010). Immersion occurs through interactions with real or simulated versions of the systems, people, teams, environments, organizations, etc. they would encounter in actual operational situations while working through realistic scenarios (or missions). Since the focus of this chapter is distributed debriefing, we are interested in simulation-based training that involves interacting with systems that collect data on the events occurring during the training and how trainees respond to them during the exercise since some forms of technology are required to facilitate learner interactions.

Debriefs for distributed simulation-based training now are de rigueur largely because distributed simulation-based training is a more prevalent learning modality, which has been driven by several factors. First, real-world missions and events have become increasingly complex. In the military domain, missions often involve joint operations between multiple US forces (e.g., Air Force, Navy, Army), allied/coalition (e.g., NATO) forces, and sometimes nongovernmental agencies (NGOs) like the Red Cross. The heterogeneous participants in such missions generally are geographically distributed, and thus it is difficult and expensive to coordinate co-located training for these groups. In some cases, it also fails to capture the actual distributed nature of the tasks being simulated. In part, this reflects broadening mission sets that include those not traditionally handled by the military (e.g., humanitarian aid). This broadening of missions and their joint/coalition emphasis also require a wider range of skills, including teamwork and coordination across organizational boundaries, and implies that one overall (often called "community" or "mass") debrief may not address all training requirements across participants, teams/groups, and organizations.

In the healthcare domain, there are events (e.g., mass casualties) that are similarly complex in requiring coordination across geographic and/or organizational boundaries, and training events need to reflect this complexity as well as appropriate debriefing opportunities and content across collaborating organizations and agencies. Second, pressures to reduce training costs are making it harder to support the expense of bringing people together for the purposes of training. For example, operating room-based team training exercises have proven to be expensive and difficult to conduct as they require a centralized facility and a team of expert clinicians to conduct the exercises, evaluate the participants, and debrief them verbally (Hippe et al. 2020). The situation is even more challenging when conducting healthcare training in resource-constrained countries overseas, including long travel distances and high costs for bringing medical instructors for onsite teaching.

Third, rapid advances in enabling technologies make it relatively easy to use existing simulators and computing systems for distributed simulation. For computer-based simulation (e.g., simulated desktop systems, VR-based simulators), the prevalence of high-speed internet and inexpensive, cloud-based computation and storage solutions make it easier and more cost-effective to bring trainees together virtually. Additionally, high levels of network reliability, bandwidth, and speed are dropping in price far more quickly than new simulation technologies are being introduced, making the basic technology infrastructure needed for distributed simulations more available. In particular, DIS (distributed interactive simulation) and HLA (high-level architecture)—the two communications standards used for almost all military simulations—can run over wide-area networks as long as those networks have adequate performance characteristics. This means that simulators that often had to be in adjacent rooms to interoperate effectively can now be separated by thousands of miles. Fourth, restrictions on bringing people together for training (e.g., COVID-19 pandemic) necessitate innovations in transforming traditional in-person learning into distributed, simulation-based learning environments.

Issues to Consider in Providing Distributed Debriefing for Simulation-Based Training

Because distributed simulation-based training involves multiple people in different locations, it necessarily will involve multiple simulators or simulated systems. Often, but not always, this also means that multiple different *types* of simulators or simulated systems (e.g., clinical systems, weapons platforms, and/or operational elements depending on the training domain) will be used. Unless all of the learners have the same training needs, there also will be an explicit (or implicit) hierarchy of training objectives for the event. Some of these training objectives will target individual roles, some across roles or for groups of roles, and some will focus on overall outcomes (i.e., the mission as a whole). Designing and implementing scenarios for distributed simulation can be more complex for more specialized training, and the debriefs obviously have crucial dependencies on the scenario. However, the important topics of training objective development and scenario design are beyond the scope of this chapter.

Because of the differences in training objectives between learners, the debriefs for distributed simulation-based training events generally reflect (at least) two levels of the basic hierarchical structure of these objectives: separate debriefs for each role, element, or platform; and a community-wide, or mass, debrief that addresses coordination across roles and overall scenario objectives. Different pieces of information about performance and the scenario itself are required for each role and the community debrief, and different expertise or perspectives are generally required to deliver the debriefs, all of which create several complexities.

First, *it is desirable that an instructor (or a designated participant-instructor) be physically present for each separate role, element, or team in the distributed simulation.* Because some important actions (or lack of actions) may not be recorded in sufficient detail by automated components of the distributed simulation for immediate post-exercise analysis, humans can provide a level of detail that can be quite valuable. For example, if participants communicate by voice and all voice communications are recorded, while language processing software may be able to transcribe these communications in near real time, it currently is not adequate to support pedagogically useful analysis rapidly. Furthermore, replay of *all* communications for post-exercise analysis requires too much time and must be synchronized with event replay during the review in order to be meaningfully linked with other actions.

Second, identifying points for debriefing across roles/elements to be addressed in the community debrief, and deciding how to characterize those issues, requires *information fusion across the distributed simulation platforms.* Those issues will involve interaction, coordination, and/or communication across platforms or roles/elements. The identification of such problems can be difficult, and the diagnosis and suggested remediation for problems will tend to be more so. This fusion must be addressed via procedures, technology, and preparation of the instructors.

Because of the hierarchical nature of the various learning objectives being trained during these events, and the need to collect and combine information across different roles and systems, more time and energy must be expended to prepare the debriefs;

this is true for debriefing at the role/element, team/platform, and community levels. Instructor-observers must share their perspectives to gain insights into events and actions that they did not observe, identify performance issues that may span across roles and levels of hierarchy, and synthesize appropriate feedback for these various learners and stakeholders. Thus, we believe that *more formal and more extensive use of computer-supported debrief preparation tools is needed* to effectively conduct debriefs for distributed simulation-based training than for traditional co-located training. This is true to a lesser, but still significant, extent even if a single distributed team, element, or platform is being trained because the distributed instructors still must share insights and build up reasonably extensive shared situation awareness about the exercise.

Computer-supported methods of rapidly moving through and analyzing performance data can be of particular value in distributed training because more cognitive processing will tend to be required by distributed instructors (or a single instructor observing multiple, distributed learners) than in co-located training. Importantly, while shared situation awareness typically may *not* be possible, even for distributed training of a single element, it is needed to understand the complete scenario timeline and the ways in which trainee performance propagates across time and roles/elements. Therefore, well-designed, computer-supported debriefing tools should assist in sharing the debriefing workload across instructors, allowing each to address what he or she has observed in detail, thus facilitating a more global awareness of performance across training participants. Moreover, the effectiveness of debriefing tools for distributed teams is highly dependent on the instructors' ability to capture, evaluate, and analyze extensive performance data during the training scenario. Therefore, it also is desirable to *incorporate computerized algorithms and methods* to automate the data collection, classification and categorization of performance data, and the subsequent generation of summary information and visualization cues to facilitate post-event debriefing and reduce the cognitive loads for both instructors and trainees (Hanoun & Nahavandi, 2018). Ideally, these analysis and debriefing technologies also would facilitate *collaboration between the distributed instructors* since they operate under considerable time and task pressure: debriefs typically take place within one hour of completing the training exercise while experiences are fresh in participants' and instructors' minds. Task pressure comes from the fact that failing to address important aspects of the training and provide feedback to learners during the debrief can result in seriously impaired learning which attenuates the purpose of the training event.

The Rest of This Chapter

In the following sections, we discuss the *functions and methods for conducting debriefs* in some detail. We then dissect the *challenges and opportunities* afforded by the increasing importance of distributed debriefs. Then, we address the *requirements* for distributed debriefs followed by a discussion of *current methods* for conducting them. We conclude the chapter with a more speculative discussion of *the future of distributed debriefs* for simulation-based training.

DEBRIEFING FUNCTIONS AND METHODS

Debriefings and AARs fulfill diagnostic, instructional, and social functions. We define these functions and then turn to a brief review of the current techniques and technologies that attempt to support computer-mediated debriefs (i.e., debriefs that leverage computer systems for data collection, performance analysis and review, collaboration, and instruction).

Functions of Debriefs

The principal function of debriefs is *instructional*: they must convey the right lesson to the right people at the right time. In order to do this, debriefs must help learners and instructors *diagnose*—identify and characterize—specific episodes of performance during the training event. That is, the debriefs should facilitate the recall of periods of *correct* and *incorrect* performance in context. Once diagnosed, episodes of correct performance can be used instructionally to reinforce those behaviors and generalize them to similar (but not identical) future circumstances. Similarly, specific instances of *incorrect* performance can be used to discourage repetition of those behaviors in the future, cue recall of correct performance knowledge, and associate it with similar circumstances. These activities should be targeted at trainees who need to learn and who have the capacity and motivation to do so. A debriefing is not just a process for gathering performance data and delivering lessons, it also helps participants to discover those lessons themselves as well as help both instructors and trainees discover performance failures (and successes) and diagnose their causes.

In addition to helping to diagnose performance and facilitate instruction for continued performance improvement, debriefs should help instructors to link performance with learning objectives to help focus their instruction. For example, a medical team in simulation-based training to learn better teamwork skills may have to perform cardiopulmonary resuscitation (CPR) as part of the simulated case. During the debrief, the learners may diagnose that their chest compressions were too shallow over the course of the event and want to focus instruction on ways to improve. However, the training objective was to learn good teamwork skills, so the debrief should help to focus diagnosis and instruction on episodes of performance that link to that objective while also allowing for identifying and improving areas of critical performance that may be ancillary to the event's purpose (e.g., improving CPR skills).

Debriefs also serve a social function. For all involved, while they are an opportunity to demonstrate and assess technical competence, and learn how to discriminate good from bad performance and diagnose its cause, they also help develop social competence by practicing how best to convey critiques of oneself with candor and of co-participants or trainees with diplomacy. In addition to facilitating positive social interactions, unfortunately debriefs also may exacerbate the inherent constraints of prevailing social structures, namely the formation of social or "status" hierarchies. "Status" is the social "ranking" of an individual relative to others. It is the fundamental basis for social hierarchy because those higher in the hierarchy are seen as more competent and legitimately owning their status, which in turn grants legitimacy to the resulting hierarchy (Galinsky et al. 2008). Research has demonstrated that people routinely use status characteristics, such as organizational role, time in service, ethnicity, race, profession, gender, and expertise (Bales et al. 1951; Berger et al. 1966; Berger et al. 1977) as the basis for generating performance expectations, consistent with culturally salient stereotypes. From a perspective of debriefing, these status characteristics could negatively influence setting performance objectives and diagnosing performance for discussion. In addition, they can drive interactions with learners and instructors that result in patterns of unequal contributions within the group (Silver et al. 1988; Silver et al. 2000), which inhibits the information-exchange process and limits the potential positive instructional impact of diversity within the group. There is evidence from more recent research to support that status characteristics confer (dis)advantage even in virtual groups (Bélisle & Bodur 2010; Principe & Langlois 2013), which has implications for debriefings of distributed simulation-based training. The good news is that strategies exist to not only mitigate the deleterious effects of status, but they can also enhance team learning outcomes and the team's ability to diagnose performance failures, which is discussed later in the chapter.

In sum, debriefing supports diagnosis of performance, recall of performance in training, understanding of expert performance, generalization to future situations, and assessment and display of competence.

Methods of Debriefs

Techniques and technologies for computer-mediated debriefs and AARs have evolved to support these functions with varying levels of effectiveness.

One method for supporting the *instructional* function of debriefs is through *replaying* the events of the training scenario. For example, replaying video footage of medical teams during training has served as a valuable source for debriefing following team training (Sawyer et al. 2016). Replay of the events encountered by the learners during the training scenario also has a strong effect on learning since serial replay eases the recall of events and learner performance because it conforms to the serial structure of episodic memory (Tulving 2002). Replay can also reinforce memory for normative sequences of events (Schank 1982), and thus help learners generalize from scenario-specific episodes of performance to similar future situations. Serial replay helps trainees recognize sequential actions that aggregate into failure (or success) as a scenario evolves, and therefore support performance diagnosis and the skills underlying diagnosis and, hopefully, prognosis. Finally, it should help to minimize hindsight bias in diagnosing performance since learners and instructors can see the recorded behaviors in context (Fanning & Gaba 2007).

However, replay often is implemented in ways that provide an overall view of the scenario without the ability to "drill-down" to individual roles or more granular information. For example, replay is often implemented as a set of icons moving over an overhead view of the training environment (e.g., vehicle icons moving over a tactical map). While this is a useful representation for cueing recall of the overall tactical state of forces and tactical actions of units, it does not help trainees understand the situation or learning environment from their perspective or that of other roles in the simulation. Replay systems rarely record and represent learners' displays, instruments, or viewpoints/perspectives, nor do they support recall of *responses* because the systems often do not record and represent the trainee's use of the simulators or simulated systems. Rare exceptions, mainly in the field of aviation training, include an F-16 distributed debriefing system developed by the Air Force Research Laboratory (AFRL) that combines a central tactical view with instrument displays on each side; an example of implementing "overlap," a technique for maintaining visual momentum in a display (Bennett & Flach 2012). Similarly, the Dismounted Infantry Virtual After-Action Review System (DIVAARS, developed by the Army Research Institute and the University of Central Florida) provides multiple viewpoints of the simulation space during replay (Goldberg et al. 2003). Another system developed to teach the operation of anesthesia machines used an augmented reality (AR)-based debrief system to provide virtual overlay information on the real-world training environment and allowed playback of recorded training experiences through a user-controlled egocentric viewpoint (Quarles et al. 2008; Quarles et al. 2013). These systems help operators to extract information across various perspectives of the scenario while maintaining overall context (Woods' visual momentum; Woods 1984), which better facilitates detailed discussions about how learners perceived the environment and the actions they took in it.

While replay is useful to help learners understand the course of an overall situation, the serial nature of many of its implementations makes it more difficult for learners to relate parts of the scenario to other parts. To help relate non-sequential scenario events, it may be useful to show multiple instances of a class of events and learners' behavioral responses, or to even relate events in one simulation to similar events in another. To do this, debriefing systems must allow for marking events so that learners and instructors can access them on-demand (also called "random access") rather than sequentially. While some systems allow instructors to "bookmark" events for future review, in our experience, they are rarely used. Debrief systems rarely help instructors identify and navigate from one instance of a class of events to the next because simulation systems are rarely instrumented with measurement systems that can categorize events and the human responses to them, as well as link them to the learning objectives of the simulation scenario. One such system (Salter, et al. 2005) categorizes events, assesses performance in those events, displays those categories and assessments, and links each instance directly to its replay. This design gives trainers random access to multiple instances of a given class of events and thus supports trainees as they attempt to generalize from instances to the larger class. The previously described AR-based system for learning anesthesia machine operations also supports random access to scenario events and learner performance. This allows both instructors and trainees to navigate between different viewpoints to visualize key events in time or to better view information that may have been previously occluded (Quarles et al. 2008; Quarles et al. 2013), as well as use the standard *replay* controls of traditional video-based debriefing platforms.

The *diagnostic function* of debriefs typically is supported by the debriefing techniques of the instructors with some scaffolding support by debriefing technologies. Debriefing techniques encourage instructors and trainees to identify and analyze strengths and shortcomings in performance. The Navy and the Air Force, for example, decompose debriefs for large simulated and live exercises into independent debriefs of small elements or packages, and then conduct overall debriefs, supported by technology, involving the entire training audience (community or mass debriefs). The element (or role) debriefs typically identify specific performance failures that support diagnosis in the subsequent community debrief. The Army has codified its diagnostic method in a set of questions that learners explore during the debrief: What was supposed to happen? What happened? What accounts for the difference? (Dixon 2000), and a set of guidelines (e.g., "Call it like you see it," "No thin skins") that encourage participants to think critically and to be candid in their review of events. Similarly, the commercial aviation community espouses debrief methods that engage learners in identifying and diagnosing performance failures. Studies of these methods and their impact on diagnostic quality are rare. However, one analysis of debriefs in commercial aviation (Dismukes et al. 2000) found that instructors often failed to engage trainees in diagnostic (or any) discussions during the debrief, and instead dominated these sessions with monologues concerning their own observations.

When techniques fail in this way, instructors have few diagnostic technologies on which to fall back. In general, debriefing systems are incapable of generating diagnoses because they do not incorporate expert behavioral models against which to compare trainee performance. Nor do most systems generally record data concerning trainee performance or compute measures that summarize that performance, attribute effects to individual performers, or relate causes to effects. That said, in the healthcare domain, there are several new and innovative systems that attempt to collect, diagnose, and summarize learner performance to facilitate more robust debriefs. A mixed reality AAR system has also been used in the training of medical procedures such as central venous access (CVA) (Lampotang et al. 2013), in which a physical human simulator is augmented with 3D virtual human anatomies such as the lungs and veins, allowing the user to visualize the needle insertion procedure inside the human body. In addition to integrating an augmented display, the operations of the instruments during the procedure are tracked by the simulator using an embedded six DoF (degree-of-freedom) magnetic sensor; this allows the trainee to observe their own data and reflect on their own performance.

Computer-based multimedia (audio, video, text, and graphs) debriefing has been shown to be useful in teaching not only medical skills, but also nontechnical skills that do not relate to medical knowledge or technical procedures, but instead encompass situation awareness and interpersonal skills. The Interpersonal Scenario Visualizer (IPSVize) debriefing tool (Raij & Lok, 2008) was developed to allow medical students to review their interactions with virtual human patients. These interactions are captured, logged, and processed by the simulator to produce spatial, temporal, and social visualizations to drive discussion during the debrief. In addition, the system also evaluates students' actions and provides feedback to help them gain insights into methods for improving their interactions with real patients.

In a web-enabled, scenario-based training program for debriefing emergency medical teams (T-TRANE, developed by Aptima, Inc. and the University of Maryland Shock Trauma Center) (Xiao et al. 2007), video segments are used for demonstrating good and poor examples of teamwork skills. Learners use these exemplars to identify instances of ineffective teamwork and discuss how the shortcomings could be resolved.

Unfortunately, video taken of simulation-based training is limited by the viewpoint of the camera, and fine details of learner actions, as well as other team members' behaviors, may be unobservable. These limitations hinder the ability of the instructors and learners to fully appraise the teamwork performance during debriefing. To counteract this constraint, a recent system collects multiple synchronized data streams to capture a multitude of intraoperative data, such as physiological parameters from both patients and healthcare professionals, and audiovisual data from an in-room wide-angle camera, laparoscope video, and wearable cameras (Goldenberg et al. 2017; OR Black Box). Additionally, automated data analysis enables the platform to generate a joint team performance report that could be used as a tool for structured postoperative multidisciplinary debriefing (van Dalen et al. 2021).

If debriefs are meant to help trainees learn expert alternatives to their incorrect performance, they must present those alternatives in some way. Few high-end simulators (e.g., flight simulators and driving simulators with physical cockpits) can generate examples of expert performance because they do not incorporate computational models of expert behavior. Thus, alternative behaviors are highlighted largely through discussion; that is, alternatives are said, not shown. While this may be sufficient in some circumstances, research has demonstrated that verbally describing expert solutions to complex problems (e.g., team planning and execution of military air missions) produces learning outcomes that are reliably inferior to describing and "playing" solutions using video review (Scherer et al. 2003) or visual animations (Shebilske et al. 2009). In the latter paper, expert solutions were generated by an optimization model. In traditional intelligent tutoring systems, these solutions are generated by heuristic or rule-based models of expertise. Recently, eye-tracking technology has been used to track the eye movements of learners during simulation-based training. These data provide objective, detailed information to use to diagnose performance, and they also can be used to compare the learners' gaze behaviors to expert-like gaze behaviors to better drive performance improvements. This use of eye-tracking data appears to be useful for improving patient safety practices such as those relating to patient identification (Henneman et al. 2014).

Finally, debriefing techniques (but seldom the technologies) support the *social processes* by which participants assess the competence of their colleagues and assert their own. As stated previously, learners are differentiated in terms of their status characteristics (gender, race, ethnicity, education, professional rank, etc.). Because status characteristics confer social (dis)advantage pursuant to cultural stereotypes (Webster & Hysom 1998), learners can be thought of as "low-status" or "high-status" *relative to other team members*. Of note, what makes stereotypes so powerful is that both high-status and low-status team members believe them (e.g., not only do men believe that men are better in math and science, but many women also hold this belief). Because of these status characteristics, learners in the simulation environment will perceive one member to be especially qualified to perform a task, and other team members will defer to that "high-status" member. As such, high-status group members often enjoy more opportunities to contribute, have more influence, and have their contributions more positively evaluated. In contrast, low-status group members will defer to high-status members and limit their contributions (Berger & Cohen 1972; Berger et al. 1977). Subsequently, lower-status team members will become less likely to offer other (sometimes critical) information to the group during the debrief because they perceive their information and insights as less valuable. This process, also known as the "burden of proof" process (Berger & Cohen 1972; Berger & Webster 2006), proceeds unless some event or some new information interrupts the process. Therefore, whether in-person or distributed, it is critical that debriefing techniques and technologies incorporate norms for equal participation (Cohen 1993), and that instructors are taught to be attuned to the structural features of the team and help direct interactions accordingly.

The functions of debriefing—instructional, diagnostic, and social—are partially supported by debriefing systems, and debriefing techniques help instructors to fill the gaps, particularly supporting diagnostic and social functions. However, instructors often have difficulty applying good debrief technique, and the increasing need for distributed debriefing may make it even more difficult for them to use these methods, as we discuss below, which provides opportunities to design more of these functions into instructional and debriefing technologies.

CHALLENGES OF DISTRIBUTED DEBRIEFS

Recent advances in technology have allowed for the development of coordinated simulation tools, which can be used to simultaneously train groups of individuals who are dispersed across several geographical locations. As outlined above, these distributed training exercises can be conducted with less cost and risk than traditional live training events, allowing diverse groups of individuals to collaboratively train whole mission exercises more frequently than was ever before possible (e.g., Dwyer, Fowlkes, Oser, & Salas 1996). While distributed training can produce more effective and routine training events, this approach to training can complicate debriefs from both technical and social standpoints. We discuss the potential challenges that distributed training can pose to debriefing within the framework of the five key functions of a debrief discussed previously, namely: diagnosis of performance, recall of performance in training, understanding of expert performance, generalization to future situations, and assessment and display of competence within the social setting of a debrief.

PERFORMANCE DIAGNOSIS

The effectiveness of a debrief hinges on the ability of the instructors to support learners in diagnosing the underlying causes of performance failure (or success), and attributing those causes directly to individual or team behavior. As noted above, serial replay capabilities are usually provided as a global representation of mission performance, which trainees observe to recall and learn the general flow of the simulated scenario. Traditionally, such replays do not include views of specific displays or instruments, which can provide the needed context of individual constraints and reasoning in diagnosing performance. When the training environment is extended to include multiple, diverse, and distributed training groups, the challenges of using this training approach become increasingly apparent.

Because distributed simulation-based training substantially reduces the logistics and costs associated with live training events, more diverse trainees (or more trainee groups) can participate. For example, a distributed training event for the US Navy might include an E2-C Hawkeye aircraft (to provide surveillance coordination), a flight of F/A-18 Hornets (to provide suppression of enemy air defenses and strike ground targets), as well as ground control and intelligence support. The participant diversity in this example, represented by two air platforms and supporting ground elements, permits complexity in the type of mission used and generates corresponding complexity in the data generated from the simulated mission and the interdependencies of actions between elements to accomplish satisfactory mission performance. With current procedures and technologies, instructors and learners often diagnose successes and failures and analyze the interdependencies of actions to support these outcomes themselves with some technological support. Because most current instructional technologies cannot capture the subtle details of actions and communications between *remotely* located elements, the diagnosis of performance likely will be hindered in these cases.

The debriefing process can be structured to encourage instructors and learners to collaboratively identify performance shortcomings and strengths, for which instructors are an essential guide in the diagnostic process. However, under a distributed training process, participants and instructors likely will not all be collocated, thus inhibiting the instructors for each role/element from observing and integrating performance information in real time across the team. In the best case, performance measures may be collected automatically through the simulators or other simulation systems into a common repository, or remotely by instructors observing performance via global views of the overall scenario. By distributing learners and instructors across locations, instructors also will face challenges in diagnosing performance, and ultimately may be less helpful in supporting performance diagnosis for trainees. The distribution of training participants also may force asynchronous communications and interactions between trainees and training sites. From a social structure and social processes standpoint, this constraint may further exacerbate limited contributions from low-status members, and these missed opportunities for unique observations across learners and roles could have potentially life-threatening implications in real-world situations.

Therefore, to support these diagnostic processes and leverage diagnosis to improve future performance, learners and instructors must be successful in recalling mission performance, comparing that performance to expert behavior, and extrapolating behaviors to future situations. Unfortunately, each of these key processes can be inhibited by the dispersion of trainees across geographical locations, as we note below.

Performance Recall, Comparison, and Extrapolation

Traditional replay techniques can provide great utility in supporting memory for the sequence of scenario events (Schank 1982) in traditional training exercises. Distributed training exercises can compromise this recall by increasing the load placed on instructor and trainee memory through two independent mechanisms. First, because distributed training exercises can involve richer interactions with heterogeneous training groups, the data and behaviors associated with these interactions increase in number and complexity. For example, communications in non-distributed training events may be limited to face-to-face voice communications, but in distributed simulation-based events, these same communications must occur over a network, and will involve communications between different types of roles and simulation systems. Second, distributed training exercises rely on simulation systems that have evolved into sophisticated data collection tools, which can exponentially increase the amount of data that are collected (Jacobs, Cornelisse, & Schavemaker-Piva 2006), and may need to be recalled by instructors. Although all these data may not be conventionally discussed in traditional debriefing processes, they may be included in distributed debriefs, which would increase memory demands and reduce the likelihood of recalling any truly granular performance data. This increased data complexity and deluge will require a different set of techniques and technologies that might include computational synthesis and drill-down capabilities into debriefing systems that support distributed training events.

The comparison of these collected data to expert alternatives for diagnostic purposes is further complicated by the complexity of the interactions between distributed learners. As the number of individuals involved in the training exercise increases, the predictability of their interactions and the ability to optimize or simulate these interactions in an understandable way to instructors and trainees also becomes complex. In addition, the grouping of similar performance data becomes an essential component to understanding performance trends and extrapolating those trends to predict future behavior. Because the distributed training environment involves integrating many data from multiple, heterogeneous data streams, this training approach can reveal rich, informative patterns in trainee behavior. As in each of the critical debriefing processes discussed thus far, the sheer quantity of data and the heterogeneity of behaviors can complicate this process beyond that encountered with traditional training events.

Assessment and Display of Competence

The final function that debriefs provide is supporting the social processes through which trainees can assess themselves and the team and provide feedback to their colleagues, as well as assert their own expertise. Whereas distributed debriefing processes can influence the functions described above by generating more complex and diverse performance data, the distribution of trainees and instructors during debriefing also constrains the social mechanism of appraisal quite differently.

Distributed debriefing, and the tools that are used to support communication during these processes, can affect social appraisal and display of competence in three key ways. First, information-sharing tools (e.g., collaborative desktops) and communication channels (e.g., video teleconference) can be useful in exchanging knowledge across disparate locations; however, these tools can still constrain the type of information that is readily shared beyond the ways that the hierarchical structure of the group may already constrain. These tools are especially useful at exchanging text-based, or pre-generated spatial content, but are not highly effective at facilitating the types of interactions that occur between learners who are collocated at a whiteboard and visualize and discuss scenario events and diagnose performance. Second, the microphones and sound quality associated with teleconferences and video teleconferences can inhibit fluid discussion and instigate misunderstanding between disparate participants, even with the most advanced systems. These communication technologies also are poor at transmitting any radio communications that occurred during the scenario and are being replayed during the debrief. Finally, these communication devices cannot capture the nonverbal communications (e.g., gestures, eye gaze, facial expressions) that can be used to effectively assess and display competence. While traditional, in-person training processes can rely on nonverbal information exchange, such as eye contact, distributed training environments strip this form of communication from the essential social interactions that occur between individual trainees, as well as between trainees and their instructors.

REQUIREMENTS FOR DISTRIBUTED DEBRIEFS

In the preceding section, we identified several challenges that distributed debriefings might, and frequently do, encounter. However, standard procedures for conducting effective distributed debriefing with collaborative technologies have not yet been defined. While a range of processes could be used to prepare and deliver debriefs in the distributed environment, the utility of these approaches depends on both the context in which they are being used and the design of the training technology to support the instructor. It is tempting to discuss the technical and procedural requirements for distributed debriefs separately. However, due to their very nature, distributed debriefs combine both technology *and* process in ways that are difficult, if not impossible, to separate. To ensure that the technologies and the processes interact seamlessly, it is important to address the design of these pieces simultaneously and collaboratively. We discuss distributed debriefing requirements below, relating them to the previously defined debrief functions: diagnosis of performance, recall of performance training, understanding of expert performance, generalization to future situations, and assessment and display of competence. Importantly, while these requirements may be met by a single technology, it is likely that an integrated set of tools and institutional processes will be needed to conduct effective distributed debriefs for simulation-based training.

Communication

Distributed training events require constant communication of information and data throughout the events, and the subsequent distributed debriefs also must support communication between the various locations involved, during both the preparation and delivery phases of the debrief. At a minimum, voice communication must be supported. Ideally, video conferencing will also be supported across all sites to allow for nonverbal communication and rapport development between participants. This is no small technological feat across multiple locations. While current technologies exist that support this requirement, there are numerous technical issues that pose some issues. Naturally, the frequency of technical issues increases with more distributed sites. Aside from technical issues, each distributed simulation event must choose and follow some basic guidelines for using these communication technologies, beginning with the start of the communication (i.e., who calls whom) and including turn-taking tips on reducing extraneous noise, and how electronic information will be shared. Communication is a critical requirement of any distributed debrief and all the previously defined functions require it.

Collaboration

Closely tied to communication is collaboration. The distributed debrief must allow sites to coordinate on the content that should be debriefed, the strategy for debriefing, and the actual delivery of the debrief. All supporting information related to the execution of the simulation exercise—performance data, feeds from simulators or simulated systems, video feeds to support replay and perspective-taking—must be shared across sites. Collaboration technologies that allow all participants to view the same information simultaneously will reduce confusion and facilitate the creation of common ground between all instructors and learners. Optimally, the collaboration technology also will allow each site to take control of the information so they can interact with the information and help illuminate any performance results they deem appropriate from *their perspective* to the rest of the participants. Here again, rules on effective use of the collaboration technology are imperative as confusion in turn-taking can quickly lead to a chaotic debriefing, both in the preparation and delivery phases. In addition, creating norms for equal participation across sites and roles will help alleviate negative influences of the status hierarchy (Cohen 1993), which will further support the collaborative environment by facilitating more equal interactions and contributions. As with communication, without collaboration mechanisms, none of the other debriefing functions can be fulfilled in a distributed fashion.

Automated Data Capture

Distributed simulation-based events generally run on a very tight schedule, and instructors are not given much time to develop their debriefs post-event (e.g., 20–30 minutes is fairly standard for a large event). Any distributed debriefing technology must facilitate rapid development, part of which is accessing available performance data and simulator feeds. Allowing instructors to view performance data specific to their element and common across all participants supports diagnosis, recall, understanding, generalization, and, ultimately, overall performance assessment. This technology should accept and process performance data and simulator feeds automatically or semi-automatically (in the case of any observer-based measures used during the event) with little direction on the part of the instructor.

Data Presentation

Any performance data and measurements collected during the exercise have two potential presentation audiences: the instructors (during debrief preparation) and the learners (during the debrief itself). How this performance data is presented strongly affects instructors' and learners' ability to diagnose, understand, and, subsequently, assess their performance during the exercise. The distributed debrief technology should allow performance data to be presented at varying levels of detail, as it relates to all learners, subsets of learners, and individual learners. Drill-down capabilities are key as are the methods by which performance data are presented (e.g., replay, on a timeline, by event, textual representations, graphs) since they also can influence the interpretation of the information presented.

Data Selection

Not all performance data is relevant for each individual learner, nor are all the data relevant to the entire group of participants. In order to facilitate diagnosis and understanding of performance, instructors must be able to easily select relevant performance data, simulator feeds or displays, communications feeds, and video (as available). Thus, distributed debriefing systems must allow instructors to identify and select key scenario events and associated performance data that are indicative of both good and poor performance. Subsequent review of performance during these key events should facilitate debriefs across sites at the role, group, and community levels.

Replay Perspective

While viewing performance data is important, replaying exercise events is equally critical. Showing these events from multiple viewpoints can greatly facilitate diagnosis, recall, understanding, generalization, and assessment of trainee performance. This is particularly important for assessing coordination and teamwork. Viewing events on a map (or overhead representation of the training environment), from a first- (or third-) person view, can provide context and perspective to participants that otherwise would be difficult to obtain. Similarly, it may be useful to display simulation system/simulator controls and interaction artifacts (e.g., gauges and instrument panels) in order to provide a more common understanding of element capabilities across trainees. The distributed debriefing system must allow instructors to replay selected exercise events from these multiple viewpoints to the extent they are available. Additionally, the technology must allow instructors to choose when and how these viewpoints are presented in order to best fit the overall structure of the debrief.

Expert Models of Performance

Viewing and analyzing performance in comparison to defined standards or models of expert performance can greatly assist instructors in diagnosing trainee performance. Discussing how the trainees' performance compared to those standards can assist in understanding what went well or poorly and why. The distributed debriefing technology should present alternative (or expert) models of performance for each role, either in the form of quantitatively modeled behavior or performance categories. When the domain knowledge of the training task has been clearly identified, the most simple and effective way to deliver expert models is through a rule-based system (Grosan & Abraham, 2011) in which the expert performance is represented by a set of rules and coded into the system to mimic the behavior of human experts under different circumstances (Qi et al. 2020). Expert performance models created using machine learning techniques are generally considered more flexible, and suitable, for capturing experts' technical skills in complex procedures than a set of fixed rules. In fact, a machine learning model trained with expert data can simulate expert performance simultaneously with the learners in the simulation-based training event, demonstrating effective feedback of learning (Rhienmora et al. 2011, Wijewickrema et al. 2018).

Flexible Delivery Style

The way in which a distributed debrief is conducted varies according to the institution sponsoring the training exercise, the domain(s) being trained, and the instructors conducting the training. It therefore is important that any distributed debriefing technology not unduly constrain or force instructors into presenting feedback to trainees in a specific style; this requirement can be tricky to fulfill. Certainly, some instructional strategies may be more effective than others, and, indeed, some instructors may be more effective than others. However, all other things being equal, instructors should be allowed to tell the performance narrative (Fiore et al. 2007) in the manner that best suits their needs. For example, it may be appropriate to conduct the debrief in a narrative style while allowing instructors to drill down to specific training objectives and aspects of performance as needed. In this way, the debrief can be tailored to each training audience in a way that maximally supports learning for that community and training environment.

Post-Exercise Review

Once the distributed training exercise is complete, the debrief should be available for offline review by each distributed site. Additionally, each learner and instructor should receive a performance report after the exercise is complete. These should facilitate a variety of post-exercise analysis activities, such as more in-depth review of each role's (or site's) performance, evaluation of the training effectiveness of the simulated scenarios, evaluation of instructor effectiveness, and evaluation of the effectiveness of the debrief.

Store Lessons Learned

Distributed exercises accomplish much more than just teaching individual learners in that particular moment of training. Each distributed simulation-based event results in a variety of lessons learned (e.g., about trainees, instructional strategies and training materials, and instructors) that should be used when planning the next distributed training event. These lessons should be reviewed by local units/departments to facilitate continued learning opportunities, and they should be used to prepare for real-world situations like those that were trained. Retaining and using institutional knowledge is a difficult process in any environment, and the distributed debriefing system must facilitate institutional learning by providing a mechanism to accumulate these lessons learned, distribute them, and use them when designing the next exercise and developing an associated performance assessment plan.

Scalable

Distributed simulation training exercises come in all shapes and sizes. At their simplest, they will involve two different learners, each located at a different location. The roles of the learners may, or may not, be similar. At the other end of the spectrum, large-scale distributed simulation events, such as the US Air Force's Virtual Flag exercise, may involve ten different sites with hundreds of trainees, thousands of training activities at multiple hierarchical levels (e.g., strategic, operational, tactical), and dozens of interconnected systems ranging from high-fidelity aircraft simulators to sophisticated computer-based simulations; adding to this complexity, some sites may host a variety of different types of simulators. Distributed debriefing solutions should consider the scale of the distributed events they intend to support so the complexity of the training is appropriately supported by the technology in both useful and usable ways.

Ease of Use

Instructors may be involved in a distributed exercises only sporadically, so the debriefing system needs to be easy to use. Time spent immediately prior to training events, particularly large-scale exercises, typically focuses on reviewing relevant techniques and procedures, becoming familiar with the training scenario being executed during the exercise, and mitigating any issues with the distributed technology. There typically is little time to become familiar with *how* the technology works. Therefore, any distributed debriefing technology must allow instructors to quickly learn (or re-learn) the system with minimal difficulty or instruction; learners also may need to use the front end of the system during the debrief so the user interface should be as intuitive to use as possible.

CURRENT TECHNIQUES FOR DEBRIEFING DISTRIBUTED TEAMS

Many existing simulation environments provide the technical building blocks upon which many of the requirements listed above can be fulfilled, and while some have been realized by existing technologies, we are unaware of any fully implemented distributed debriefing system that addresses all these requirements.

State of the Art in Distributed Debriefing

Today's distributed simulation training has come a long way. Both the US Navy and the US Air Force regularly conduct complex virtual training exercises that involve a variety of simulated platforms and locations (e.g., the Navy's Operation Brimstone and the Air Forces' Virtual Flag Exercises). Research and development efforts across all the US military services continue to develop additional simulation environments for use locally and in a distributed manner. In addition, the development of specifications and protocols, like the Distributed Debrief Control Architecture (DDCA) (SISO 2016) and the Distributed After-Action Playback and Review (DAAPR) (Streit 2020), hold the promise for better integration and improved interoperability for debriefing systems that support distributed simulation-based training. However, these frameworks mostly support allowing existing debrief systems to operate together and share information, and those current systems do not support all the necessary functions for distributed debriefs discussed in this chapter; in particular, the specifications mostly support the interoperation of those systems' scenario and video replay functions.

Large-Scale Distributed Simulation Training Exercises

During large-scale virtual training exercises hosted by the US Air Force and US Navy, the distributed debriefs heavily rely on common tools found throughout the uniformed services; for example, standard video teleconferencing applications are used to connect sites during briefs and debriefs. Performance data collected during the event generally are presented to all trainees as a slideshow during a community (or mass) debrief; elements and smaller units sometimes will conduct a short debrief (sometimes called a "hot wash") as they put together their thoughts and lessons learned to be presented at the mass debrief. During the community debrief, networked collaboration software is used to share slides between sites and save them to commonly accessible shared network drives. The diagnosis of performance and delivery of the debriefs are typically left up to the individual instructors for each unit, although some overall guidance on areas of good and poor performance may be provided by an internal assessment team. Finally, the slides largely contain textual information with some still images, although generally the images are not diagnostic, but rather serve either to provide graphical reminders to participants or to break up the monotony of the text-heavy slides.

This method of developing and conducting distributed debriefs certainly has advantages. The use of commonly available software applications and collaboration tools minimizes the maintenance required by each site and alleviates the need to learn another application by already busy instructors. Additionally, the free-form nature of the slide show allows instructors to add whatever content is desired. On the other hand, these commonly available tools do not allow instructors to take full advantage of the plethora of simulation data available to them. However, even if they had these data, they would not have time to analyze and integrate the lessons learned into the debrief given typical time frames. A technology focused more on providing instructors with immediate access to data, quick analytics, and summary feedback regarding the trainees' performance may help them best use available data while still developing the debriefs rapidly. This innovation would allow them to address more instructional points than they can currently.

Small-Scale Distributed Simulation Training Exercises

Smaller-scale, distributed training events are experimenting with distributed debriefing technologies. The US Navy's DDSBE AAR (developed by Aptima, Inc.) and US Air Force Lab's (AFRL) Distributed Mission Training Collaborative Briefing and Debriefing System are two such examples; they are primarily used to conduct distributed debriefs across two sites (Salter et al. 2005). Both systems support communication and collaboration between sites, the replay of events using an interactive tactical map, and the view of various instruments from individual simulators. In addition, the DDSBE AAR collects and presents formal performance data collected during the training exercise, and instructors can select specific performance data points for discussion and presentation during the debrief, which synchronized with other data sources and shared across sites. In addition to these systems, MAK Technologies has developed several applications that enable data logging (via MAK Data Logger) across simulators connected to HLA and DIS protocols (via VR-Link) as well as visualization software to enable replay of captured data (via VR-Vantage Stealth). However, it is unclear how well these systems instantiate the requirements discussed in this chapter to fully realize the promise of distributed debriefs for simulation-based training.

These smaller-scale distributed debriefing tools certainly show promise for becoming a scalable solution that also could be used during large-scale distributed training exercises. However, additional work needs to be done to ensure that collaboration methods are scalable across multiple sites; incorporate multiple viewpoints during replay; ensure formal performance data are collected and available for all trainees; and support data and analytics use after the exercise.

A relatively new concept in simulation-based medical education is *telesimulation* (TS), in which telecommunication and simulation resources are used to provide education, training, and assessment for learners at distributed locations. With telesimulation, the internet is used to connect simulators between an instructor and trainees in different locations. In an early study, researchers were able to assess the effectiveness of TS in teaching the Fundamentals of Laparoscopic Surgery (FLS) to surgeons in Africa with instructors in Canada (Okrainec et al. 2010). Since then, there has been a rapidly growing interest in applying TS to other medical areas such as emergency medicine and anesthesia (McCoy et al. 2017). Of course, with telesimulation comes *teledebriefing*, which involves an instructor in a remote location using cameras, microphones, and basic videoconferencing software to provide feedback to learners while observing them conduct the simulation in real time (Ahmed et al. 2014). Teledebriefing with telesimulation could help reduce costs and eliminate the barriers of both distance and time (Honda & McCoy 2019), thus making it an ideal tool for delivering debriefs to geographically separate healthcare teams. A more recent study has demonstrated the feasibility and effectiveness of providing real-time training in a mass casualty incident (via telesimulation) with integrated debriefing (via teledebriefing) to healthcare providers overseas using Google Glass (McCoy et al. 2019).

SUMMARY

Currently, most developers of distributed debriefs focus either on the technology or on researching academic issues surrounding debriefing processes for distributed learners. Little information is publicly available describing efforts to combine these two strands in meaningful ways. In order to be truly successful, distributed training exercises must conduct debriefs that rigorously and formally promote learning. We believe that, by considering the current challenges to these issues and the requirements presented above, distributed debriefing technologies and techniques can begin to meet this lofty goal.

ACKNOWLEDGMENTS

We would like to acknowledge Xinwen Zhang and Samuel Alfred for their contribution to the literature review for this updated chapter.

REFERENCES

Ahmed, R., King Gardner, A., Atkinson, S. S., & Gable, B. (2014). Teledebriefing: Connecting learners to faculty members. *Clinical Teacher* 11(4):270–273.

Bales, R. F., Strodtbeck, F. L., Mills, T. M., & Roseborough, M. E. (1951). Channels of communication in small groups. *American Sociological Review* 16(4):461.

Bélisle, J., & Bodur, H. O. (2010). Avatars as information: Perception of consumers based on their Avatars in virtual worlds. *Psychology and Marketing* 27(8):741–765.

Bennett, K. B., & Flach, J. M. (2012). Visual momentum redux. *International Journal of Man-Machine Studies* 70(6):399–414.

Berger, J. M., Cohen, B. P., & Zelditch, M. (1966). Status characteristics and expectation states. In J. Berger, M. Zelditch, & B. Anderson (Eds.), *Sociological Theories in Progress* (pp. 29–46). New York: Houghton Mifflin.

Berger, J., Cohen, B. P., & Zelditch, M. (1972). Status characteristics and social interaction. *American Sociological Review* 37(3):241–255.

Berger, J. M., Fisek, M. H., Norman, R. Z., & Zelditch, M. (1977). *Status Characteristics and Social Interaction: An Expectation States Approach*. New York: Elsevier Scientific Publishing Company.

Berger, J., & Webster, M. (2006). Expectations, status, and behavior. In P. J. Burke (Ed.), *Contemporary Social Psychological Theories* (pp. 268–300). Stanford, CA: Stanford University Press.

Cohen, E. E. (1993). From theory to practice: The development of an applied research program. In J. Berger & M. Zelditch (Eds.), *Theoretical Research Programs: Studies in the Growth of Theory* (pp. 385–415). Palo Alto, CA: Stanford University Press.

Dismukes, R. K., McDonnell, L. K., & Jobe, K. K. (2000). Facilitating LOFT debriefings: Instructor techniques and crew participation. *International Journal of Aviation Psychology*, 10:35.

Dixon, N. M. (2000). *Common Knowledge: How Companies Thrive by Sharing What They Know*. Cambridge, MA: Harvard University Press.

Dwyer, D. J., Fowlkes, J., Oser, R. L., & Salas, E. (1996). Case study results using distributed interactive simulation for close air support training. *Proceedings of the 7th International Training Equipment Conference* (pp. 371–380). Arlington, VA: ITEC Ltd.

Fanning, R. M., & Gaba, D. M. (2007). The role of debriefing in simulation-based learning. *Simulation in Healthcare* 2(2):115–25. https://doi.org/10.1097/SIH.0b013e3180315539. PMID: 19088616.

Fiore, S. M., Johnston, J., & McDaniel, R. (2007). Narrative theory and distributed training: Using the narrative form for debriefing distributed simulation-based exercises. In S. M. Fiore & E. Salas (Eds.), *Toward a Science of Distributed Learning* (pp. 119–145). Washington, DC: American Psychological Association.

Galinsky, A. D., Magee, J. C., Gruenfeld, D. H., Whitson, J. A., & Liljenquist, K. A. (2008). Power reduces the press of the situation: Implications for creativity, conformity, and dissonance. *Journal of Personality and Social Psychology* 95(6):1450–1466.

Goldberg, S. L., Knerr, B. W., & Grosse, J. (2003). Training dismounted combatants in virtual environments. Paper presented at RTO HFM Symposium on Advanced Technologies for Military Training. Genoa, Italy. 13–15 Oct 2003. Accessed on DTIC https://apps.dtic.mil/sti/pdfs/ADA428918.pdf on 28 August 2021.

Goldenberg, M. G., Jung, J., & Grantcharov, T. P. (2017). Using data to enhance performance and improve quality and safety in surgery. *JAMA Surgery* 152(10):972–973.

Grosan, C., & Abraham, A. (2011). Rule-based expert systems. *Intelligent Systems Reference Library* 17:149–185.

Hanoun, S., & Nahavandi, S. (2018). Current and future methodologies of after action review in simulation-based training. *12th Annual IEEE International Systems Conference, SysCon 2018 – Proceedings*.

Henneman, E. A., Cunningham, H., Fisher, D. L., Plotkin, K., Nathanson, B. H., Roche, J. P., … Henneman, P. L. (2014). Eye tracking as a debriefing mechanism in the simulated setting improves patient safety practices. *Dimensions of Critical Care Nursing* 33(3):129–135.

Hippe, D. S., Umoren, R. A., McGee, A., Bucher, S. L., Bresnahan, B. W. (2020). A targeted systematic review of cost analyses for implementation of simulation-based education in healthcare. *SAGE Open Medicine* 8:1–9.

Honda, R., & McCoy, C. E. (2019). Teledebriefing in medical simulation. In *StatPearls*. Retrieved from https://www.ncbi.nlm.nih.gov/books/NBK546584/

Issenberg, S. B., McGaghie, W. C., Petrusa, E. R., Lee Gordon, D., & Scalese, R. J. (2005). Features and uses of high-fidelity medical simulations that lead to effective learning: A BEME systematic review. *Medical Teacher* 27(1):10–28. https://doi.org/10.1080/01421590500046924. PMID: 16147767.

Jacobs, L., Cornelisse, E., & Schavemaker-Piva, O. (2006). Innovative debrief solutions for mission training and simulation: Making fighter pilots training more effective. *Proceedings of the Interservice/Industry Training, Simulation, and Education Conference (I/ ITSEC).* Orlando, FL.

Kozlowski, S. W. J., & Klein, K. J. (2000). A multilevel approach to theory and research in organizations: Contextual, temporal, and emergent processes. In K. J. Klein & S. W. J Kozlowski (Eds.), *Multilevel Theory, Research, and Methods in Organizations: Foundations, Extensions, and New Directions* (pp. 3–90). San Francisco, CA: Jossey-Bass.

Larnpotang, S., Lizdas, D., Rajon, D., Luria, I., Gravenstein, N., Bisht, Y., ... Robinson, A. (2013). Mixed simulators: Augmented physical simulators with virtual underlays. *Proceedings – IEEE Virtual Reality.*

Lateef, F. (2010). Simulation-based learning: Just like the real thing. *Journal of Emergencies, Trauma, and Shock* 3(4):348–352. https://doi.org/10.4103/0974-2700.70743

McCoy, C. E., Sayegh, J., Alrabah, R., & Yarris, L. M. (2017). Telesimulation: An Innovative Tool for Health Professions Education. *AEM Education and Training* 1(2):132–136.

McCoy, C. E., Alrabah, R., Weichmann, W., Langdorf, M. I., Ricks, C., Chakravarthy, B., Lotfipour, S. (2019). Feasibility of telesimulation and Google glass for mass casualty triage education and training. *Western Journal of Emergency Medicine* 20(3):512.

Okrainec, A., Henao, O., & Azzie, G. (2010). Telesimulation: An effective method for teaching the fundamentals of laparoscopic surgery in resource-restricted countries. *Surgical Endoscopy* 24:417–422.

Principe, C. P., & Langlois, J. H. (2013). Children and adults use attractiveness as a social cue in real people and avatars. *Journal of Experimental Child Psychology* 115(3):590–597.

Qi, D., Ryason, A., Milef, N., Alfred, S., Abu-Nuwar, M. R., Kappus, M., De, S., & Jones, D. B. (2020). Virtual reality operating room with AI guidance: Design and validation of a fire scenario. *Surgical Endoscopy* 35:779–786.

Quarles, J., Lampotang, S., Fischler, I., Fishwick, P., & Lok, B. (2008). Collocated AAR: Augmenting after action review with mixed reality. *Proceedings – 7th IEEE International Symposium on Mixed and Augmented Reality* 2008, *ISMAR 2008.*

Quarles, J., Lampotang, S., Fischler, I., Fishwick, P., & Lok, B. (2013). Experiences in mixed reality-based collocated after action review. *Virtual Reality* 17:239–252.

Raij, A. B., & Lok, B. C. (2008). IPSViz: An after-action review tool for human-virtual human experiences. *Proceedings – IEEE Virtual Reality.*

Rhienmora, P., Haddawy, P., Suebnukarn, S., & Dailey, M. N. (2011). Intelligent dental training simulator with objective skill assessment and feedback. *Artificial Intelligence in Medicine* 52(2):115–121.

Salter, W. J., Hoch, S., & Freeman, J. (2005). Human factors challenges in after-action reviews in distributed simulation-based training. *Proceedings of the Human Factors and Ergonomics Society 49th Annual Meeting.*

Sawyer, T., Eppich, W., Brett-Fleegler, M., Grant, V., & Cheng, A. (2016). More than one way to debrief. *Simulation in Healthcare* 11(3):209–217. https://doi.org/10.1097/SIH.0000000000000148

Schank, R. C. (1982). *Dynamic Memory: A Theory of Reminding and Learning in Computers and People.* Cambridge: Cambridge University Press.

Scherer, L. A., Chang, M. C., Meredith, J. W., & Battistella, F. D. (2003). Videotape review leads to rapid and sustained learning. *The American Journal of Surgery* 185(6):516–20. https://doi.org/10.1016/s0002-9610(03)00062-x. PMID: 12781877.

Shebilske, W., Gildea, K., Freeman, J., & Levchuk, G. (2009). Optimizing instructional strategies: A benchmarked experiential system for training. *Theoretical Issues in Ergonomics Science* 10(3):267–278.

Silver, S. D., Cohen, B., & Rainwater, J. (1988). Group structure and information exchange in innovative problem solving. *Adv Gr Process* 5:169–194.

Silver, S. D., Troyer, L., & Cohen, B. P. (2000). Effects of status on the exchange of information in team decision-making: When team building isn't enough. *Advances in Interdisciplinary Studies of Work Teams* 7:21–51.

Simulation Interoperability Standards Organization (SISO). (2016). Standard for Distributed Debrief Control Architecture (SISO-STD-015-2016). Orlando, FL: SISO, Inc.

Streit, A. (2020). DDAPR: The dapper way to debrief together. Paper presented at International Training Technology Exhibition and Conference (IT2EC) 2020. London UK.

Tulving, E. (2002). Episodic memory: From mind to brain. *Annual Review of Psychology* 53: 1–25.

van Dalen, A., Jansen, M., van Haperen, M., van Dieren, S., Buskens, C. J., Nieveen van Dijkum, E., Bemelman, W. A., Grantcharov, T. P., & Schijven, M. P. (2021). Implementing structured team debriefing using a Black Box in the operating room: Surveying team satisfaction. *Surgical Endoscopy* 35(3):1406–1419. https://doi.org/10.1007/s00464-020-07526-3

Webster, M., & Hysom, S. J. (1998). Creating status characteristics. *American Sociological Review* 63(3):351–378.

Wijewickrema, S., Ma, X., Piromchai, P., Briggs, R., Bailey, J., Kennedy, G., & O'Leary, S. (2018). Providing automated real-time technical feedback for virtual reality based surgical training: Is the simpler the better? *Artificial Intelligence in Education*. London.

Woods, D. D. (1984). Visual momentum: A concept to improve the cognitive coupling of person and computer. *International Journal of Man-Machine Studies* 21:229–244.

Xiao, Y., Schimpff, S., Mackenzie, C., Merrell, R., Entin, E., Voigt, R., & Jarrell, B. (2007). Video technology to advance safety in the operating room and perioperative environment. *Surgical Innovation* 14(1):52–61.

9 Performance Assessment in Simulation

Steve Hall, Michael Brannick, and John L. Kleber

CONTENTS

Performance assessment is a key element in simulation. In a training context, performance assessment will lay the foundation for feedback to the pilot or flight crew, and in a research context, performance assessment is typically the key to assessing the impact of various factors of interest, such as training or equipment design. There is no single way to measure performance and practical issues typically limit the type and amount of performance data that can be collected during a simulation session.

DOI: 10.1201/9781003401360-9

For convenience we classify measures along two dimensions. First, we describe measures as being either subjective or objective. Subjective measures are provided directly by human judges. For example, an instructor might rate a crew *satisfactory* on a paper-and-pencil scale of Mission Analysis. Objective measures are provided by simulators as a result of recordings or calculations. For example, performance might be defined in terms of deviation from a desired flight path or flight parameter (e.g., airspeed, heading) or in terms of external pilot behaviors such as pushing a button. For the second dimension, we describe measures as being either qualitative or quantitative. By qualitative, we mean measures that are categorical in nature. For example, one might either simply pass or fail a maneuver. For a different example, a pilot's behavior might be allocated by an instructor to a category such as *assertiveness* or *decision-making.* Quantitative measures, on the other hand, indicate magnitude. For example, altitude and vertical speed are both quantitative. We also consider numerical ratings made by judges to be quantitative if the ratings indicate degree, so that a pilot given a rating of 4 is indicated by an instructor as being more proficient than another pilot given a rating of 3 under similar circumstances.

The choice of performance assessment technique should be driven by the purpose of the simulation and practical limitations. Training scenarios may require feedback for specific flight crew behaviors. Such feedback may include both subjective appraisals of quality of behavior and objective information about the frequency or timeliness of behaviors. Research endeavors typically seek to examine the impact of specific factors on some outcome, such as flight technical error (FTE) or perceived workload. The former can be efficiently quantified with objective data, while the latter is typically assessed with subjective workload survey instruments.

The purpose of this chapter is to present both subjective and objective measures of performance that are commonly used in simulation. A thorough discussion of measurement reliability accompanies the subjective measurement discussion, whereas the objective section emphasizes logistical and practical issues associated with automated methods of performance assessment.

While this chapter uses flight training and flight simulation as the context of interest, the principles and guidelines presented in this chapter can be applied to any number of simulated training events where performance measurement is of interest.

SUBJECTIVE METHODS OF PERFORMANCE MEASUREMENT

In general, humans are excellent at sensing and perceiving information. We can also become very adept at knowing what to perceive, that is, we learn what is (and is not) worthy of attention in a given situation. In many cases, if we want to evaluate human performance, there is no other choice but to use expert judgment. For example, if we want to evaluate the quality of coordination of two pilots working together in a cockpit, we will most likely be forced to rely on the judgment of an instructor pilot. For these reasons and others, humans are often used as performance measurement devices. Not surprisingly, the most commonly used measures of work performance are based on human judgment (Landy & Farr, 1983).

Purpose of Performance Measures

Simulators are typically used to train and evaluate skilled performance. For example, a flight simulator can be used to teach navigation skills. Performance measures, therefore, should support the training and evaluation functions of simulators. Numerical (quantitative) evaluations of proficiency are typically of primary interest in simulators. Such measures are useful in evaluating and documenting individual skill levels for certification or for adequacy of preparation in dealing with the real (not simulated) task. Although numerical evaluations are not very useful in providing developmental feedback in training, they are useful in evaluating the quality of a training program. That is, evaluations of proficiency can be used in program evaluation research.

Special Properties of Performance Measures in Simulators

Often the individual using the simulator (the target of evaluation) is aware of the evaluation, and the evaluation has consequences for the target (e.g., certification). This can cause a type of bias known as the Hawthorne effect (McCambridge et al., 2014). In such circumstances, the performance measure should be considered one of maximal performance rather than typical performance. Because of the consequence of the measurement and because the performance measure is related to skills that the target considers important, such evaluations can be threatening or anxiety-provoking for the target. In some cases, the apprehension that results from being evaluated can interfere with performance of the task, thus resulting in a poorer evaluation.

Not only does the simulator encourage maximal performance but it does so for a limited time. Typically, the simulator is used to teach or evaluate specific skills. The instructor and student will use the simulator for a time brief enough to allow them to maintain their attention on that specific task. This means that ratings evaluating simulator performance are less taxing for memory than typical performance appraisal ratings, which often cover job performance for a year. Further, for a given session, the experience in the simulator is designed to target one or more skills, such as navigation or coordination. Such training and evaluation design considerations help both the judge and target focus on a limited range of behaviors compared to typical job performance ratings.

There are both advantages and disadvantages to the use of humans as judges of performance in simulators. From the standpoint of measurement, one of the main drawbacks to using human judges is that such judges may disagree with one another (see Guion, 2011). Kenny (1991) developed a general model of consensus and accuracy of ratings of interpersonal perception. The factors in the model included (a) the amount of information given to the judges, (b) the degree to which the judges attend to the same behaviors, (c) the degree to which the judges make similar inferences or interpretations of the same behaviors, (d) reliability or consistency of the target (ratee) behavior, (e) degree to which the judges base their ratings on irrelevant behavior, and (f) degree of communication of impressions of the target prior to rating.

Of Kenny's factors, items (b), (c), and (d) appear the most important for performance ratings in simulators. We discuss factors (b) and (c) here, and factor (d) later in the chapter. The amount of information given to different judges (factor a) tends to be similar for a given scenario using a given simulator. For example, two judges might watch the same recording of pilots flying a simulator. Even though two judges watch the same participants in the same simulator at the same time, they may disagree in their evaluation of what they saw for several reasons. The judges may attend to different behaviors, so that one sees something that another does not (factor b). For example, only one of two instructors may notice that the copilot has become lost. The judges may interpret the same behavior to mean different things (factor c). Again, one judge may find a behavior overly confrontational, but another judge may find the same behavior to be appropriately assertive. When judges record behaviors, make ratings, or otherwise produce evaluations, they may do so in idiosyncratic ways. One judge might assign an assertive behavior to an interpersonal dimension, but another might classify the same behavior as belonging to decision-making, flexibility, or some other dimension (for general discussions of performance ratings, see Landy and Farr, 1980, 1983).

DEFINING AND ASSESSING RELIABILITY

There is a very substantial literature on the reliability of measurement (e.g., Crocker & Algina, 1986; Nunnally & Bernstein, 1994; Wigdor & Green, 1991; Coulacoglou & Saklofske, 2017). There is a somewhat more manageable literature on the reliability of judges (e.g., Shrout & Fleiss, 1979; Hallgren, 2012). Because judges commonly disagree, anyone who uses judges to evaluate performance in simulators should conduct a study to compute one or more indices that quantify how well the judges agree with one another. In this section, we illustrate the more commonly used indices of how well judges agree with one another. We provide recommendations for choosing suitable estimators for the most commonly occurring situations in practice. The choice of disagreement index depends primarily on two issues: (a) whether the judgments are qualitative or quantitative, and (b) whether differences in means across judges are important or meaningful.

Data Requirements

To study how well judges agree with one another, we have to collect data. Ideally, we should have a large, representative sample of judges and a large, representative sample of targets (pilots, teams, technicians, or whoever gets judged). Judges and targets should be crossed so that each judge sees each and every target and provides the judgment of interest for each (e.g., whether a pilot passes a specific test or an evaluation of the degree to which a team showed good coordination). The benefit of such a study is that you will actually learn what you want to know, that is, your evaluation of how well judges agree will be accurate. Unfortunately, such data collection can be rather expensive. Practical constraints may force a less informative design.

TABLE 9.1
Ratings from Three Judges on Ten Targets

	Judge		
Target	1	2	3
1	5	5	5
2	4	3	3
3	3	3	3
4	5	4	5
5	2	1	3
6	4	3	3
7	2	2	2
8	1	1	2
9	5	4	5
10	4	4	5
M	3.5	3.0	3.6
SD	1.43	1.33	1.26

At a minimum, we have to have at least two judges and some number of targets; the minimum possible number of targets is two, but some larger number is really needed for the calculations to be meaningful: say ten targets for the sake of argument (see Flack et al., 1988; Sim & Wright, 2005, for choosing the number of targets). Table 9.1 shows hypothetical data for three judges (represented as columns) on ten targets (represented as rows). The judges have recorded their judgments in the form of numbers ranging from 1 to 5. These judgments can be thought of either as categorical (polite, assertive, etc.) or quantitative (e.g., 1 = poor to 5 = excellent) for purposes of illustration.

Qualitative versus Quantitative

Qualitative ratings are categorical. An example of a categorical rating is one in which the judge merely indicates whether a pilot passes or fails a simulated task. Sorting teams of pilots on the basis of their working style into groups with labels such as *cooperative*, *confrontational*, *rational*, and *polite* would be another example. Qualitative ratings are labels applied to performance that either indicate group membership or falling above or below some threshold to indicate passing or failing in a task.

Quantitative ratings indicate the magnitude or degree of something. Ratings made on a scale from "poor" to "excellent" indicate increasing proficiency and can be assigned numbers (e.g., 1 to 5) that correspond to degree of proficiency. Some argue that judges' ratings are not measures in the same sense as measures provided by thermometers or airspeed indicators (for more detail on the issues, see Annett, 2002). Regardless of one's position on the issue, we feel that it is useful to act as if ratings are quantitative measures because the results of doing so are helpful in practice.

A Qualitative Index

If the judgments are categorical (qualitative), then disagreement can be quantified using percentage agreement and other indices of association. Although percentage agreement is easy to compute and (on the face of it) easy to interpret, it can be misleading. Suppose that one judge assigns "pass" to 100% of the targets, and another judge assigns "pass" to 80% of the targets. Then percent agreement will be 80%, which looks good on the face of it. However, because there is no variance in the first judge's ratings, there is no statistical association between the two sets of judges' ratings. The statistic we recommend for the analysis of categorical judgments is called *Cohen's kappa* or just *kappa* for short (Cohen, 1960). Kappa adjusts percentage agreement for chance agreement, so that the agreement is reduced if a large amount of it would be expected by chance.

Suppose that the data in Table 9.1 were categorical. To compute kappa, we would first compute a contingency table that shows the agreement in assignment to categories (see Table 9.2). Note that for the first target, all three judges agreed that the target was a "5"; so "3" is recorded in the fifth column. For the second target, two of the judges called it a "3" and one called it a "4," so a "2" is written in the third column and a "1" is written in the third column and a "1" is written in the fourth column. The rest of the rows proceed in a similar manner.

The formula for kappa is

$$K = \frac{P(A) - P(E)}{1 - P(E)},$$

where $P(A)$ is the proportion of times that judges agree, and $P(E)$ is the proportion of times that we would expect the judges to agree by chance (Siegel & Castellan,

TABLE 9.2
Contingency Table

	Category					
Target	1	2	3	4	5	S_i
1					3	1
2			2	1		0.33
3			3			1
4				1	2	0.33
5	1	1	1			0
6			2	1		0.33
7		3				1
8	2	1				0.33
9				1	2	0.33
10				2	1	0.33
C_j	3	5	8	6	8	
p_j	0.1	0.167	0.267	0.2	0.267	

1988, p. 285). Note that, in general, we have k judges, N targets, and m categories. In our example, we have $k = 3$ judges and $N = 10$ targets for a total of kN, or 30 judgments. To compute chance agreement, we hypothesize that row frequencies will be proportional to column totals. We find column totals by adding across the rows for each column. The totals are shown in the second-to-last row of Table 9.2, labeled C_j.

To find the proportion of judgments in each category, we divide the column totals by the total number of judgments, that is, C_j/kN. The result is shown in the last row of Table 9.2, labeled p_j. The proportion of agreement expected by chance is

$$P(E) = \sum\nolimits_1^j p_j^2$$

$$= 0.1^2 + 0.167^2 + 0.267^2 + 0.2^2 + 0.267^2$$

$$= 0.22.$$

Now we need to compute the proportion of agreement, $P(A)$. One way to do so is to first compute agreement for each target, Sj. We can do so with the following equation:

$$S_1 = \frac{1}{k(k-1)} \sum\nolimits_1^m n_{1j}(n_{1j} - 1)$$

$$=1/(3)(2)[0 + 0 + 0 + 0 + (3)(2)]$$

$$= 6/6 = 1.$$

The observed proportion of agreement among the judges is average agreement over targets:

$$P(A) = \frac{1}{N} \sum\nolimits_1^N S_i$$

$$P(A) = \frac{1 + 0.33 + 1 + \cdots + 0.33 + 0.33}{10} = 0.50.$$

Finally, our value of kappa is

$$K = \frac{0.5 - 0.22}{1 - 0.22} = 0.36,$$

which is a rather modest level of agreement.

Quantitative Indices

Is the Difference in Means Meaningful?

There are several indices of interjudge reliability that we can use when the ratings are quantitative. If the difference in means among judges is not meaningful or not important, we can use the correlation coefficient or a certain type of intraclass coefficient (the fixed case). On the other hand, when the difference in means among judges is meaningful, we can use another intraclass coefficient (the random case). We will define the indices and explain the reasons for the choices among indices as we go along.

Correlation Coefficient

Suppose that the data in Table 9.1 are quantitative and indicate instructor ratings of the level of proficiency of targets in completing a task in a simulator. One index of the degree to which judges agree about the relative standing of targets is the correlation coefficient, sometimes called Pearson's r (e.g., Guion, 2011, p. 240):

$$r = \frac{\sum z_X z_Y}{N},$$

where

$$z_X = \frac{X - M_X}{\mathrm{SD}_X},$$

and N is the number of pairs (targets), X is the raw score given by a judge, M is the mean rating for a judge, and SD is the standard deviation (sample, not population estimate) of the judge's ratings. The correlation coefficient is computed once for each pair of judges and indicates the degree to which the judges' scores rise and fall together across targets.

The correlations among the three judges' scores are shown in Table 9.3. All three correlations are quite large, indicating substantial agreement among the judges. Notice, however, that, as shown in Table 9.1, Judges 1 and 3 tended to give higher scores than Judge 2. The correlation can be quite high even though the judges have very different mean ratings. The ratings from a very lenient judge and a very severe judge will be highly correlated so long as they both tend to rate the same targets relatively high and relatively low, even though they do not agree on the specific numbers that are assigned to a target. Note that the correlations appear to indicate higher agreement among the judges than did kappa. This is because, if the judgments are categorical, any disagreement is as substantial as any other disagreement. (If

TABLE 9.3
Correlation Matrix

Judge	1	2	3
1	1		
2	0.93	1	
3	0.86	0.86	1

it misses, an inch is as good as a mile.) The correlation, however, essentially gives credit for being close.

For research purposes, the correlation coefficient is often used to show the degree of association between two variables. In such a context, a difference in means is often of no importance. However, in applied contexts, we typically collect data to make decisions about people. In such a circumstance, differences in means across judges become very important. For example, one judge may pass or certify a performance that another judge would fail. Clearly, disagreements of this nature would be important.

Intraclass Correlations

Intraclass correlations (ICCs) can estimate how well judges agree with one another while taking mean differences into account. There are several different ICCs that can be computed. All of them are related to the analysis of variance (ANOVA). We will illustrate the use of two of these. To compute the ICCs, we first need results from an ANOVA in which the ratings are the dependent variable, and the judge and target are the independent variables. Table 9.4 shows the way in which the data would be input to a computer, and Table 9.5 shows the ANOVA results. Notice that we have labeled one sum of squares BMS for *between targets mean square*, another is labeled JMS for *between judges mean square*, and the last is labeled EMS for *error mean square*. In this design, there is only one observation per cell, so the interaction and error terms are not separately estimable (Shrout & Fleiss, 1979).

Shrout and Fleiss (1979) described the computation of two general classes of ICCs, namely, random and fixed. The typical ANOVA interpretation of random and fixed effects would be that random ICCs consider judges to be sampled from some larger population, whereas the fixed ICCs consider the judges in the study to be sampled from only the judges of interest. However, the main difference between the two formulas is the way in which mean differences in judges are handled. The random ICCs reduce the index of agreement for differences in means across judges, but the fixed ICCs do not. Therefore, the choice of ICCs is better informed by the way in which data will be collected and used in practice. If, in the actual use of the ratings (that is, when the simulator is used for actual training, performance evaluation, etc., and the judges' evaluations actually count), the same judge or judges evaluate all targets, then a fixed ICC should be used. On the other hand, if different judges evaluate different targets, then a random ICC should be used. In most practical applications, there are multiple judges, and each target is rated by only one judge. Therefore, the random ICC will usually apply in practice.

In our current study, we have three judges (the jargon is that $k = 3$). We can estimate the reliability of a single judge, or we can estimate the reliability of the average of all three judges. We can do this for both the random- and fixed-effects cases. The computations are illustrated in Table 9.6 for all four possibilities (random versus fixed case, and one versus three judges). Notice that reliability estimates are larger for the fixed effects case than for the random effects case. This is because the JMS term appears in the denominator of the random effects estimates but not for the estimates in the fixed effects estimates. The JMS term is the estimate of variance due to

TABLE 9.4
Data Layout for ANOVA and ICC Computation

Rating	Judge	Target
5.00	1	1
4.00	1	2
3.00	1	3
5.00	1	4
2.00	1	5
4.00	1	6
2.00	1	7
1.00	1	8
5.00	1	9
4.00	1	10
5.00	2	1
3.00	2	2
3.00	2	3
4.00	2	4
1.00	2	5
3.00	2	6
2.00	2	7
1.00	2	8
4.00	2	9
4.00	2	10
5.00	3	1
3.00	3	2
3.00	3	3
5.00	3	4
3.00	3	5
3.00	3	6
2.00	3	7
2.00	3	8
5.00	3	9
5.00	3	10

TABLE 9.5
ANOVA Summary Table

Source	Sum of Squares	Degrees of Freedom	Mean Square	Label
Judge	2.07	2	1.03	JMS
Target	44.97	9	5.00	BMS
Judge*Target	3.93	18	0.22	EMS

TABLE 9.6
Computation of Intraclass Correlations

Formulas	Numbers
One random judge	
$\frac{BMS - EMS}{BMS + (k-1)EMS + k(JMS - EMS)n}$	$\frac{5-0.22}{5+2(0.22)+(3(1.03-0.22)/10)} = 0.84$
One fixed judge	
$\frac{BMS - EMS}{BMS + (k-1)EMS}$	$\frac{5-0.22}{5+(2)0.22} = 0.88$
All k (three) random judges	
$\frac{BMS - EMS}{BMS + (JMS - EMS)n}$	$\frac{5-0.22}{5+(1.03-0.22)10} = 0.94$
All k (three) fixed judges	
$\frac{BMS - EMS}{BMS}$	$\frac{5-0.22}{5} = 0.96$

differences in means among the judges. The difference in results between the fixed and random cases will depend on the size of the differences in mean ratings from the different judges.

Number of Judges

Note also in Table 9.6 that increasing the number of judges from one to three increases the reliability estimate. As we noted previously, the typical case in practice is for a single judge to rate each target. That single judge is usually not the same for all targets. In such an instance, the reliability estimate that would apply is the estimate for one random judge. In our study, the value for that estimate was 0.84, which is respectable for some purposes. However, we might want a better reliability if the rating has serious consequences for target (e.g., if the rating causes the target to lose time on the job). We can use a variant of the Spearman-Brown formula to estimate the number of judges we would need to obtain any given level of reliability. The formula we need is (Shrout & Fleiss, 1979)

$$m = \frac{\rho^*(1-\rho_L)}{\rho_L(1-\rho^*)},$$

where ρ_L is our estimate of one (random, in our case) judge, ρ^* is our aspiration level, and m is the resulting number of judges. We have to round off m to the next highest integer, as judges cannot be expressed in fractions. Suppose we wanted to achieve a reliability of 0.90; we would find that

$$m = \frac{0.9(1-0.84)}{0.84(1-0.9)} = 1.71 \cong 2.$$

Therefore, we would need two (random) judges to assure a reliability of at least 0.90.

To recapitulate the distinction between fixed and random judges in practical applications, suppose we have a pool of five instructors, of whom two will be available to rate each crew on each simulation, but the same two instructors will not always be paired; the random judges formula would then apply because differences in the judges' means would influence the ratings. On the other hand, suppose we have only two instructors available, and these two must evaluate each and every crew. In that case, the fixed judges formula would apply because differences in means among the judges would not influence the ratings.

There is also a third possibility, that is, there is some absolute standard that is based on the ratings (e.g., there is a numerical scale from 1 to 5, with a passing point of 3). In such a case, the calibration of the judges becomes of interest. However, we are unaware of well-developed psychometric approaches to evaluating such a calibration. In such instances, calibration would hinge upon having a gold standard of performance, and these are rarely available in practice. If such instances were widely available, then human judgment would probably be unnecessary.

Other Designs for Assessing Agreement among Judges

For studying interjudge reliability, we advocate that each judge evaluates each and every target. Such designs present logistic challenges for researchers, particularly in terms of getting multiple judges together. It is often possible to make recordings (e.g., videotape) of targets' responses to simulations to ease the burden of gathering multiple judges at once. Once recorded, the target can be evaluated more or less at the judge's leisure. More complex designs in which some nesting of targets within judges can also be used (see Crocker & Algina, 1986; Cronbach et al., 1972; DeShon, 2002; Shavelson & Webb, 1991). However, you will probably have to hire a statistician to analyze the data and compute the reliability estimate.

There are also methods that can be used when each judge sees only a single target. Such methods include ICC(1) (Shrout & Fleiss, 1979) and r_{wg} (James et al.,1984). However, we recommend against using such methods because they have serious flaws as indices of agreement between judges. ICC(1) can be used in situations where judges can be grouped in some way and differences can be compared across groups; essentially, this amounts to using ANOVA between subjects to estimate reliability. The problem with this design is that we do not know if the differences come from the targets or the judges (or both) because different judges see different targets. The r_{wg} method compares a distribution of judges' ratings having variance against a distribution where scores are distributed uniformly. However, this is not a method that provides reliability estimates (Kozlowski & Hattrup, 1992).

Enhancing Reliability

Reliability of measurement is essential for the measures to be useful in training or performance evaluation (e.g., Baker & Salas, 1992, 1997). The main approaches to improving the reliability of subjective performance measures are (1) increasing the number of judges, (2) changing the task, and (3) training the judges. We have already described how to estimate the number of judges needed to obtain any desired level of reliability. Unfortunately, increasing the number of judges is often prohibitively expensive. A single judge may be all that is available.

Changing the Task

An excellent discussion of rating scales and formats can be found in Guion (2011, pp. 449–465). As Guion noted, the consensus of researchers in this area is that format effects due to appearance are small. That is, it makes little difference whether the scales are shown horizontally or vertically, whether the scales have five or seven response categories, or whether numbers or words are used to label the response options. This is not to say that careful scale development is not important. It is important because it specifies the task (the content of the items) ultimately given to the judge. Reliability among judges can be improved by making the judges' task easier.

One way to do this is to make the ratings more easily observable, quantitative, and behavioral. That is, relatively concrete, observable behaviors are easier to evaluate or record than relatively abstract concepts that must be inferred from subtle patterns of behavior. It is easier to count incidents of team members shouting at one another than it is to infer the degree of hostility being felt in the same team. Although reliability is increased by making the judges' task simpler, there is often a price to be paid for this simplicity. Often the simpler ratings are deficient, meaning that they do not fully capture the construct that was intended. Hostility can be expressed in many different ways, of which shouting at team members is only one. If the judges consider only shouting, then their reliability (agreement) should be good. However, the resulting measure will be an index specifically of shouting and will be a deficient measure of hostility.

Rater Training

Rater training is often a good method to use to improve agreement between judges. There are many kinds of rater training available. Probably the current favorite is frame of reference training (Bernardin & Beatty, 1984; Roch et al., 2012). Good training programs involve training in attention (teaching the set of observed behaviors relevant to the construct of interest) and standards for evaluation (Bernardin & Buckley, 1981). For example, suppose we are interested in crew decision-making. During a simulated mission, the crew encounters an equipment problem (say a boost pump failure). How does the crew handle this problem? Training might include attending to how long it takes the crew to break out the checklist, whether they skip steps in the checklist, whether there is concurrence among crew members on a specific step, and so forth. Training on the evaluative part might include what behaviors indicate satisfactory, above-average, and below-average performance on the problem. We recommend training that is specific to the simulation and evaluation form being used. Generic training, or training aimed at reducing common rating errors, is not likely to improve the reliability of judges' ratings.

Table 9.7 shows some possible formats for items dealing with decision-making. Some of the items are very simple checklist items such that the judge merely records whether the behavior was observed. Another item asks the judge how the crew did on the event (boost pump failure) as a whole. The final item asks about

TABLE 9.7
Example Rating Scales

Checklist Items **Check the Appropriate Box for Each Item**	**Yes**	**No**
1. Checklist out within 30 seconds of boost pump failure	[ ]	[ ]
2. Checklist complete within 2 minutes of start of checklist	[ ]	[ ]
3. Problem fuse correctly identified	[ ]	[ ]
Global evaluations Circle the number that indicates your opinion	1 = Poor 2 = Below average 3 = Average 4 = Above average 5 = Excellent	
1. Overall handling of boost pump failure	1 2 3 4 5	
2. Decision-making	1 2 3 4 5	

decision-making in general. Notice how the judge's cognitive work increases as the generality of the item increases. For the specific checklist items, the judge just has to attend to a specific behavior and record what happens. For the boost pump item, the behaviors are circumscribed by the boost pump event, but the judge must evaluate the proficiency of the crew, which involves comparing the behaviors of the crew to some standards of quality of performance. The final item asks the judge to recall and integrate behaviors from multiple events into an overall evaluation of decision-making. In addition to the boost pump failure, there may be other events built into the scenario. The judge has to remember how the crew handled the other events and somehow aggregate multiple behaviors before rendering an overall judgment. As we mentioned previously, as scales become more specific, reliability increases, but so does deficiency.

SPECIAL PROBLEMS WITH SIMULATORS

The Gouge

Many kinds of training and evaluation involving simulators require one or more scenarios that are developed to evaluate or teach specific things. Once developed, the scenarios are typically used for a period of months or even a year or more. As people experience the scenarios, they may tell others what they encountered. After a period of time, those encountering the simulation are not at all surprised by what happens during the simulation. When some people can prepare in advance, but others cannot, the measures obtained will most likely mean different things for the two groups because they are being measured under different conditions. This can be a problem, particularly after the scenario becomes well known.

One potential solution to such a problem is to develop rapidly reconfigurable event sets. The idea is for the instructor or simulator operator to quickly change the scenario in such a way that the participants will not know what to expect. The benefit is that the participants cannot avoid a proper evaluation by preparing for a specific event. There are disadvantages to such an approach as well. Time and effort are required to develop multiple scenarios intended to tap similar skills and behaviors. Perhaps more seriously, there is some doubt about whether different events can provide comparable information. For example, in the assessment center literature, exercises that appear to tap similar skills produce scores that are not highly correlated with one another. It is difficult to guess how participants will respond to an event built into a scenario. There is no empirical basis to guide the choice of events, so a successful outcome for rapidly reconfigurable event sets is uncertain.

Instructor Attitudes

Psychologists typically have an inductive approach to decision-making. They like to collect the relevant behaviors, evaluate them, and come to a decision about a person. So, for example, if psychologists have to certify a person as competent to complete a task, they will first analyze the task to determine what behaviors need to be completed. Then they will determine conditions for successful performance, that is, define performance standards. They will create the simulation to allow for the proper conditions and behaviors, and then watch the participant in the simulation. They will record and evaluate the participant's behaviors. If the behaviors meet the predetermined standard, they will certify the person. The recorded data are the basis of the decision.

Instructors, like managers, often do not share the psychologist's mindset. To many, the rating forms do not serve as a source of evidence from which a decision can be made. Rather, the forms serve as documentation of a decision that has already been rendered. Some instructor pilots, for example, feel that the important thing is their word or decision about whether a pilot is competent to fly. To them, the forms are a waste of time; all that is necessary is a "yes" or "no" from the instructor pilot. Such instructors will complete the forms if they are forced to, but one should not expect highly reliable data from such an individual.

We recommend that subjective performance assessment forms be developed as much as possible in cooperation with the judges who will be using them. If the judges find them overly burdensome, difficult to understand, or irrelevant to the purpose of the simulation, they will not be motivated to use them properly. Without the cooperation of the judge, reliability of the ratings will be poor.

OBJECTIVE METHODS OF PERFORMANCE MEASUREMENT

Some may see objective measures[1] of performance as a relatively simple way to assess "true" pilot ability. Objective measures may seem simpler to use and easier to validate, but such is not always the case. There are real limits to the utility of objective data in assessing pilot performance, and the costs associated with getting objective data are sometimes prohibitive. On the other hand, objective measures can provide valuable information about pilot performance for certain flight maneuvers; furthermore, the quantitative performance measurement process can be perfectly replicated across pilots, sessions, researchers, simulation platforms, and even in real aircraft. There are situations where subjective measures of performance are not feasible or precise enough for the research at hand. Certain aspects of flight performance, such as "stick and rudder" skill, are arguably best measured with data provided directly by the simulator. Certain tasks, such as instrument approaches, lend themselves to assessment with objective data. This is not to say that subjective performance assessment is of no value in such situations; on the contrary, both measurement approaches should be used when possible, as the two approaches will each provide unique information about different aspects of flight performance.

Ironically, using objective measures of flight performance is sometimes more difficult than using subjective measures. Even though there are benefits to using objective data (e.g., reliability, precision, and standardization), there is a potentially high cost involved in obtaining and analyzing objective data. The difficulty involved in using objective measures of performance will vary depending on the flight tasks being assessed—whether or not additional parameters such as team functioning are being assessed—and the simulation platform and software being used. End users of objective performance measures may disagree on the actual meaning of the outcome data or how to use such measures most effectively. In applied situations, pilots, instructors, and managers may be reluctant to use or even record objective measures of performance. The bottom line is that using objective measures in flight performance research is not easy, not simple, and not necessarily superior to the use of subjective methods, but they can certainly add value in many situations. The focus of this section is on the objective measurement of FTE during specific flight maneuvers. Other aspects of pilot behavior and performance, such as workload or situational awareness, are not considered.

AUTOMATED DATA COLLECTION SYSTEMS

Many simulation platforms have the capacity for automated data collection. Most simulation packages, either directly or through a third-party add-on, can output flight parameter information at some sampling rate. Output formats vary from text files to graphs of the aircraft's position through time. Text file output is usually done so that each line of data represents a given set of flight parameters at a given point in time. Data collection rates are usually flexible and are sometimes set as high as 60 samples per second. Some systems allow the user to specify the parameters to be saved, while others dump a predefined set of parameters. In some cases, instantaneous rate-of-change data can be collected, allowing the researcher to examine flight performance in terms of "smoothness."

Systems vary in terms of what data can be collected, but usually both flight parameter data (e.g., position, airspeed, and course deviation indicator [CDI] needle deflection) and pilot input data (e.g., control surface input and button presses) can be collected. Also, the data are usually time stamped, thus allowing the researcher to coordinate specific events and activities across data collection platforms. These time stamps facilitate synchronizing flight data with other sources of data such as cockpit video and audio recordings and eye-tracking information. A word of caution to the researcher: not all data collection programs will use the same reference clock for time stamping purposes!

Data collection software is readily available for the Elite and Microsoft Flight Simulator (MSFS) software packages. Elite sells a data collection module separate from the simulation software, while FlightRecorder is freely available from the internet for use with MSFS. Similarly, Frasca's latest simulators are configured from the factory to collect flight parameter data. Surprisingly, some of the more advanced simulation systems that we have worked with do not readily make such data available, but usually these data can be accessed with some minor programming. In some situations, physically getting the data from the simulator is easier than getting permission from the manufacturer to access the data in the first place.

Most data collection programs are designed to record raw data, which often results in a rather lengthy data file. To date, we have not seen a commercially available product that will scientifically analyze flight performance, but we have seen custom-made packages that analyze FTE in real time given specific flight criteria. Unfortunately, these packages are not readily available, leaving most researchers with the lengthy data files saved by the commonly available data collection packages.

FLIGHT TECHNICAL ERROR

Objective data can be used to generate various measures of FTE. FTE quantifies the degree to which the position or orientation of the aircraft deviates from some ideal state. In other words, FTE metrics focus on the pilot's ability to make the aircraft attain some predetermined goal specified by an external agency; hence, FTE is probably best used as a measure of "stick and rudder" skill.

FTE and pilot performance are certainly not synonymous. There is much more to safely operating an aircraft than FTE can quantify. On the other hand, a pilot's ability to maintain precise control of the aircraft is a prerequisite for successful operations.

Data feeds from simulators are ideal for FTE computations, but deriving meaningful FTE-based performance metrics requires a great deal of planning. The following sections detail various forms of FTE measures and their potential uses.

Deviation-Based Metrics

Deviation metrics are the backbone of most FTE measures. As the name implies, a deviation metric compares a given flight parameter to a specific value for that parameter. For example, the pilot may be instructed to maintain an altitude of 5,000 feet, and actual altitude data collected at some sampling rate are compared to this target value. The difference between the two values is the raw outcome of interest. A similar metric can be computed using positional data. The pilot is instructed to maintain a flight path, specified via Global Positioning System (GPS) or perhaps a

VOR (very high frequency omnirange) radial, and the actual position of the aircraft is compared to the target flight path.

Root Mean Square Error (RMSE)

Deviation data must be aggregated in some way to create a single outcome metric of interest. One of the most common methods of aggregation is to compute the RMSE. It works by first squaring each deviation, averaging these squared deviations, and taking the square root of the average to return to the original units. By doing so, the polarity of the deviations is eliminated, and gross deviations are exaggerated. If an observed flight path has no deviations from the desired flight path, RMSE will equal zero. Given this lower bound of zero, the distribution of RMSE will not be normal and violate a basic assumption for most parametric statistic procedures. To address this issue, RMSE data can be transformed using the natural logarithm function, and this will usually result in a distribution of RMSE data that appears normal (see the data in Figure 9.1).

The data in the figure were collected during a flight simulation study where pilot performance was assessed as a function of keeping the aircraft centered on the glide slope during a simulated instrument approach. Deviation from the glide slope was measured using the deviation (in dots) of the glide slope needle on the CDI, where a value of zero dots indicates perfect alignment with the glide slope. The raw data are presented in the left panel of the figure and clearly show the skewed nature of RMSE (this skewed shape can be reproduced with virtually any RMSE data). On the right, the data were transformed using the natural log function, producing a more normal distribution. This allows the use of ANOVA and other general linear models to be applied to the performance data. A problem with such transformations is that

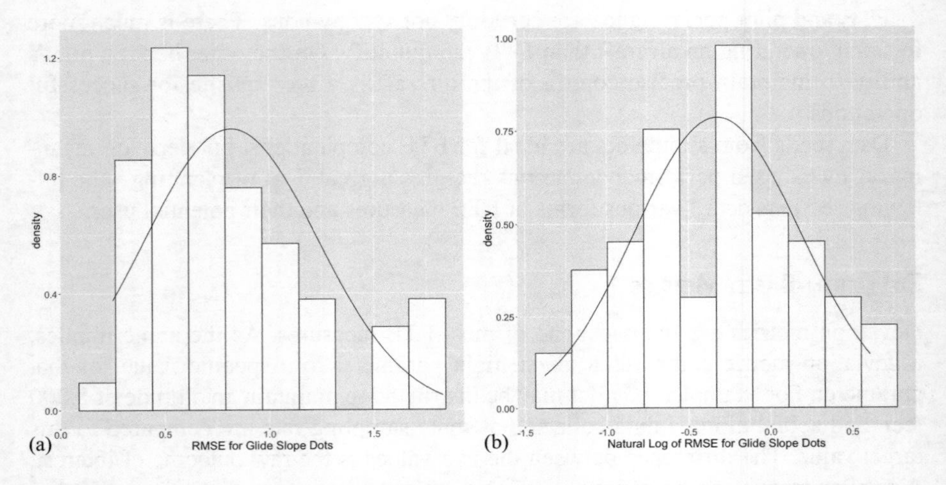

FIGURE 9.1 Illustration of raw RMSE data (left panel) and transformed RMSE data (right panel).

they obscure the meaning of the data so that the transformed outcome measures are no longer directly interpretable. This is especially problematic for deviation-based metrics expressing FTE in terms of average deviation and so forth; but non-transformed RMSE values have limited real-world interpretability and are not negatively impacted through transformations.

RMSE cannot be interpreted as a simple average deviation because the square transformation gives more weight to gross deviations. Similarly, RMSE values cannot be simply compared to prescribed performance standards, such as the Federal Aviation Administration (FAA) practical test standards (PTS) to determine whether some criterion level of performance was met. On the other hand, RMSE is a very sensitive measure of flight performance and is more likely than other indices of performance to show differences across different instrumentation systems, training programs, or other manipulated scenarios.

Changes in performance can be assessed using RMSE data but observed differences in RMSE values across groups or treatments are not directly interpretable. This issue can be addressed by reporting effect sizes as opposed to just reporting means and *F* tests (Cohen, 1994). Researchers should consider using either Cohen's *d*, which expresses group mean differences in terms of pooled standard deviation units, or omega squared, which is an estimate of variance accounted for in the outcome metric by some manipulated variable.

Number of Deviations and Time-Outside Standard

Rantanen and Talleur (2001) discussed several other deviation-based performance metrics. The number of deviations (ND) outside of a tolerance indicates the pilot's ability to control the velocity of the aircraft during a tracking task. It is like counting the number of times a car runs off a racetrack. Low ND values are desired but may be misleading because the aircraft could stray outside tolerance only once but stay outside of the desired flight path for the duration of the flight. As such, ND value must be interpreted considering the amount of time spent outside the tolerance, that is, time in deviation (TD). TD values add information beyond RMSE and ND, and low TD values reflect higher levels of performance.

Time within FAA Practical Test Standard

The notion of objective criterion performance measurement can be taken a step further by comparing the amount of deviation to some preset standard of performance. For example, a researcher could use FAA practical test standards to define acceptable and unacceptable levels of deviation. The time within standard (TWS) metric is conceptually similar to the TD metric discussed earlier, the main difference being that TWS focuses on the amount of time spent within a standard, and TD focuses on the amount of time spent outside the standard. The goal of the TWS metric is to quantify performance relative to known and accepted standards in such a way that the metric is conceptually friendly and directly interpretable.

The FAA has dictated standards of performance for the various pilot ratings and endorsements for various phases of flight. For example, the ATP (airline transport pilot) standard for an instrument approach is ¼-scale deflection on the localizer and glide slope, and +/– 5 knots on airspeed. In contrast, the standard for an instrument approach under the instrument rating test standards is ¾-scale deflection on the localizer and glide slope, and +/10 knots on airspeed. These criteria were designed to be applied by instructor pilots but can easily be applied to objectively measured flight data. The end results are measurements of the proportion of time spent within either the ATP or instrument rating standards, whether or not a specific landing approach met the ATP or instrument rating standard, and, at the group level, the proportion of pilots in a sample that performed at the ATP or instrument rating level.

TWS data can be used in many ways. For example, the TWS for a single maneuver being performed by a single pilot can be established and used to provide easily understood feedback to the pilot (e.g., 75% of the last approach was within ATP standards). Such data could be compared across trials to establish the learning curve for a particular pilot (see Mengelkoch et al., 1971) or to evaluate changes in performance over long periods of time (see Taylor et al., 2007). To enhance the quality of the feedback, TWS data can be compartmentalized across various flight parameters, such as glide slope tracking, localizer tracking, airspeed, etc.

TWS data can also be used to estimate the proportion of a specific pilot population that can fly to a specified FAA standard by quantifying maneuver performance in a binary fashion. A pilot who completes a maneuver with a TWS score less than 1 would be failed on that maneuver. Once a group of pilots has been tested and scored on that maneuver, it is easy to compute the proportion of the sample that was able to successfully complete the maneuver. These data can be generalized to some broader population overall, assuming that random sampling procedures are followed. Such estimates can be compared across treatment groups to determine the real-world impact of a system on average pilot performance. This kind of treatment effect is directly interpretable and will likely be preferred by program sponsors and the aviation sector in general. In either use, the advantage of TWS over RMSE is that the participant and end users (including program sponsors) can examine the TWS numbers to determine whether the impact of a treatment or intervention is practically meaningful.

Although the TWS metric has the benefit of being easily and directly interpreted, it does have the drawback of being prone to ceiling effects. This is especially likely when the standard being applied is an easy standard or if the pilots under evaluation are high performers. In such cases, a more sensitive measure, such as RMSE, will be required to differentiate the performances. On the other hand, if performance in some area by some population of pilots is already at a high level, various interventions designed to improve pilot performance cannot have a practical positive impact. Another drawback to the TWS metric is that it is not a sensitive measure of performance, meaning that small improvements in performance or heterogeneous improvements in performance are not likely to be detected when using this metric even if the improvement in performance is real. Where TWS is likely to detect differences across groups of pilots or across system implementations is when experimental conditions designed to increase workload and/or stress are added to the study design. Under "normal" flight operations, most recreational and almost all professional pilots can obtain perfect TWS scores. Maintaining perfect TWS during stressful situations such as troubleshooting in-flight emergencies, managing heavy traffic, or flying through inclement weather is much less likely. Thus, two treatment conditions can be compared by examining the drop in TWS scores as flight conditions move from easy to difficult. Any intervention that allows pilots to maintain higher TWS scores during difficult flight scenarios can arguably be deemed as being operationally superior (all other factors, such as cost, being equal).

The researcher may choose to examine specific phases of flight (e.g., takeoff, cruise, approach, and landing) or performance during specific maneuvers (e.g., turns about a point, stalls, and level flight). Objective measurement techniques based on real-time data dumping allow the researcher to examine specific phases of flight or even segment-specific maneuvers to decompose performance across the various components of a maneuver or phase of flight. It is important to remember that different phases of flight require different KSAOs (knowledge, skills, abilities, and other characteristics) to complete successfully, and knowing that a pilot is able to fly an instrument approach well does not necessarily mean that he or she will perform well in other phases of flight. Similarly, changes in training or aircraft instrumentation may impact performance differently across different maneuvers or phases of flight. Systems designed to enhance performance under IFR (instrument flight rules) conditions may not enhance performance under VFR (visual flight rules) conditions.

NON-FTE MEASURES

Not all aspects of pilot performance can be directly measured via positional or orientation data. There are several pilot tasks that can be measured via automated data collection mechanisms that can provide additional information about pilot performance.

Rates of Change

Most simulators can provide rate of change data associated with roll, pitch, and altitude. Some aircraft operators may wish to establish criteria about how fast the pilot changes the orientation of the aircraft to enhance passenger comfort, or more aptly, reduce the likelihood that passengers become airsick. The maximum rate of change can be extracted for a given maneuver and compared to some maximum allowed rate. If the pilot exceeds the maximum allowed rate, the pilot will fail that maneuver. If performance is being scored on a point system, points might be deducted for exceeding the maximum rate.

Operators may also wish to evaluate vertical velocity at touchdown in the interest of equipment longevity and passenger comfort. Harsh landings (i.e., touchdowns with a high vertical speed component) are not only uncomfortable for passengers but can also cause structural damage to the aircraft. Some operators use onboard flight recorders to evaluate real-time flight data following incidents in flight (such as passenger injury or aircraft damage). These same protocols can be used in simulation-based pilot performance evaluation.

Control Input

Some flight scenarios require quick and decisive action by the pilot, such as a specific button or switch activation, throttle adjustment, or yoke input. Other research projects might wish to investigate the frequency or magnitude of surface controls or the sequence of control activations. Again, most simulators can provide such data. Switch and button activation are usually recorded in terms of button or switch position for a given time frame, whereas control surface and power plant settings are typically given as numbers on some scale.

The key to using such data is to construct scenarios that require specific inputs relative to specific events. The analyst must know when the event was initiated, and the response made by the pilot. Control input data can also be used to flag segments in a flight scenario. For example, the researcher may be interested in FTE on approach once flaps are dropped to 10°.

Switch activation data are especially useful when used to evaluate human interface design for communication and navigation equipment. Such evaluations can be performed using the "virtual" devices included as part of the software, where the pilot uses a keyboard or mouse to control the virtual device. More sophisticated simulators will use physical switches to control the function of virtual equipment. Similarly, physical mock-ups of radio stacks are commercially available, allowing the pilot to interface with a set of physical communications equipment. If funding is not a problem, the researcher can procure a specific avionics package and interface the package into the flight simulation software. This usually requires the construction of both hardware and software interfaces to allow the equipment to communicate with the software.

SUMMARY

There are a variety of ways to define and measure pilot performance. The desired usage of the data should drive the choice of measurement. Aircraft operators and pilot training centers tend to prefer subjective measures provided by SME (subject matter experts) evaluations of performance, whereas researchers tend to prefer objective measures. Each has its strengths and weaknesses, and the optimal choice is not always clear.

Regardless of the method chosen, steps should be taken to ensure that the measure is psychometrically sound. Whenever human judges provide ratings, disagreements among judges are very likely, and efforts must be made to understand the magnitude of such disagreements, that is, we must estimate reliability. If the magnitude of such disagreements is large, then action must be taken to improve reliability. Although reliability is seldom a problem with objective measures (at least when it comes to the calibration of the machine; the pilot's consistency of performance may be another matter), the researcher should be careful to properly interpret observed differences. It is easy to overstate the practical significance of observed differences in FTE measures; we suggest that the researcher translate any observed differences in performance to real-world consequences (e.g., reduced risk and enhanced passenger comfort).

NOTE

1. The phrase "objective measures" in this chapter refers specifically to measurement based on flight data provided by the simulator.

REFERENCES

Annett, J. (2002). Subjective rating scales: Science or art? *Ergonomics*, *45*, 966–987.

Baker, D. P., & Salas, E. (1992). Principles for measuring teamwork skills. *Human Factors*, *34*, 469–475.

Baker, D. P., & Salas, E. (1997). Principles for measuring teamwork: A summary and look toward the future. In M. T. Brannick, E. Salas, & C. Prince (Eds.), *Team performance assessment and measurement: Theory, methods, and applications* (pp. 343–368). Mahwah, NJ: Erlbaum.

Bernardin, H. J., & Beatty, R. W. (1984). *Performance appraisals: Assessing human behavior at work*. Boston: Kent Publishing.

Bernardin, H. J., & Buckley, M. R. (1981). Strategies in rater training. *Academy of Management Review*, *6*, 205–212.

Cohen, J. (1960). A coefficient of agreement for nominal scales. *Educational and Psychological Measurement*, *20*, 37–46.

Cohen, J. (1994). The earth is round ($p < .05$). *American Psychologist*, *49*, 997–1003.

Coulacoglou, C., & Saklofske, D. H. (2017). *Psychometrics and psychological assessment: Principles and applications*. Cambridge, MA: Academic Press.

Crocker, L., & Algina, J. (1986). *Introduction to classical & modern test theory*. New York: Holt, Rinehart, & Winston.

Cronbach, L. J., Gleser, G. C., Nanda, H., & Rajaratnam, N. (1972). *The dependability of behavioral measurements*. New York: Wiley.

DeShon, R. P. (2002). Generalizability theory. In F. Drasgow & N. Schmitt (Eds.), *Measuring and analyzing behavior in organizations: Advances in measurement and data analysis* (pp. 189–202). San Francisco, CA: Jossey-Bass.

Flack, V. F., Afifi, A. A., Lachenbruch, P. A., & Schouten, H. J. (1988). Sample size determinations for the two rater kappa statistic. *Psychometrika*, *53*, 321–325.

Guion, R. M. (2011). *Assessment, measurement, and prediction for personnel decisions* (2nd ed.). New York: Routledge.

Hallgren, K. A. (2012). Computing inter-rater reliability for observational data: An overview and tutorial. *Tutorials in Quantitative Methods for Psychology*, *8*(1), 23.

James, L. R., Demaree, R. G., & Wolf, G. (1984). Estimating within-group interrater reliability with and without response bias. *Journal of Applied Psychology*, *69*, 85–98.

Kenny, D. A. (1991). A general model of consensus and accuracy in interpersonal perception. *Psychological Review*, *98*, 155–163.

Kozlowski, S. W. J., & Hattrup, K. (1992). A disagreement about within-group agreement: Disentangling issues of consistency versus consensus. *Journal of Applied Psychology*, 77, 161–167.

Landy, F. J., & Farr, J. L. (1980). Performance rating. *Psychological Bulletin*, *87*, 72–107.

Landy, F. J., & Farr, J. L. (1983). *The measurement of work performance: Methods theory and applications*. New York: Academic Press.

McCambridge, J., Witton, J., & Elbourne, D. R. (2014). Systematic review of the Hawthorne effect: New concepts are needed to study research participation effects. *Journal of Clinical Epidemiology*, *67*(3), 267–277. https://doi.org/10.1016/j.jclinepi.2013.08.015

Mengelkoch, R. F., Adams, J. A., & Gainer, C. A. (1971). *The forgetting of instrument flight skills as a function of the initial level of proficiency* (Report No. NAVTRA DEVCEN 71-16-18), Port Washington, NY: U.S. Naval Training Center.

Nunnally, J. C., & Bernstein, I. H. (1994). *Psychometric theory*. New York: McGraw-Hill.

Rantangen, E. M., & Talleur, D. A. (2001). Measurement of pilot performance during instrument flight using flight data recorders. *International Journal of Aviation Research and Development*, *1*(2), 89–102.

Roch, S. G., Woehr, D. J., Mishra, V., & Kieszczynska, U. (2012). Rater training revisited: An updated meta-analytic review of frame-of-reference training. *Journal of Occupational and Organizational Psychology*, *85*(2), 370–395.

Shavelson, R. J., & Webb, N. M. (1991). *Generalizability theory: A primer*. Newbury Park, CA: Sage.

Shrout, P. E., & Fliess, J. L. (1979). Intraclass correlations: Uses in assessing rater reliability. *Psychological Bulletin*, *86*, 420–428.

Siegel, S., & Castellan Jr., N. J. (1988). *Nonparametric statistics for the behavioral sciences*. New York: McGraw-Hill.

Sim, J., & Wright, C. C. (2005). The kappa statistic in reliability studies: Use, interpretation, and sample size requirements. *Physical Therapy*, *85*(3), 257. https://doi.org/10.1093/ptj/85.3.257

Taylor, J. L., Kennedy, Q., Noda, A., & Yesavage, J. A. (2007). Pilot age and expertise predict flight simulator performance: A 3-year longitudinal study. *Neurology*, *68*(9), 648–654.

Wigdor, A. K., & Green B. F. (Eds.). (1991). *Performance assessment for the workplace* (Vol. I). Washington, DC: National Academy Press.

10 Performance Measurement Issues and Guidelines for Adaptive, Simulation-Based Training

Phillip M. Mangos and Joan H. Johnston

CONTENTS

INTRODUCTION

In our original chapter we sought to advance discussion of applying psychometric principles to learning assessment in adaptive simulation-based training (SBT) (Mangos & Johnston, 2009). We discussed how the training potential of simulation

DOI: 10.1201/9781003401360-10

systems is increased by focusing on their capability for learning assessment and that important training outcomes can be achieved by treating an SBT system as a tool for generating customized training content around a core of embedded, potentially high-fidelity assessment "items." Specifically, more accurate diagnoses about the causes of suboptimal performance would allow a higher proportion of training time focused on correcting unique skill deficiencies; thus achieving desired performance levels in less time, and improving generalization and maintenance of trained skills to job settings. Our discussion led to a set of guidelines for the development and evaluation of performance measures in adaptive SBT based on a theory-driven, confirmatory performance measurement framework. We explored the essential characteristics of performance measures within the context of this framework that included supporting a confirmatory strategy for measuring and evaluating performance in adaptive training contexts, and sound inferences regarding hypothesized relationships among training objectives, performance episodes, and outcomes. We discussed several dimensions along which performance measures can vary, described desirable characteristics of measures along each dimension, and introduced a set of principles articulating how individual performance measures can be used to support adaptive training. This chapter updates our initial effort by briefly discussing research trends and advances in the psychometrics of assessment, and provides exemplars of adaptive SBT implementation and updated guidelines. It is beyond the scope of this chapter to fully explore these topics; therefore, comprehensive references are provided for further exploration.

RESEARCH ADVANCES

Progress on adaptive SBT capabilities could not have been achieved without the 40+ years of research and development invested by the US Department of Defense. Two notable programs of research that began in the 2000s focused on rapid advancement of adaptive learning assessments, psychometric accuracy of assessments, and non-proprietary, government-owned assessment authoring software. In 2003, the Defense Advanced Research Programs Agency (DARPA) initiated the DARPA games for a training program (DARWARS) with the goal of creating a vast on-demand, multi-user, online military adaptive SBT environment (Chatham, 2007; O'Neil, Baker, Wainess, Chen, Mislevy, & Kyllonen, 2004). Ultimately, DARWARS succeeded in advancing intelligent tutoring system technologies and game-based learning (Bell, Johnston, Freeman, & Rody, 2004; Chatham, 2007; Smith & Bowers, 2019). The DARWARS Ambush! trainer prototype enabled the US Army to embed its Games for Training (GfT) curriculum across Soldier training sites (Roberts & Diller, 2014). The DARWARS tactical language trainer transitioned to operational deployment as the Virtual Culture Awareness Trainer (VCAT), which we describe in more detail later in the chapter (Johnson, 2010; Johnson & Lester, 2016).

Starting in 2009, the US Army launched a decade-long program to develop adaptive learning principles (see Durlach and Lesgold (2012) and Spain, Priest, and Murphy (2012)), and a software architecture—the Generalized Intelligent Framework for Tutoring (GIFT) —to instantiate learning principles, accelerate production, and reduce the cost of adaptive instruction (Sottilare, 2013). Significant accomplishments include access to free, government-owned, online/cloud-based software for designing, authoring, and implementing adaptive instruction; and authoring low-cost individualized, adaptive SBT, and extensive documentation of research findings through annual symposiums, expert workshops, and design guidelines publications (see gif-tutoring.org). Currently, the Army is extending the GIFT architecture to build adaptive SBT for teams (Johnston, Sottilare, Sinatra, & Burke, 2018). Challenges include identifying team competencies and assessments, developing learning assessments that provide a comprehensive and unbiased picture of how a team performed, and devising methods and technologies for reducing and converting raw performance data into meaningful feedback for the team (Johnston et al., 2018).

With continued expansion and maturation of learning technologies, the widespread use of sophisticated adaptive SBT for multi-team systems is assured; nevertheless, there are major barriers to implementation. A recent treatise on human–computer interface grand challenges includes a lengthy discussion on the need for research to focus on advancing adaptive learning design processes, metrics, and evaluation tools that accurately assess student learning progress (Stephanidis, Salvendy, Antona, Chen, Dong, Duffy, Fang, et al., 2019).

ADAPTIVE SBT IMPLEMENTATION

The implementation of adaptive simulations has greatly expanded and extends well beyond traditional K-12 math and physics courses to more complex skills such as military combat (Domeshek, Ramachandran, Jensen, Ludwig, Ong, & Stottler, 2019; McCarthy, 2008), cross-cultural interactions and foreign language (Johnson, 2010; Johnson & Lester, 2016), and medical skills (see also, Maheu-Cadotte, Cossette, Dubé, Fontaine, Lavallée, Mailhot, et al., 2020). As noted above, the US military has made a major contribution to adaptive simulation development because it recognized the return on investment in delivering more effective training to its active and reserve military 24-7 (Fletcher, 2009; Kulik & Fletcher, 2016; Ma, Adesope, Nesbit, & Liu, 2014). By the 1990s, the US Navy was able to transition early advances in adaptive SBT to develop and maintain critically important sailor skills needed to operate computerized systems in combat ships. Three examples are the Radar System Controller Intelligent Training Aid (RITA ITA), the Anti-submarine Warfare/Anti-surface Warfare Tactical Air Controllers (ASTAC) ITA, and the Tactical Action Officer Intelligent Tutoring System (TAO-ITS).

The RITA ITA was designed to develop operator skills in conducting standard procedural tasks and responding to information in the radar's visual display, including recognizing and managing data presentation such as chaff, jamming, and clutter (McCarthy, 2008; McCarthy, Pacheco, Banta, Wayne, & Coleman, 1994). McCarthy et al. (1994) found a majority of sailors reported the RITA ITA improved these skills and expressed a desire that it be incorporated into their curriculum. Following implementation, the Navy reported they were able to increase the throughput of trainees without additional instructors and improve on-the-job performance (McCarthy et al., 1994).

Following the success of RITA ITA, the Navy developed an ASTAC ITA to address a high sailor dropout rate (greater than 25%) due in part to an outdated training system. An ASTAC provides air control for fixed and rotary-wing aircraft on board Navy ships and the ASTAC ITA was developed to increase practice opportunities on these tasks (McCarthy, Wayne, & Deters, 2013). McCarthy et al. (2013) conducted a study indicating that the Navy's implementation of several cost-savings measures that included the ASTAC ITA likely resulted in a cost avoidance of almost $1 million per year and a lower ASTAC dropout rate.

The underlying assessment architecture used in both ITAs is the ExpertTrain™ system (McCarthy et al., 2013). Key simulation events are the learning objectives; the expert performance model assesses trainee responses to the event, compares them to expected actions, updates the learner model, and triggers an instructional response that determines whether to provide feedback to the learner. The ExpertTrain™ learner model was attributed to Murray's (1991) endorsement-based modeling approach that enables combining different types of behavioral assessments of varying reliability. For example, the ASTAC ITA collects trainee test responses that include event-based computer inputs, speech acts, and practical tests with the simulation after training. The ITA creates a repository of behavioral evidence for each learning objective that enables selecting an array of data to determine mastery of learning objectives, and for adapting the training environment to move ahead.

About the time the RITA ITA was under development, the Navy had also determined that better training was needed to improve preparing its Tactical Action Officers (TAO) for combat decision making. It invested in the TAO ITS, which is an adaptive team-training system that uses intelligent agents to play the role of simulated crew members (Domeshek et al., 2019). The tutor evaluates trainee performance in real time, infers whether they used proper tactical principles, and provides real-time feedback and coaching. A learning management system enables instructors to review, evaluate and remediate trainees, provide detailed feedback, and grade performance. More recently, the TAO ITS was integrated into the Northrup Grumman developed PC-based, Open architecture for Reconfigurable Training Systems (PORTS) that runs the Combined Tactical Training and Analysis System. The enhancement to TAO ITS in PORTS is a speech-enabled graphical user interface that enables trainees to verbally command and receive information with simulated crew members. The ITS architecture includes expert, curriculum, tutor and student models, with an assessment method that combines data from training conditions (e.g., tactical actions made by the TAO and other friendly and hostile actors), event sequences in the simulation, and student responses to identify gaps in knowledge and skills (Domeshek et al., 2019).

By the mid-2000s, adaptive SBT was expanded to solve training needs resulting from the large numbers of service members being deployed and redeployed to Iraq and Afghanistan. For example, as noted earlier, the DARPA DARWARs program enabled advances in authoring, intelligent agents, and natural language processing technologies to enable rapid development, testing, and deployment of the DARWARS tactical language trainer which uses interactive avatars to train local culture and customs across a variety of geographic locations (Johnson, 2010; Johnson & Lester,

2016). The Department of Defense Language and National Security Education Office now hosts the VCAT as an operational training readiness tool.

The DARWARS program had a significant impact on moving gaming platforms into mainstream training venues, with VCAT being an early example of what is now termed Serious Games (SGs) for training (Interservice/Industry Simulation, Training, and Education Conference, 2021). Consequently, by the mid-2000s, the SGs industry had established a niche in adaptive learning environments (Laamarti, Eid, & Saddik, 2014) and since then has seen dramatic growth across most public and private sector industries. Clement (2021) reported the SGs commercial market value had risen to an estimated US$3.5 billion in 2018, and was projected to be $24 billion by 2024. Recent meta-analyses have found the effectiveness of SGs is about the same as other training methods (Johnson, Deterding, Kuhn, Staneva, Stoyanov, & Hides, 2016; Maheu-Cadotte et al., 2020). Many examples of SGs can be found at the Interservice/Industry Simulation, Training, and Education Conference (I/ITSEC) (2021) website for SGs. The I/ITSEC SGs Challenge was established to improve training effectiveness by encouraging developers to incorporate such learning principles as measurable, challenging objectives, adaptive assessments, gameplay dynamics to increase user engagement, positive and negative feedback on progress, and making them easily accessible online 24/7 (Cannon-Bowers & Bowers, 2010). We propose that a key enabler of SG success is the focus on training a small set of well-defined, observable task skills which control the cost of developing assessment and measurement. In summary, the growth in commercially viable SGs continues unabated and will likely increase exponentially with the expansion of intelligent agent technologies and access to big data. Continued success, however, requires ensuring the learning assessments used are valid and reliable.

A CONFIRMATORY PERFORMANCE MEASUREMENT FRAMEWORK FOR ADAPTIVE SBT

The trends summarized previously—the use of performance information for diagnosis, use of simulation for distributed training, and increased automation and customization of training content—present a number of challenges for the development of performance measures. The hardware and software technologies contributing to these trends have evolved at a rapid pace. As a result, trainers have the opportunity to extract vast amounts of raw data from a simulation, employ data mining algorithms to seek out consistencies in the data, or rapidly reconfigure training scenarios based on an arbitrary notion of the trainee's skill deficiencies. In other words, it is very easy for the trainer to set up an ad hoc training strategy with accompanying ad hoc performance measures and little guidance in terms of defining training objectives, identifying critical skills to be trained, developing valid performance measures that convey individual differences in the critical skills (and allow changes in the skills to be modeled over the course of training), and evaluating training effectiveness. More generally, there is a distinct requirement to integrate useful and valid assessment architectures within these learning technologies. As exemplified by the GIFT architecture and multiple SBT and tutoring technologies that use it, assessment provides a foundational utility for virtually every learning application within these systems. A deeper understanding of *how* these symptoms accelerate student learning reveals the critical distinction between performance *measurement* (i.e., nonevaluative description of learning relevant behaviors), *assessment* (i.e., use of performance information to make inferences regarding the latent knowledge, skills, abilities, or other personal characteristics that drive observable performance), and *diagnosis* (i.e., identifying the root causes of suboptimal performance).

In an effort to provide such guidance, we propose a comprehensive measurement framework useful for guiding the development and evaluation of performance measures in adaptive SBT contexts (see Figure 10.1). Specifically, we define the criteria for a performance measurement framework and use it to provide specific principles and guidelines for the development and use of performance measures. The essential characteristic of performance measures conforming to this framework is that they support a confirmatory strategy for measuring and evaluating performance in adaptive training contexts. It is confirmatory in that it is a closed-loop system of predictions and observations that enable sound inferences regarding hypothesized relationships among training objectives, performance episodes, and outcomes. This highlights the need for a process-based approach to training planning, execution, and evaluation where learner performance is measured continuously and training outputs directly lead to revised inputs.

Similar representations of the training process, its phases, and the role of performance measures have been proposed in other research and development efforts. For

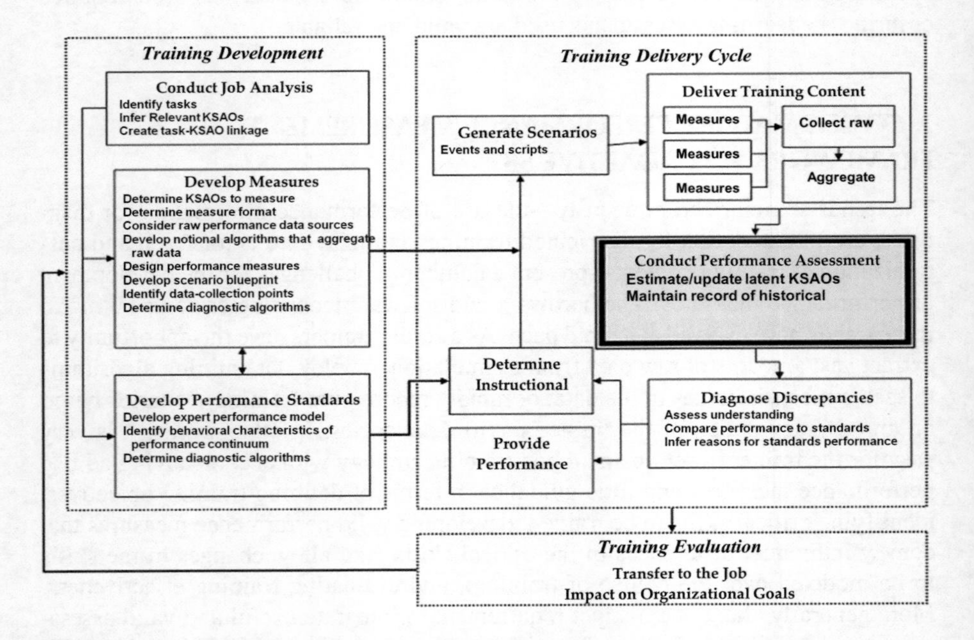

FIGURE 10.1 Performance measurement framework for simulation-based training.

example, Zachary, Cannon-Bowers, Bilazarian, Krecker, Lardieri, and Burns (1999) proposed a model of embedded training systems that describes the cycle through which historical training data and job performance standards are combined to create performance objectives; training objectives are used to script scenario events and derive performance measures; and performance measures are used as the basis for performance diagnosis, feedback, and revision of trainee performance history. Our model augments and extends this framework, specifically:

(1) Knowledge, skills, abilities, and other personal characteristics (KSAOs) necessary for effective job performance are derived from job/task analytic data, rated on independent dimensions (e.g., importance and criticality), and linked empirically to specific tasks.
(2) Critical KSAOs are used to define performance standards and drive the development of performance measures embedded within scenario content.
(3) Performance standards are used to determine instructional objectives, which in turn are used to script and generate training scenarios with embedded performance measures.
(4) Observed performance on the embedded measures provides estimates of the underlying KSAOs represented by the measures.
(5) Performance standards and KSAO estimates jointly determine performance feedback and subsequent training objectives.
(6) Revised training objectives drive the generation of new training content targeting deficient KSAOs.
(7) Performance measures obtained throughout training are used to evaluate the training system itself, continuously improving it with innovative instructional methodologies with the ultimate goal of maximizing training transfer.

A number of features of the framework should be noted. First, a feedback loop is integrated into the training delivery cycle portion of the model that emphasizes performance measures are error-prone indicators of latent KSAOs and the training delivery cycle reflects an ideographic approach by iteratively assessing relevant KSAOs and strategically inserting scenario elements designed to evoke learning throughout training. We propose that KSAOs can be estimated indirectly via repeated performance measurement under changing scenario conditions. The confirmatory process of repeatedly exposing trainees to varying, graded training content follows the logic of adaptive testing and allows for repeated, increasingly accurate measurements of latent KSAOs (Embretson & Reise, 2013). The resulting KSAO measurements drive subsequent generation of scenario content tailored to the learner's deficient KSAOs and permit the possibility of modeling performance changes throughout training. Second, the model emphasizes the complementary roles of deliberate practice and performance feedback. Performance feedback provides a mechanism for highlighting specific gaps in the learner's knowledge structure, correcting misconceptions, and guiding students through effective solutions to specific problems. However, we believe that a complementary mechanism should serve the express purpose of generating training content that targets specific skill deficiencies. Adaptive SBT can provide such a capability by allowing for focused, deliberate practice within a realistic simulation environment, a necessary capability to support expert performance (Ericsson, Hoffman, Kozbelt, & Williams, 2018). Although the focus of our discussion is on performance measurement issues in adaptive training, it should be noted that the same issues are relevant for the development of tailored feedback.

Third, the model assumes that training content will be customized to target the KSAOs of a single trainee. This limitation stems from the fact that the theory and methods necessary to develop tailored training are still in early stages of development. Thus, it would be difficult to generate training content that is simultaneously tailored to the needs of multiple individuals, each with unique skill deficiencies, performing interdependent tasks in a team setting. Although research has pointed to the potential use of intelligent agents to simulate the actions of teammates resulting in a simulated team environment and allowing practice of critical teamwork behaviors (Zachary, Santorelli, Lyons, Bergondy, & Johnston, 2001).

DIMENSIONS AND ESSENTIAL CHARACTERISTICS OF PERFORMANCE MEASURES

As adaptive SBT continues to rapidly mature, so does the need to identify and develop desirable characteristics of performance measures, which includes psychometric quality (DiCerbo, Shute, & Kim, 2019; Katz, LaMar, Spain, Zapata-Rivera, Baird, & Greiff, 2017). To address this gap, we identified six dimensions—validity, criterion relevance, reliability, measure invariance, objectivity and intrusiveness, and diagnosticity—along which performance measures in adaptive SBT can vary. The idea is for the SBT to create high levels of each of these dimensions. This provides a basis for identifying desirable characteristics of performance measures and recommendations for improving measurement practices.

VALIDITY

Performance measurement refers to the process through which behaviors observed within a job or training environment are translated into a comprehensive summary snapshot of an individual's performance in a particular setting, providing the foundation for subsequent performance evaluation, management, or prediction. Performance measures are commonly referred to as criteria because they provide a basis for arriving at evaluative judgments about individual job performance and often serve as the dependent variable (or criterion) in validity research. The concept of a set of interrelated, observable behaviors that are relevant for accomplishing higher-order job or training objectives is consistent with the notion of a psychological construct, a conceptual term used heuristically to articulate, describe, and predict a set of related, covarying behaviors associated with a phenomenon of theoretical interest (e.g., intelligence, personality, anxiety, and expertise) (Binning & Barrett, 1989; Cronbach & Meehl, 1955; Edwards & Bagozzi, 2000). A measure, in contrast, is an observed score recorded using some measurement method (e.g., self-report, interview, observation, and objective measurement) that represents an empirical analog of a construct (Edwards & Bagozzi, 2000).

Discussions of the difference between constructs and measures in the context of personnel selection, training, and performance appraisal, though limited, have distinguished between an ideal, hypothetical criterion construct (i.e., the domain of all behaviors important for attaining organizational, job, or training objectives) and the actual measures used to representatively sample this domain of behaviors (Binning & Barrett, 1989; Borman, Grossman, Bryant, & Dorio, 2017). Stated more precisely, the behavioral domain comprising the hypothetical criterion construct can be considered an expression of the collective KSAOs important for organizational, job, or training effectiveness, often derived from job/task analyses, and formalized in explicit training or learning objectives. However, despite the performance criterion/measure distinction, there is no explicit unifying model useful for articulating the meaning of the validity of a performance measure. Validity studies have emphasized the predictive, criterion-related validity of a selection system or training intervention, that is, the degree to which performance on a selection test or during training predicts subsequent real-world performance. Less attention has been paid, however, to what is meant by the validity of the criterion measure in and of itself. Discussions of a measure's construct validity often invoke the idea of a nomological network which is an expected pattern of relationships among constructs and between constructs and empirical observations (e.g., measures). Traditionally, evidence for a measure's construct validity has been provided in the form of significant correlations, with measures purporting to measure the same or similar construct (i.e., convergent validity), and nonsignificant correlations, with measures purporting to measure dissimilar constructs (i.e., discriminant validity). Such evidence can be derived using analytical techniques such as the multimethod-multitrait matrix (Campbell & Fiske, 1959), and the resulting pattern of relationships has been termed the *nomological network* for the target measure. The traditional approach to construct validity has focused primarily on the degree of empirical support for a nomological network and the resulting quality of the inferences that may be drawn from it (Cronbach & Meehl, 1955). However, reliance on a nomological network as a foundation for assessing the validity of a measure introduces a potential problem that may be especially pronounced with respect to the validity of performance measures, specifically, that the meaning of a construct and its measures emerge as a result of their configuration within the nomological network (Borsboom, Mellenbergh, & van Heerden, 2004). This introduces the tautological fallacy of using the network to implicitly define the constructs of which it is composed strictly in terms of their relationships with each other and without reference to theoretical terms (Borsboom et al.).

An alternate view of validity, proposed by Borsboom and colleagues (2004), poses two criteria for a valid measure: the construct exists in the real world and variations in the construct cause analogous variations in the measure. This view transfers the locus of evidence for validity from the observed relationships among measures to the response processes that convey the causal effect of a psychological construct on its measure (i.e., substantive validity; Messick, 1995). Consider measures used in the natural sciences—a common thermometer may be considered a "valid" measure of temperature because thermal energy exists in the ambient environment and transfers the energy to the thermometer and causes the mercury to rise to a degree that depends completely on the amount of energy transferred. No allusion to a nomological network is necessary because evidence for the validity of the instrument lies entirely in the causal sequence of events linking variations in the construct with variations in the measurement instrument.

This "back to basics" view that validity basically reflects how much a measure is synchronized with its intended, invisible, underlying construct is especially relevant to and necessary for the development of performance measures for adaptive SBT. It corresponds with the notion of evidence-based measurement being used increasingly in the context of SBT and ITS assessment. First, this conception of validity requires a detailed theory of response processes with respect to a set of KSAOs. That is, one must explicitly describe the behaviors associated with a given KSAO, and the specific behaviors that are indicative of various levels of effectiveness with respect to that KSAO. This constraint is often satisfied in SBT and ITS research, because both require the articulation of an expert performance model that describes the ideal behavioral patterns to be expected with respect to critical KSAOs vis-à-vis the training scenario content. In fact, many of the adaptive SBT applications described above operate off an expert model in which optimal sequences or patterns of responses are used as a reference model against which trainee performance is compared. Thus, by adopting a detailed theory of response processes, users can maximize the diagnostic potential of adaptive SBT by articulating responses to scenario content that are associated with various effectiveness levels of specific KSAOs. Second, it may be difficult to "build" nomological networks for performance measures, that is, to predict relationships between the target measures and other measures that target similar and dissimilar constructs, which are useful for construct validity inferences. We believe this to be especially true because, ideally, performance measures should be developed to target specific, nonredundant sets of behaviors such that no two measures assess the same constructs. In such cases, a performance measure's nomological network would be composed exclusively of other measures with which one would not expect significant relationships. Third, it would be difficult to examine patterns of covariance among performance measures under dynamic scenario conditions because this could confound inferences regarding the validity of the individual performance measures and the effects of scenario modifications on changes in targeted KSAOs. Finally, researchers have noted the potential utility of response processes to support validation of measures assessing complex, multidimensional constructs (Ployhart, 2006). In order to understand whether a performance measure captures the latent construct it is designed to measure, one must understand the response processes associated with the underlying construct. This is especially true for complex constructs such as situational awareness, multitasking performance, dynamic workload management, and team communications assessed in adaptive SBT systems.

A validation strategy that considers the response processes associated with relevant KSAOs is likely to support the development and refinement of performance measures consistent with the proposed measurement framework. Valid performance measures accurately represent the latent KSAOs driving observable training performance by capturing essential KSAO-specific response processes. This allows for comparison of observed performance against theory-driven benchmarks of expert performance, accurate estimation of latent KSAOs, and fine-grained customization of subsequent training content.

Criterion Relevance

Another critical issue is the degree to which performance measures comprehensively, yet efficiently, capture the domain of behaviors they represent. This notion invokes the concepts of criterion relevance, deficiency, and contamination (Borman, Grossman, Bryant, & Dorio, 2017). Criterion relevance refers to the idea that performance measures should correspond to the actual performance demands of the training situation. Measures should accurately assess only the KSAOs they represent and not other irrelevant sources of variance. Two potential problems that may compromise criterion relevance are criterion contamination and deficiency. Criterion contamination refers to the degree to which a performance measure taps variance unrelated to performance demands. For example, contamination might occur if one used an automated speech recognition capability to assess the quality of team communications but the program required users to memorize keywords not typically used in the training scenario in question. In this case, the measure tapped an irrelevant construct (i.e., working memory) rather than the intended construct (i.e., communication quality). Criterion deficiency occurs when a performance measure fails to sample important training behaviors. Deficiency might occur if one intended to measure the quality of the trainee's communications during a simulation but the scenario content was designed such that too few or only a limited variety of team communication opportunities were offered. This can cause critical problems with transfer of training.

Although criterion relevance is implicitly defined as the absence of criterion contamination and deficiency, an additional consequence is enhanced parsimony and efficiency of measurement. A measurement strategy conforming to the proposed framework will include a suite of uncorrelated measures, each measuring a unique, relevant aspect of performance, that collectively provide a comprehensive, representative sample of the hypothetical criterion domain. Additionally, it is possible to maximize criterion relevance by ensuring that training objectives are based on thorough, accurate job analytic results. A comprehensive job analysis methodology will provide the blueprint detailing which KSAOs should be included in the training context, by providing empirical estimates of KSAO importance and task–KSAO linkages. Consequently, training objectives based on thorough job analysis results will help ensure that only relevant criteria are included in training scenarios.

Reliability

Adoption of a perspective in which performance measures are characterized as error-prone surrogates of a hypothetical performance domain encourages examination of the sources of unsystematic variance. In the language of classical test theory, validity usually refers to measurement accuracy, whereas reliability refers to measurement consistency (e.g., across test items, measurement intervals, or raters). The two concepts usually go hand in hand, emphasizing the notion that reliability is a necessary, but not sufficient, condition for validity. The logic underlying this notion is that measurement error in the form of unreliability must be minimized to assess the true magnitude of the statistical relationship between variables when provided as evidence for validity (Crocker & Algina, 1986; Nunnally, 1978). A related concept, method variance, refers to measurement error associated with using different methods to measure a construct, often in the context of construct validation (e.g., multitrait-multimethod matrices; Bagozzi & Yi, 1990; Coovert, Craiger, & Teachout, 1997).

In the context of adaptive, simulation-based performance measurement, traditional, classical test theory notions of measurement error and reliability assessment take on a new meaning. Most importantly, method variance means something different when measuring performance in adaptive training contexts, given the fact that individuals are observed in continuously changing measurement contexts, because the performance construct itself is changing. That is, the latent KSAOs represented by performance measures are expected to improve (hopefully) over the course of training, driving subsequent scenario modifications. So, both the latent trait and the environment/intervention affecting the latent trait are changing simultaneously. Thus, the traditional conception of reliability as consistency across performance measures or measurement opportunities may not be sufficient as an indicator of measurement error in the context of adaptive training.

As with any performance measurement system, reliability is a prerequisite to validity. Classical test theory notions of internal consistency as reliability may be useful in adaptive training contexts if one could establish the dimensionality of the performance measures. However, this may be limited by the fact that the performance constructs of interest in adaptive training contexts are often complex and multidimensional. Additionally, test–retest reliability cannot be meaningfully computed when the latent construct influencing test scores changes over time (as with adaptive training). It is possible, however, to develop and test covariance structures (e.g., latent growth curve models) that model performance changes over time and allow for the inclusion of scenario events as time-varying covariates (Bollen & Curran, 2006). Such analyses may provide useful insight into how reliably performance measures are functioning across time and contexts. However, such analyses have intense statistical power requirements (highlighting the need for adequate numbers of subjects) and can be employed only when there are repeated opportunities to observe performance (reiterating the requirement for multiple observational opportunities).

Measure Invariance

Measurement invariance relates to the need for and "apples-to-apples" comparison between measured skills or cognitive attributes and task requirements. According to the proposed measurement framework, one must observe a trainee's behavior across measurement contexts in order to make meaningful, reliable inferences about his or her latent KSAOs. Given that adaptive training entails continuous modification of the training environment, an essential property of performance measures is the ability to assess critical training behaviors across a range of varying scenario content. Measure invariance refers to the degree to which performance measures retain their essential measurement properties and thus can be used to make meaningful comparisons across observations and under transient scenario contexts. Psychological measures usually contain multiple assessment items varying in content and along critical psychometric parameters (e.g., difficulty, discrimination). Consistent with the notion of a psychological construct as a theory of behavioral consistency over varying contexts (Cronbach & Meehl, 1955), the specific pattern of responses across items allows observation of behavioral consistency and, consequently, inference of the individual's standing on the latent construct (Embretson, 2006).

As Embretson (2006) has noted, in response to the ongoing search for nonarbitrary metrics in psychological research, a number of innovative psychometric theories and methodologies have emerged from the psychological and educational assessment domains. It is possible to draw from these theories and methods to develop performance measures that provide meaningful information about the performance construct across changing task or scenario conditions. A notable advantage of assessment methods such as computerized adaptive testing is the use of psychometric models derived from modern test theory (i.e., item response theory) to produce trait score estimates that do not depend on population distributions (Embretson & Reise, 2013). In contrast to traditional, normative assessment strategies, in which test scores are norm referenced via classical test theory analyses (Crocker & Algina, 1986), trait scores derived from adaptive tests gain meaning by direct comparison of estimated trait levels to item parameters (e.g., the item's difficulty level). This enhances diagnosticity, retains desirable psychometric properties of the test while reducing the length of test administrations, and allows meaningful performance comparisons across individuals who received different item sets. With the resulting measurement correspondence, we can make better and more accurate decisions about what training or assessment content is going to provide the most appropriate level of challenge given the person's currently measured skill level.

The same standards are relevant for adaptive SBT. Use of normative performance scores (e.g., mean performance across a scenario) does not allow direct observation of what specific elements of a scenario influenced performance, and thus prevents unequivocal inferences of the individual's standing on the latent KSAO's driving performance. That is, it does not allow one to measure which specific aspects of the scenario posed the greatest level of difficulty for the trainee. The logic of modern test theory demands that test items be scaled according to their difficulty. Analogously, one may scale scenario content according to its difficulty level by defining the minimum level of the underlying skill the trainee must possess to successfully "pass" a scenario event. Taking this logic one step further, it is also possible to create scenario content to target a very specific, exact cognitive skills or personal attributes needed for successful performance. This reflects the modern test theory concept of dimensionality. Knowing which dimensions are being measured, and to what degree they are being challenged in the scenario provides a foundation for more accurate and meaningful sequencing of scenario content around the trainee's iteratively measured skill level. Subsequently, performance measures can be constructed around statistical models that allow aggregation of individual performance observations across performance episodes of varying difficulty. However, for performance measures to be useful in such a context, it would be necessary for them to sensitively capture critical behaviors across the spectrum of difficulty levels and across a wide range of scenario content. Similar endeavors have been attempted, with some evidence of success, to model learner performance in complex problem-solving domains, using item response theory and other probabilistic models, such as Bayesian networks (Embretson, 1997, 1998; Levy & Mislevy, 2004; Mangos, Campbell, Lineberry, & Bolton, 2012; Mislevy, 1995; Mislevy & Wilson, 1996; Pirolli & Wilson, 1998). Such a process, if applied to adaptive SBT, would allow for a fine-grained analysis of how individuals respond to specific elements that comprise a scenario, address how behavioral responses correspond to specific skill sets or changes within an individual's knowledge structure, and support scenario generation that adaptively challenges deficient knowledge and skills.

Objectivity and Intrusiveness

Simulation-based training performance measures can differ according to their level of objectivity. Such differences may depend on characteristics of the simulation environment, which can range from realistic, immersive, and automated to artificial and contrived. The level of realism and automation inherent within the simulation can influence whether assessments can be performed passively from within the simulation environment or whether an external intervention is required to observe and record performance. The resulting degree of objectivity has implications for the quality of the inferences to be made regarding the latent KSAOs underlying observable performance. Whereas objective measures afford direct observation and measurement of specific behaviors useful for assessing and evaluating performance, subjective measures introduce an additional source of error variance in the form of rater error, potentially undermining the quality of the inferences to be drawn regarding an individual's performance (Borman, Grossman, Bryant, & Dorio, 2017; Landy & Farr, 1980). Additionally, obtrusive measures can create a source of criterion contamination by distracting the trainee's attention away from scenario content in order to attend to the performance measure. An example of this may be seen in typical cases of aviation or unmanned systems training where instructors sit by students as they perform the simulation, and asked situation awareness-related questions as the trainee is attempting to fly the aircraft.

It is possible to create opportunities for direct, naturalistic performance measurement when measures are embedded directly into the simulation environment. Virtual, augmented, and mixed-reality simulations continuously push the envelope with respect to the level of physical fidelity, enabling increasingly realistic representations of real-world scenarios and high levels of presence (Mangos et al., 2012). Automated performance measures embedded in such systems are capable of passively recording critical behaviors without disrupting the training exercise. Often, it is possible to glean data in either raw or aggregated form directly from the simulation environment. This capability has been enhanced by a trend toward common data standards imposed by frameworks such as GIFT and ADL. Such data, when aggregated in a meaningful way, can be used to form direct, unobtrusive performance measures. However, subjective measures are often needed to assess constructs such as situational awareness that are difficult to assess purely with raw performance data. In such cases, a combination of objective and subjective measures can be used to provide a more comprehensive portrayal of performance effectiveness. However, this could introduce a number of additional measurement challenges, including the identification of highly skilled subject matter experts (SMEs) to serve as raters, intensive SME training to ensure accurate, reliable ratings of behaviors that often reflect highly specialized skills, difficulties in attending to all relevant performance information, criterion deficiency, and interrater agreement measurement. Use of multiple raters combined with intensive rater training, behaviorally based performance measures, and stringent interrater agreement and reliability criteria can help mitigate such difficulties.

Diagnosticity

Simulation-based training measures can differ further according to their capabilities for translating raw performance information into assessment information useful for diagnosing skill deficits. In some training contexts, a single performance measure can be used as an indicator of the trainee's standing on the latent KSAO it represents (e.g., number of algebra problems completed correctly as a measure of mathematical ability). Such measures are termed reflective in that they reflect or represent the manifestation of a single construct. Often, however, we are interested in constructs that represent the composite of multiple component variables. Measures for such constructs are termed formative, given that the construct is formed or induced by its measures (Edwards & Bagozzi, 2000). Formative measures are increasingly common in SBT environments because they are commonly used to target more complex constructs, such as situational awareness, teamwork skills, multitasking ability, and communication effectiveness—constructs formed by aggregating measures of more basic constructs (e.g., working memory performance on a single task dimension, and attention to a single visual/auditory stimulus).

A critical issue with respect to diagnosticity is the extent to which performance measures, whether reflective or formative, provide insights regarding why an individual is performing at a suboptimal level. Measures that allow comparison of patterns of responses to a theoretical model of response processes, that is, measures conforming to the Borsboom et al. (2004) validity framework outlined earlier, are likely to be useful from a diagnostic standpoint. Such measures support inferences regarding an individual's skill deficiencies by virtue of the observed pattern of responses. Thus, a measure that is "valid" in terms of its ability to capture KSAO-specific response processes is also likely to be diagnostic. A second issue with respect to diagnosticity is the fact that the use of purely aggregate measures without attention to the composite measures of which they are formed could confound diagnostic inferences. Aggregate measures are useful in adaptive training contexts by providing a summary index of the complex variety of behaviors that occur within the simulation. However, the individual measures of which an aggregate measure is composed can reflect unique constructs or rater perspectives. Aggregation treats meaningful variance associated with unique perspectives or constructs as measurement error, introducing a form of aggregation bias (James, 1982; Morgeson & Campion, 2000; Sanchez & Levine, 2000) and rendering difficult the drawing of inferences regarding why a deficient score on an individual performance measure was observed. Thus, aggregate measures may not provide the diagnostic precision necessary to customize training content to target skill deficiencies. In such cases, aggregate measures may be more useful for providing performance feedback to trainees, whereas narrower, individual performance measures may be necessary for structuring adaptive training content.

An especially promising measurement solution related to the issue of diagnosticity is the branch of psychometrics related to cognitive diagnostic assessment. This field goes beyond the usual emphasis of what level of the trait is being shown by a test respondent to diagnosing the actual, specific cognitive deficits and misconceptions that lead to an incorrect response. This methodology is closely related to the concepts of evidence-based measurement and principled, structured problem-solving. These models provide a mechanism for drilling down and isolating an individual's skill deficiencies based on unique patterns of responses to strategically sequenced test items (Henson, Templin, & Willse, 2009). With the right combination of test items, these models effectively show the probability of having a specific deficiency (e.g., conceptual misunderstanding) with respect to the skill or knowledge domain being measured. For example, they have significant utility in identifying whether a student has reached specific developmental cognitive milestones, such as the acquisition of basic mathematical properties for addition and subtraction, based on specific patterns of incorrect responses when solving math problems. The utility of cognitive diagnostic models in isolating trainee skill deficiencies highlights their potential effectiveness when applied to the measurement of advanced, executive-level cognitive constructs (i.e., superordinate processes necessary to orchestrate subordinate mental processes, such as attention and working memory). In more advanced military task domains, such as command and control missions, a failure to attend to cues at the right time may be the result of a number of executive cognitive skill deficiencies, including fatigue-related vigilance decrement, channelized attention, or tunnel vision, overwhelming task-switching costs, or general mental overload. This renders the models readily extensible to diagnosing failures in advance cognitive skills within the context of military mission tasking, providing a basis for future training scenario adaptation.

MEASUREMENT PRINCIPLES FOR ADAPTIVE TRAINING

Table 10.1 summarizes the critical dimensions of performance measures, the associated challenges with respect to the proposed confirmatory measurement framework, and the desirable characteristics of measures along each dimension. The primary

TABLE 10.1
Strategies for Addressing Performance Measurement Challenges

Performance Measure Dimension	Challenges	Measurement Strategies
Validity: The degree to which a performance measure accurately represents a performance construct in the real world (i.e., the latent KSAOs driving observable performance), and variations in the construct cause analogous variations in the measure (i.e., essential KSAO-specific response processes) (Borsboom et al., 2004)	Articulate a model of response processes and ensure measures are capable of capturing critical responses as specified by the model	1. Ensure measures are sensitive to subtle variations in the latent KSAOs driving observable performance, capturing behavioral patterns corresponding to different levels of the latent KSAOs 2. Employ specific performance benchmarks derived from theory-driven expectations about "expert" or optimal performance (e.g., expert performance model)
Criterion relevance: The degree to which the domain of behavior is captured by performance measures	Minimize criterion contamination and deficiency as potential sources of measurement error	1. Ensure measures comprehensively, accurately, and parsimoniously sample the criterion domain2. Map measures directly to specific training objectives guided by real-world performance demands (e.g., derived from job/task analyses)
Reliability: The degree to which measurement is consistent across context, raters, and time. Method variance: variance that is due to using different methods, e.g., observation, multiple raters, self-report	Hypothesize a model of behavioral consistency across scenario contextsModel changes in the latent KSAOs over time	1. Ensure repeated performance measurements across multiple observational opportunities2. Assess internal consistency reliability when the dimensionality of the measures can be established3. Assess reliability within a latent growth curve modeling approach that incorporates the presence or absence of scenario content as time-varying covariates; ensure adequate levels of statistical power

(*Continued*)

TABLE 10.1 (CONTINUED)
Strategies for Addressing Performance Measurement Challenges

Performance Measure Dimension	Challenges	Measurement Strategies
Measure invariance: The degree to which performance measures provide invariant measurements across transient scenario contexts	Develop performance measures that provide meaningful, interpretable metrics across performance contexts	1. Use modern test theory as a psychometric framework to scale scenario content according to its difficulty level2. Ensure performance measures are sensitive to critical behaviors across a wide range of scenario content and difficulty levels
Measure objectivity: The degree to which the measure allows direct versus indirect observation and measurement of specific behaviors. Intrusiveness: The degree to which the trainee is made aware of the measurement process	Employ a measurement strategy that does not interfere with performanceIntegrate objective/subjective measures with different loci of assessment into a coherent suite of measures	1. Use automated, embedded performance measures based on raw or aggregated data that can be obtained directly from simulation environment2. Combine subjective measures, when it is necessary to use them, with objective measures3. Employ multiple raters, rater training, behaviorally based performance measures, and stringent interrater agreement and reliability criteria when subjective rater judgments are necessary
Diagnosticity: The process of translating raw performance data into assessment information	Meaningfully aggregate raw performance data into a summary index useful for performance assessment and subsequent diagnosis and scenario modifications	1. Adopt a validity model that considers response processes2. Ensure that performance measures have the necessary precision to make diagnostic inferences; use aggregate measures mainly for performance feedback, and narrower, individual measures as a basis for structuring training content

challenge with respect to validity is to articulate a model of response processes and ensure that the measures are capable of capturing critical responses as specified by the model. With respect to criterion relevance, the challenge is to minimize criterion contamination and deficiency as potential sources of measurement error. For reliability, the challenge is to hypothesize a model of behavioral consistency across scenario contexts and to model changes in the latent KSAOs over time. For measurement invariance, the challenge is to develop performance measures that provide meaningful, interpretable metrics across performance contexts. For objectivity and intrusiveness, the challenge is to employ a measurement strategy that does not interfere with performance and to integrate objective measures with different loci of assessment into a coherent suite of measures. For the diagnosticity dimension, the challenge is to meaningfully aggregate raw performance data into a summary index useful for performance assessment and subsequent diagnosis and scenario modifications. In addition to the specific guidelines provided for each dimension, we offer an additional set of more general principles relevant to the effective application of performance measures in adaptive training contexts, which are as follows: ensure that performance measure development is guided by a sound theoretical framework, identify and exploit measurement affordances of the adaptive training environment, and consider training evaluation strategies early in the performance measure development process.

PRINCIPLE 1: ENSURE THAT PERFORMANCE MEASURE DEVELOPMENT IS GUIDED BY SOUND THEORY

We believe that the primary ingredient for sound performance measures is grounding in a sound theoretical framework. Substantial theory development in the areas of learning, skill acquisition, practice, cognitive modeling, and psychometrics has resulted in robust, detailed theories useful for guiding training designs. The decision to customize training content or feedback delivery, automate aspects of scenario generation or performance measurement, or vary the pacing, content, or sequencing of training content should have a clearly defined theoretical rationale drawn from these lines of research. This body of knowledge will be useful for articulating how specific training interventions will influence immediate learning and performance as well as long-term retention and transfer performance. Individualized instruction in the form of adaptive training and intelligent tutoring systems has a growing theoretical and empirical research base supporting its effectiveness. A major stimulus for this research base emerged from theorizing on the concept of aptitude-treatment interactions (ATIs), which suggest that instructional interventions are effective to the extent they meet the individual needs of the learner (Cronbach & Snow, 1981; Snow & Lohman, 1984, 1993). Research on ATIs has revealed that the effectiveness of instruction is influenced by specific individual differences, endorsing an idealized model for instruction in which instructional events are customized to challenge, accommodate, or adapt to a given learner's unique skills (Snow, 1994; Snow & Lohman, 1993). This line of research provides a useful theoretical lens for developing adaptive training content and performance measures. The notion that individuals experience the training environment differently depending on latent ability levels suggests that it is possible to systematically measure the latent abilities underlying performance, model their changes throughout training, and customize training to these evolving ability levels using adaptive instruction. Thus, it is possible to consider adaptive training as a tool for inducing a continuous sequence of ATIs throughout training, thereby providing a consistently high level of challenge without compromising motivation or overwhelming the trainee.

The research literature on ATIs emphasizes characteristics of instructional conditions that could differentially influence learning, depending on skills, abilities, and personal attributes unique to the learner. Indeed, an early application of ATIs was to inform selection for training or instruction, an application based on the tenuous assumption that the latent skills and abilities underlying training performance remained stable during training. ATI research effectively cast training performance as a between-subjects phenomenon—some training interventions are effective for a subpopulation of individuals with certain levels of a requisite skill, whereas other interventions are more effective for other, more or less skilled subpopulations. However, a critical element of adaptive training, as implied by the cyclical representation of training delivery in the proposed measurement framework, is the ability to customize instructional content, and thus model the effects of instruction on a single individual over time. This can complement ATI research by providing the within-subjects perspective needed to describe and predict how an individual learner's performance (as an indicator of the latent KSAOs that are being targeted during training) varies throughout training (Alliger & Katzman, 1997).

Research addressing individual patterns of progress through distinct cognitive stages during the learning process can provide additional insight into the mechanisms underlying ATIs (Embretson, 1997; Schoenfeld, Smith, & Arcavi, 1993). Often, the performance changes that occur when learning a complex task do not reflect a single, unitary "instance" of learning. Instead, individuals frequently experience a series of learning events in which they demonstrate effective problem-solving after experiencing impasses that emerge in the task environment (Annett, 1991; Van Lehn, 1996). A learning event corresponds to the discovery of a problem's solution after the learner experienced errors or difficulties. Impasses signal faults in the learner's knowledge structures, prompting the learner to divert attention from problem-solving to the discovery of new knowledge and questions about domain knowledge itself (Van Lehn). Paralleling the ATI literature, research has indicated that the nature and timing of learning events depend on idiosyncratic experiences with impasses and on levels of stable individual differences in the requisite KSAOs relevant for learning (Ackerman, 1987; Campbell & DiBello, 1996; Ohlsson, 1996; Snow, 1994).

These theoretical notions hint enticingly at the possibility of structuring adaptive training content to control the occurrence of learning events or even instigate a sequence of learning events throughout training (Mangos et al., 2012). It is possible to recast the concept of ATIs and learning events as deliberate outcomes of training rather than as chance phenomena confined to basic cognitive research. However, a critical contingency in applying such promising theoretical concepts to training design is that performance measures must be designed to capture these elusive phenomena, both of which represent, ultimately, subjective experiences (Schoenfeld et al., 1993; Van Lehn, 1996). This emphasizes the key issue that performance measurement is instrumental for the translation of theory into training design. Valid performance measures will be sensitive to response processes indicative of targeted KSAOs, and models of response processes can only be constructed within the framework of a specific theory of learning.

PRINCIPLE 2: CONSIDER AND EXPLOIT MEASUREMENT AFFORDANCES

Considerable variability exists in the types of performance domains for which adaptive training has been developed and, consequently, in the scenario-generation methods and simulation content used to represent these domains. As stated earlier, whereas a general challenge of job performance measurement efforts has been to identify and exploit objective performance measurement opportunities, the challenge for simulation-based performance measurement has been to reduce the abundance of objective data into meaningful diagnostic patterns. This challenge is complicated further by the need to articulate and test relationships among training interventions, the psychological constructs targeted during training, and observable performance.

The event-based training approach offered a promising solution to these challenges by incorporating "trigger events" into scenario content (Dwyer, Fowlkes, Oser, & Lane, 1997; Fowlkes, Dwyer, Oser, & Salas, 1998). Responses to these events served as indicators of the individual's standing on the relevant skill being trained. This logic provides an interesting parallel to the domain of computerized assessment. As mentioned earlier, computerized adaptive testing is an assessment method that provides iterative estimation of the targeted KSAO (Olson-Buchanan & Drasgow, 1999). Computerized adaptive testing uses the individual's responses to initial test items to provide hypothetical estimates of the underlying ability level. Subsequent items are selected on the basis of the likelihood of their providing additional diagnostic information about the underlying ability, considering both initial ability estimates and item parameters (e.g., item difficulty and discrimination) (Embretson & Reise, 2013). This form of testing relies on the logic that assessments represent a form of experimentation in which test items (representing the independent variable) elicit cumulative information about some underlying trait (the dependent variable) that influences test behavior (Embretson & Reise).

This perspective is equally applicable to SBT research. Initial formulation of the event-based approach treated scenario content as an instrument for embedding individual trigger events (Cannon-Bowers & Bowers, 2009). In the context of adaptive training, however, it is possible to reconceptualize the scenario as a palette for generating a continuous stream of simulation-based assessment content useful for iterative estimation of latent skills and subsequent scenario generation. Simulation-based training scenarios are often scripted with the primary objective of realistically recreating real-world tasks or problems. Assessment content is often an afterthought in such a model, and trainers are left to force assessment opportunities out of the resulting scenario content. However, one can maximize the assessment potential of the SBT environment by considering its measurement affordances in advance, and by developing scenario content around these affordances.

Several additional lines of assessment research may provide specific guidance for exploiting the measurement affordances and realizing the assessment potential of the simulation environment. One assessment method—situational judgment tests (SJTs)—provides descriptions of problem situations likely to be experienced in the task environment along with potential solutions ranging in effectiveness as response options. SJTs purport to measure the trainees' expectations of the effectiveness of different performance options, given realistic task cues, essentially treating the quality of these expectations as an indicator of expertise (Chan & Schmitt, 2002). The parallels between SJTs and SBT are obvious; indeed, SJTs were developed as a low-fidelity alternative to more sophisticated assessments at a time when limited computing and simulation capabilities prevented higher-fidelity assessments. It is possible now, however, to draw from SJT methods to develop simulation-based versions of SJT items to make use of the measurement opportunities afforded by SBT.

An additional area of research, on the measurement construct of a "time window," provides useful guidance for measuring and assessing performance in light of the opportunities for trainee actions offered by the simulation task (Rothrock, 2001). A time window is a measurement construct useful for decomposing simulation content according to which specific activities can be performed within a given period of time. For example, in the air defense warfare domain, the presence of task cues (e.g., three unknown radar tracks on the radar screen) and the actions of one team member (e.g., the Air Intercept Coordinator illuminating one track as hostile) can engage a window of opportunity for another team member to perform a variety of actions differing in their effectiveness (e.g., ignore or engage the track). The time window defines the time period bounded by the emergence of cues or operator actions that constrain performance and the execution of some action by the target performer. The development of the time window as a formal measurement construct is based on the premise that a functional relationship exists between action constraints and time availability, a notion grounded further in the theory of situated cognition (Hutchins, 1995; Lave, 1988). By explicitly defining operational and time constraints on performance, the time window construct is likely to be a useful tool for reducing vast amounts of objective simulation data and allowing useful inferences on the meaning of performance in light of task constraints, supporting the confirmatory measurement framework described previously.

A third line of inquiry revolves around research on mathematical modeling of human performance (Campbell & Bolton, 2005; Campbell, Buff, & Bolton, 2006; Dorsey & Coovert, 2003). This research focuses on the development of formal mathematical models of human behavior with respect to situational cues and action affordances of the simulation environment. Typically, the mathematical model specifies the relationship between terms reflecting aspects of the environment (e.g., presence or absence of specific cues) and some aspect of performance (e.g., decision making). Recent efforts have compared the effectiveness of various mathematical modeling techniques (e.g., fuzzy logic, multiple regression) and have applied mathematical modeling specifically to the development of customized feedback in SBT (Campbell & Bolton, 2005; Dorsey & Coovert, 2003). Use of mathematical modeling techniques to drive adaptive scenario generation and performance measurement would be a natural extension for this research. Specifically, because mathematical models reflect explicit, quantitative hypotheses about performance under different situational cues, they provide a mechanism for developing expert performance models as a basis for assessing and diagnosing individual performance.

The requirement for a closed loop, confirmatory measurement strategy is especially relevant considering the proliferation of primarily data-driven machine learning models and algorithms in SBT applications (Oswalt & Cooley, 2019). As illustrated with the event-based approach training, preplanned events, and measurement hooks can create the foundation for a confirmatory strategy. With advances in artificial intelligence and machine learning, there is an inherent risk in relying too heavily on data-driven recommendations for training content adaptations, or inferences about underlying trainee's skill levels. Without a theoretical framework to drive expectations about what data outputs to expect within the greater context of the training situation, this risk can lead to arbitrary decisions that are too specific to the particular context.

PRINCIPLE 3: ENSURE USEFULNESS OF MEASURES FOR EVALUATING TRAINING EFFECTIVENESS

A final guideline concerns the training and transfer validity of the SBT system as a whole. Ideally, performance measures will allow assessment of performance vis-à-vis specific training objectives, which should reflect real-world performance demands and be derived from job/task analyses. However, a potential limiting factor in the usefulness of performance measures for evaluating training effectiveness is the disconnect between the strategies used to assess SBT performance and those used to assess on-the-job performance. The former often employs finer-grained, objective measures, whereas the latter often uses broader, subjective measures (e.g., supervisor or multisource ratings). The resulting difference in the levels of analysis could limit the magnitude of the training validity coefficients that were observed to be useful for evaluating training effectiveness.

Use of performance measures consistent with the measurement framework described earlier (e.g., invariant to changing scenario contexts) may prove useful as a foundation for evaluating training effectiveness. Performance assessed using scenario-invariant measures takes on evaluative meaning only after considering the difficulty of the scenario content in which performance was observed. Thus, this framework allows for experimental manipulation of the difficulty levels of various situational elements to address how these elements influence immediate and long-term learning and performance. For example, assessment research suggests that highly discriminating test items with moderate (e.g., 50%) difficulty levels give the most diagnostic information about a person's actual trait level. The proposed framework allows for analogous research to address how situational parameters influence performance in the context of training, as well as training evaluation research to support inferences regarding long-term outcomes.

SUMMARY AND CONCLUSIONS

The purpose of this chapter is to describe the role of performance measurement in adaptive SBT contexts in light of emerging technological and methodological innovations. We have presented criteria for a confirmatory measurement framework to emphasize the necessity of sound performance measurement as the foundation for automated, adaptive training content and feedback delivery in SBT. Additionally, we have identified a number of dimensions along which performance measures can vary and the desirable characteristics of performance measures along each dimension to support the criteria for confirmatory performance measurement. Performance measures that conform to this framework are likely to provide high utility as a result of their diagnosticity, objectivity, unintrusiveness, comprehensiveness, and efficiency. However, perhaps the greatest advantage of implementing such measures is the ability to draw sound, causal inferences regarding relationships among scenario content, the latent psychological constructs targeted during training, and observable performance. The use of "valid" performance measures—that is, measures capable of transmitting the causal influence of the latent performance construct on observable performance—provides a necessary foundation for adaptive scenario modifications useful for iteratively assessing and correcting deficient skills as they change over the course of training. One effect of the rapid evolution of SBT systems has been the tendency to resort to ad hoc measurement strategies to reduce the large amounts of objective performance data resulting from these systems. We believe that this measurement framework provides a useful set of specific, quantifiable standards to counteract this trend, providing a key mechanism for improving long-term learning and retention.

REFERENCES

Ackerman, P. L., 1987. Individual differences in skill learning: An integration of psychometric and information processing perspectives. *Psychological Bulletin*, *102*, 3–27.

Alliger, G. M., & Katzman, S., 1997. When training affects variability: Beyond the assessment of mean differences in training evaluation. In J. K. Ford, S. W. J. Kozlowski, K. Kraiger, E. Salas, & M. S. Teachout (Eds.), *Improving training effectiveness in work organizations* (pp. 223–246). Mahwah, NJ: Lawrence Erlbaum Associates.

Annett, J., 1991. Skill acquisition. In J. E. Morrison (Ed.), *Training for performance: Principles for applied human learning* (pp. 13–52). Chichester: John Wiley and Sons.

Bagozzi, R. P., & Yi, Y., 1990. Assessing method variance in multitrait-multimethod matrices: The case of self-reported affect and perceptions at work. *Journal of Applied Psychology*, *75*, 547–560.

Bell, B., Johnston, J., Freeman, J., & Rody, F., 2004. STRATA: DARWARS for deployable, on-demand aircrew training. In the *Proceedings of the Interservice/Industry Training, Simulation & Education Conference* (pp. 1–9). Arlington, VA: National Training & Simulation Association.

Binning, J. F., & Barrett, G. V., 1989. Validity of personnel decisions: A conceptual analysis of the inferential and evidential bases. *Journal of Applied Psychology*, *74*, 478–494.

Bollen, K. A., & Curran, P. J., 2006. *Latent curve models: A structural equation perspective.* Hoboken, NJ: John Wiley and Sons.

Borman, W. C., Grossman, M. R., Bryant, R. H., & Dorio, J., 2017. The measurement of task performance as criteria in selection research. In J. L. Farr, N. T. Tippins, W. C. Borman, D. Chan, M. D. Coovert, R. Jacobs, P. R. Jeanneret, J. F. Kehoe, F. Lievens, S. M. McPhail, K. R. Murphy, R. E. Ployhart, E. D. Pulakos, D. H. Reynolds, A. M.

Ryan, N. Schmitt, & B. Schneider (Eds.), *Handbook of employee selection* (2nd Ed., pp. 429–447). Routledge.

Borsboom, D., Mellenbergh, G. J., & van Heerden, J., 2004. The concept of validity. *Psychological Review, 111*, 1061–1071.

Campbell, G. E., & Bolton, A. E., 2005. HBR validation: Integrating lessons learned from multiple academic disciplines, applied communities, and the AMBR project. In K. A. Gluck, & R. W. Pew (Eds.), *Modeling human behavior with integrated cognitive architectures: Comparison, evaluation, and validation* (pp. 365–396). Mahwah, NJ: Lawrence Erlbaum Associates.

Campbell, G. E., Buff, W. L., & Bolton, A. E., 2006. Viewing training through a fuzzy lens. In A. Kirlik (Ed.), *Adaptation in human-technology interaction: Methods, models, and measures* (pp. 149–162). Oxford: Oxford University Press.

Campbell, R. L., & DiBello, L., 1996. Studying human expertise: Beyond the binary paradigm. *Journal of Theoretical and Experimental Artificial Intelligence, 8*, 277–293.

Campbell, D. T., & Fiske, D. W., 1959. Convergent and discriminant validation by the multitrait-multimethod matrix. *Psychological Bulletin, 56*, 81–105.

Cannon-Bowers, J., & Bowers, C., 2009. Synthetic learning environments: On developing a science of simulation, games, and virtual worlds for training. In S. W. J. Kozlowski & E. Salas (Eds.), *Learning, training, and development in organizations* (pp. 250–282). Routledge.

Cannon-Bowers, J. A., & Bowers, C., 2010. *Serious game design and development: Technologies for training and learning*. Hershey, PA: IGI Global.

Chan, D., & Schmitt, N., 2002. Situational judgment and job performance. *Human Performance, 15*, 233–254.

Chatham, R. E., 2007. Games for training. *Communications of the ACM, 50*(7), 36–43.

Clement, J., 2021. *Serious games market revenue worldwide 2018–2024*. Statista. https://www.statista.com/statistics/733616/game-based-learning-industry-revenue-world/

Coovert, M. D., Craiger, J. P., & Teachout, M. S., 1997. Effectiveness of the direct product versus confirmatory factor model for reflecting the structure of multimethod-multirater job performance data. *Journal of Applied Psychology, 82*, 271–280.

Crocker, L., & Algina, J., 1986. *Introduction to classical and modern test theory*. New York: Holt, Rinehart, and Winston.

Cronbach, L. J., & Meehl, P. E., 1955. Construct validity in psychological tests. *Psychological Bulletin, 52*, 281–302.

Cronbach, L. J., & Snow, R. E., 1981. *Aptitudes and instructional methods: A handbook for research on interactions*. New York: Irvington.

DiCerbo, K., Shute, V., & Kim, Y. J., 2019. The future of assessment in technology rich environments: Psychometric considerations. Learning, design, and technology: An international compendium of theory, research, practice, and policy. In M. J. Spector, B. B. Lockee, & M. D. Childress (Eds.), *Learning, design, and technology: An international compendium of theory, research, practice, and policy* (pp. 1–21). New York: Springer International Publishing.

Domeshek, E., Ramachandran, S., Jensen, R., Ludwig, J., Ong, J., & Stottler, D., 2019. Lessons from building diverse adaptive instructional systems (AIS). In R. Sottilare, & J. Schwarz (Eds.), *Adaptive Instructional Systems: First International Conference* (pp. 62–75). Cham, Switzerland: Springer. https://www.stottlerhenke.com/solutions/education-and-training/intelligent-simulation-based-tutoring-nets-efficiencies-training-tactical-action-officers/

Dorsey, D. W., & Coovert, M. D., 2003. Mathematical modeling of decision making: A soft and fuzzy approach to capturing hard decisions [Special issue]. *Human Factors, 45*, 117–135.

Durlach, P. J., & Lesgold, A. M. (Eds.), 2012. *Adaptive technologies for training and education*. Cambridge: Cambridge University Press.

Dwyer, D. J., Fowlkes, J. E., Oser, R. L., & Lane, N. E., 1997. Team performance measurement in distributed environments: The TARGETs methodology. In M. T. Brannick, E. Salas, & C. Prince (Eds.), *Team performance assessment and measurement: Theory, methods, and applications* (pp. 137–154). Hillsdale, NJ: Lawrence Erlbaum Associates.

Edwards, J. R., & Bagozzi, R. P., 2000. On the nature and direction of relationships between constructs and measures. *Psychological Methods*, *5*, 155–174.

Embretson, S. E., 1997. Multicomponent item response models. In W. J. Van der Linden, & R. K. Hambleton (Eds.), *Handbook of modern item response theory* (pp. 305–322). New York: Springer-Verlag.

Embretson, S. E., 1998. A cognitive design system approach for generating valid tests: Approaches to abstract reasoning. *Psychological Methods*, *3*, 300–396.

Embretson, S. E., 2006. The continued search for nonarbitrary metrics in psychology. *American Psychologist*, *61*, 50–55.

Embretson, S. E., & Reise, S. P., 2013. *Item response theory*. London, UK: Psychology Press.

Ericsson, K. A., Hoffman, R. R., Kozbelt, A., & Williams, A. M. (Eds.), 2018. *Cambridge handbooks in psychology. The Cambridge handbook of expertise and expert performance*. Cambridge: Cambridge University Press.

Fletcher, J. D., 2009. Education and training technology in the military. *Science*, *323*(5910), 72–75.

Fowlkes, J. E., Dwyer, D. J., Oser, R. L., & Salas, E., 1998. Event-based approach to training (EBAT). *International Journal of Aviation Psychology*, *8*, 209–221.

Henson, R. A., Templin, J. L., & Willse, J. T., 2009. Defining a family of cognitive diagnosis models using log-linear models with latent variables. *Psychometrika*, *74*(2), 191.

Hutchins, E., 1995. *Cognition in the wild*. Cambridge, MA: MIT Press.

Interservice/Industry Simulation, Training, and Education Conference, 2021. *Serious games showcase and challenge: Serious games*. Arlington, VA: National Simulation Training Association. http://sgschallenge.com/serious-games

James, L. R., 1982. Aggregation bias in estimates of perceptual agreement. *Journal of Applied Psychology*, *67*, 219–229.

Johnson, W. L., 2010. Serious use of a serious game for language learning. *International Journal of Artificial Intelligence in Education*, *20*(2), 175–195.

Johnson, D., Deterding, S., Kuhn, K. A., Staneva, A., Stoyanov, S., & Hides, L., 2016. Gamification for health and wellbeing: A systematic review of the literature. *Internet Interventions*, *6*, 89–106.

Johnson, W. L., & Lester, J. C., 2016. Face-to-face interaction with pedagogical agents, twenty years later. *International Journal of Artificial Intelligence in Education*, *26*(1), 25–36.

Johnston, J., Sottilare, R., Sinatra, A. M., & Burke, C. S. (Eds.), 2018. *Building intelligent tutoring systems for teams: What matters*. Bingley: Emerald Publishing Ltd.

Katz, I. R., LaMar, M. M., Spain, R., Zapata-Rivera, J. D., Baird, J. A., & Greiff, S., 2017. Validity issues and concerns for technology-based performance assessments. In R. Sottilare, A. Graesser, X. Hu, & G. Goodwin (Eds.), *Design recommendations for intelligent tutoring system: Assessment methods* (Vol. 5, pp. 209–224). Aberdeen Proving Grounds, MD: U.S. Army Research Laboratory.

Kulik, J. A., & Fletcher, J. D., 2016. Effectiveness of intelligent tutoring systems: A meta-analytic review. *Review of Educational Research*, *86*(1), 42–78.

Laamarti, F., Eid, M., & El Saddik, A., 2014. An overview of serious games. *International Journal of Computer Games Technology*. https://doi.org/10.1155/2014/358152

Landy, F. J., & Farr, J. L., 1980. Performance rating. *Psychological Bulletin*, *87*, 72–107.

Lave, J., 1988. *Cognition in practice: Mind, mathematics and culture in everyday life*. Cambridge: Cambridge University Press.
Levy, R., & Mislevy, R. J., 2004. Specifying and refining a measurement model for a computer-based interactive assessment. *International Journal of Testing, 4*, 333–369.
Ma, W., Adesope, O. O., Nesbit, J. C., & Liu, Q., 2014. Intelligent tutoring systems and learning outcomes: A meta-analysis. *Journal of Educational Psychology, 106*(4), 901.
Maheu-Cadotte, M. A., Cossette, S., Dubé, V., Fontaine, G., Lavallée, A., Lavoie, P., Mailhot, T., & et al., 2020. Efficacy of serious games in healthcare professions education: A systematic review and meta-analysis. *Simulation in healthcare: Journal of the Society for Simulation in Healthcare*. https://journals.lww.com/simulationinhealthcare/Abstract/9000/Efficacy_of_Serious_Games_in_Healthcare.99412.aspx
Mangos, P. M., Campbell, G., Lineberry, M., & Bolton, A., 2012. Emergent assessment opportunities: A foundation for configuring adaptive training environments. In P. J. Durlach, & A. M. Lesgold (Eds.), *Adaptive technologies for training and education* (pp. 222–235). Cambridge: Cambridge University Press.
Mangos, P. M., & Johnston, J. H., 2009. Performance measurement issues and guidelines for adaptive, simulation-based training. In D. A. Vincenzi, J. A. Wise, M. Mouloua, & P. A. Hancock (Eds.), *Human factors in simulation and training* (pp. 301–320). Boca Raton: CRC Press.
McCarthy, J. E., 2008. Military applications of adaptive training technology. In M. D. Lytras, D. Gasevic, & W. Huang (Eds.), *Technology enhanced learning: Best practices* (Vol 4, pp. 304–347). Hershey, PA: IGI Global.
McCarthy, J. E., Pacheco, S., Banta, H. G., Wayne, J. L., & Coleman, D. S., 1994, November. The radar system controller intelligent training aid. In the *Proceedings of the 16th Interservice/Industry Training Systems and Education Conference* (pp. 1–10). Arlington, VA: National Training & Simulation Association.
McCarthy, J. E., Wayne, J. L., & Deters, B. J., 2013. Supporting hybrid courses with closed-loop adaptive training technology. In A. Peña-Ayala (Ed.), *Intelligent and adaptive educational-learning systems: Achievements and trends* (pp. 315–337). Berlin, Heidelberg: Springer.
Messick, S., 1995. Validity of psychological assessment. *American Psychologist, 50*, 741–749.
Mislevy, R., 1995. Probability-based inference in cognitive diagnosis. In P. Nichols, S. Chipman, & R. Brennan (Eds.), *Cognitively diagnostic assessment* (pp. 43–71). Hillsdale, NJ: Lawrence Erlbaum Associates.
Mislevy, R. J., & Wilson, M., 1996. Marginal maximum likelihood estimation for a psychometric model of discontinuous development. *Psychometrika, 61*, 41–71.
Morgeson, F. P., & Campion, M. A., 2000. Accuracy in job analysis: Toward an inference-based model. *Journal of Organizational Behavior, 21*, 819–827.
Murray, W. R., 1991. *An endorsement-based approach to student modeling for planner-controlled intelligent tutoring systems* (AL-TP-1 991-0030). Brooks Air Force Base, TX: Armstrong Laboratory.
Nunnally, J. C., 1978. *Psychometric theory*. New York: McGraw-Hill.
Ohlsson, S., 1996. Learning from performance errors. *Psychological Review, 103*, 241–262.
Olson-Buchanan, J. B., & Drasgow, F., 1999. Beyond bells and whistles: An introduction to computerized assessment. In J. B. Olson-Buchanan, & F. Drasgow (Eds.), *Innovations in computerized assessment* (pp. 1–6). Mahwah, NJ: Lawrence Erlbaum Associates.
O'Neil, H. F., Baker, E. L., Wainess, R., Chen, C., Mislevy, R., & Kyllonen, P., 2004. *Plan for the assessment and evaluation of individual and team proficiencies developed by the DARWARS Environments*. Sherman Oaks, CA: Advance Design Information.

Oswalt, I., & Cooley, T. (2019). Simulation-based training's incorporation of machine learning. *Proceedings of the 2019 Interservice/Industry Training, Simulation, and Education Conference (I/ITSEC)*. Arlington, VA: National Training and Simulation Association.

Pirolli, P., & Wilson, M., 1998. A theory of the measurement of knowledge content, access, and learning. *Psychological Review, 105*, 58–82.

Ployhart, R. E., 2006. The predictor response process model. In J. A. Weekly, & R. E. Ployhart (Eds.), *Situational judgment tests: Theory, measurement, and application* (pp. 83–106). Mahwah, NJ: Lawrence Erlbaum Associates.

Roberts, B., & Diller, D., 2014. Development methods. In T. Hussain, & S. Coleman (Eds.), *Design and development of training games: Practical guidelines from a multidisciplinary perspective* (pp. 464–475). Cambridge: Cambridge University Press.

Rothrock, L., 2001. Using time windows to evaluate operator performance. *International Journal of Cognitive Ergonomics, 5*, 1–21.

Sanchez, J. I., & Levine, E. L., 2000. Accuracy or consequential validity: Which is the better standard for job analysis data? *Journal of Organizational Behavior, 21*, 809–818.

Schoenfeld, A. H., Smith, J. P., & Arcavi, A., 1993. Learning: The microgenetic analysis of one student's evolving understanding of a complex subject matter domain. In R. Glaser (Ed.), *Advances in instructional psychology* (Vol. 4, pp. 55–177). Hillsdale, NJ: Lawrence Erlbaum Associates.

Smith, P. A., & Bowers, C., 2019. Serious games advancing the technology of engaging information. In M. Khosrow-Pour (Ed.), *Advanced methodologies and technologies in media and communications* (pp. 153–164). Hershey, PA: IGI Global.

Snow, R. E., 1994. Abilities in academic tasks. In R. J. Sternberg, & R. K. Wagner (Eds.), *Mind in context: Interactionist perspectives on human intelligence* (pp. 3–37). Cambridge: Cambridge University Press.

Snow, R. E., & Lohman, D. F., 1984. Toward a theory of cognitive aptitude for learning from instruction. *Journal of Educational Psychology, 76*, 347–376.

Snow, R. E., & Lohman, D. F. (1993). Cognitive psychology, new test design, and new test theory: An introduction. In N. Fredericksen, R. J. Mislevy, & I. I. Bejar (Eds.), *Test theory for a new generation of tests* (pp. 1–17). Hillsdale, NJ: Erlbaum.

Sottilare, R., 2013. *Special report: Adaptive intelligent tutoring system research in support of the army learning model research outline* (ARL-SR-0284). Aberdeen Proving Ground, MD: Army Research Laboratory.

Spain, R. D., Priest, H. A., & Murphy, J. S., 2012. Current trends in adaptive training with military applications: An introduction. *Military Psychology, 24*(2), 87–95.

Stephanidis, C., Salvendy, G., Antona, M., Chen, J. Y. C., Dong, J., Duffy, V. G., Fang, X., & et al., 2019. Seven HCI grand challenges. *International Journal of Human–Computer Interaction, 35*(14), 1229–1269.

Van Lehn, K., 1996. Cognitive skill acquisition. *Annual Review of Psychology, 47*, 513–53.

Zachary, W., Cannon-Bowers, J., Bilazarian, P., Krecker, D., Lardieri, P., & Burns, J., 1999. The Advanced Embedded Training System (AETS): An intelligent embedded tutoring system for tactical team training. *International Journal of Artificial Intelligence in Education, 10*, 257–277.

Zachary, W., Santorelli, T., Lyons, D., Bergondy, M., & Johnston, J. H., 2001. Using a Community of intelligent synthetic entities to support operational team training. In the *Proceedings of the Computer Generated Forces Conference* (pp. 215–224). Orlando, FL.

11 Scoring Simulations with Artificial Intelligence

Carter Gibson, Nick Koenig,
Joshua Andrews, and Michael Geden

CONTENTS

The world is currently living through what some have called the fourth industrial revolution (Schwab, 2017). In this framework, the first three revolutions related to water and steam power, electric power, and electronics and information technology to automate production, respectively. This fourth industrial revolution is characterized by technologies that blur the lines among the physical, digital, and biological spheres and include such advances as the Internet of Things, 3D printing, nanotechnology, quantum computing, and, most importantly for this chapter, artificial intelligence (AI). Use of AI has increased in the last decade and is a large driver of innovation (Rust & Huang, 2014). In obvious and less obvious ways, AI is already impacting many areas of daily life (Poola, 2017). AI gets attention for high-profile uses, such as in autonomous vehicles or how it's proving more accurate than expert

DOI: 10.1201/9781003401360-11

radiologists (e.g., Hosny et al., 2018; Schwarting, Alonso-Mora, & Rus, 2018). AI is also being used in smaller ways to subtly improve areas of modern life, such as in unlocking a phone with facial recognition, giving grammar advice for writing, filtering emails as spam, giving users personalized ads when browsing the web, or helping banks identify fraud (e.g., Ryman-Tubb, Krause, & Garn, 2018). Already ubiquitous, AI, and the fourth industrial revolution more broadly, promises to impact almost every field and job, including simulations and training.

Crucial to understanding how AI is changing the field of simulations and training is first examining where simulations started. Simulations are built to mimic or reproduce a specific context. The typical goal is to train or measure in an environment at lower risk than learning on the job. For example, it's not best practice for a pilot to learn how to perform a difficult maneuver or a surgeon to try out a new technique in the high-stakes context of their actual work. Perhaps an organization simply wants to train a leader to be a better communicator or give higher-quality performance appraisals. Simulations allow for structured, safe, and deliberate practice in a lower-stakes environment to develop skills that will transfer to the higher-stakes circumstances in the workplace.

Simulations exist across a wide continuum, from highly realistic and technical (e.g., a flight simulator that accurately reproduces all of the controls in a plane) to more conceptually representative, like a paper-and-pencil activity. At a high level, the concept of fidelity can be demonstrated by how closely a simulation can recreate the appearance and potential dynamics of the simulated scenario. A realistic flight simulator would be high fidelity, whereas the paper-and-pencil task would be lower fidelity. Though some have criticized the concept of fidelity as poorly defined (e.g., Norman et al., 2018), the term is still useful for framing thinking about simulations in the field. High-fidelity simulations may create colossal amounts of data. In the context of a flight simulator, the computer could record what inputs are made, how quickly they're made, how much pressure is applied, where the individual in the simulation is looking, and vital signs of the participant. With many of these issues, a novice may not be able to comprehend the importance of so many measures, but experts are able to take all of these inputs and provide specific feedback, advice, or general conclusions that can improve the performance of the participant in the exercise. On the other end of the fidelity spectrum, a training exercise could have a leader going through a developmental assessment center, where they work through an in-basket task writing emails, solving problems, or organizing their calendar (Motowidlo, Dunnette, & Carter, 1990). Again, a large amount of data is being created and historically has relied on expert human judgment to determine the quality of performance across the range of constructs being measured.

How can simulation performance be accurately scored across varying levels of fidelity? High-fidelity simulations may have a large number of variables with varying degrees of importance, but performance in low-fidelity simulations may still be difficult to measure and quantify empirically. For example, how do you quantify the outcomes of the simulation designed to score a proficiency exam or determine who among a group of applicants should be hired for a position? While much research has gone into measurement, less work exists to combine these sources of data to predict important criteria (Sydell et al., 2013). We introduce the concept of fidelity to show that AI can be useful across a wide range of types of simulations, regardless of the type of data that is generated.

The issue of what to do with all of the data generated by simulations and how to score them in reliable and valid ways are where AI has the potential to significantly impact how simulations are used. And while AI is changing the field, these changes were not unanticipated. That is, scholars have pointed to this future long before technology had the models and processing power to bring them about. Scholars pointed to two major predictive improvements in the area of scoring simulations: (1) combining information across item types and assessment experiences, and (2) leveraging the power of increasingly large sample sizes (and data sources) (Sydell et al., 2013). AI is following through with these promises and has significant implications for training and simulations, specifically because of its ability to automate scoring of data sources that were previously impossible to score by machines (or even, sometimes, by humans) and through more accurate models of scoring.

Of course, like many new technologies, AI won't destroy and replace what came before it, but rather provides a new tool. This chapter discusses what AI can do, what it can't, and offers suggestions for users to start incorporating it into their own work. We believe AI will change the way simulations are used to select and train individuals by allowing simulations to be more easily scaled, automating previously manual scoring approaches, and helping experts design better and more predictive weighting schemes to create more optimal scoring models of complex behaviors.

ARTIFICIAL INTELLIGENCE AND REPRODUCING EXPERT RATINGS

Automation within the field of human factors has been an area of interest for years. Since their invention, computers have become exponentially more powerful and cheaper following a pattern called Moore's Law (Moore, 1965). Several books within the field of human factors have been dedicated to the subject (e.g., Parasuraman & Mouloua, 1996; Mouloua & Hancock, 2020), and hardware and software are consistently making more complex automation possible. Mosier and Manzey (2020) discuss automated decision support systems (DSSs) and the value these systems provide in reducing user bias across a variety of industries. But what if it was possible to use expert human judgment to train a system and remove the experts from the loop?

In many disciplines, expert human judgment is leveraged for decision-making in complex tasks. From doctors reviewing patient MRIs to assessors evaluating assessment center candidates, expert judgment has been shown to outperform novice judgment (Salkowski & Russ, 2018; Schleicher et al., 1999). Wickens et al. (2016) described decision-making as including the following key features: uncertainty, time, and expertise. These features are certainly present in many high-stakes environments where decisions must be made. The complexity of human decision-making is ripe ground for AI as advances in hardware and software make replicating complex human decisions more feasible.

While AI from the 1950s to the 1980s often involved explicitly programming symbolic representations of logic and decision-making into the computer, coined

"Good Old-Fashioned Artificial Intelligence" (Haugeland, 1985, pg. 112), more recent conceptualizations of AI have leveraged the idea of machine learning and the understanding that the software can program itself if given enough data. The availability of more processing hardware at lower costs has allowed for more and more model parameters and the introduction of deep learning. Model parameters, which are defined as variables internal to the model that are estimated using data, are extremely important to all of machine learning. An example of a model parameter in a linear regression is a beta weight, which is estimated by optimizing for best fit. In the field of natural language processing, parameters in the order of magnitude of several thousand via bag-of-words were considered large just a decade or two ago. Now, we have architectures like BERT with 345 million parameters (Devlin et al., 2019) and, more recently, GPT-3 and its 175 billion parameters (Brown et al., 2020). These algorithms consist of an input layer, where data is fed into the model, several hidden layers, and an output layer where a specific prediction is made. This is considered deep learning because the neural network has several hidden layers.

The increase in the size of the parameter space leads to more and more complex representations of the data via the layers and internal neurons' ability to extract very specific subsets of information from the input data. These complex algorithms make it possible to replicate human judgment on natural language-based tasks, such as evaluating the quality of an essay, a written job simulation, or even a job candidate's interview response. This approach can be leveraged for automating simulations and systems where trained professionals currently need to evaluate natural language-based responses and make decisions. The need to automate the understanding of human language exists across several domains. In the medical field, this technology has been used to take clinical notes and predict hospital readmission (Huang, Altosaar, & Ranganath, 2020) and patient diagnoses from electronic health records (Franz, Shrestha, & Paudel, 2020). In the field of human resources, natural language processing has been used to automate job analysis (Mracek et al., 2021) and the scoring of work simulations (Tonidandel et al., 2020). In the following pages, we will outline a process for developing an algorithm that can replicate trained subject matter experts when it comes to evaluating work simulations and interview responses based on the written or spoken English language.

Assessment centers, in-person or virtual, often have several writing exercises in the form of in-baskets that require participants to respond to an email from a peer, boss, or customer. These unstructured text responses can then be evaluated for competencies relevant to success in the role. Extracting scores from these responses requires no small investment in resources. The responses can be lengthy and accurate ratings require review by trained evaluators, knowledge workers who receive substantial compensation for their expertise. In addition, the process itself is extremely repetitive, which can lead to a vigilance decrement (Thomson et al., 2015) extremely common in such work. Transformer-based NLP algorithms are most effective for use with such long-form responses. Simple chat simulations or short answer responses would likely not benefit from the added complexity these algorithms provide.

TRADITIONAL APPROACH TO SCORING OPEN-ENDED CONTENT: RATER TRAINING

The traditional method for scoring open-ended content is an expensive process in terms of both time and money. Almost by definition, the SMEs qualified to provide ratings on a complex subject are going to be both busy and expensive. Using them to rate and evaluate a large sample of any product will be challenging, whether in an academic or applied context. In the context of an organization, maintaining a stable group of trained judges can be challenging as people leave the organization or their original role. Furthermore, if ratings of a particular product are needed in a timely fashion, it may be difficult to get quick work from a judge. Plus, several of the steps require the judges to coordinate and discuss a shared frame of reference.

AI isn't going to remove humans, or in this case experts, on a given topic, but rather change their role. Using AI won't be as simple as just applying an algorithm and solving a problem; the tools described in this chapter still require significant effort to ensure appropriate data is being fed into the system. If the end goal is a program that can rate a work product, such as a writing sample, a large pool of writing samples as well as expert ratings of these samples would be needed. The way these expert ratings are collected, even for the purposes of building an AI model, is going to look a lot like it has traditionally. A large pool of literature dating back several decades exists on how to train raters most effectively (e.g., Bernardin & Buckley, 1981), and all of these proven steps still need to be followed. At its core, what data scientists are trying to do is translate a large amount of text or other data into a useful and reliable score in a standardized way. Once this training is completed, a set of judges with a shared mental model of the constructs being assessed will have been created. Deep learning can then be used to recreate these expert ratings, and ultimately be able to rate new writing samples independent of human judges in a reliable and valid way.

While many approaches could be used, frame-of-reference training is perhaps one of the most popular (Roch et al., 2012). Conceptually, the goal of this training is to get all judges onto the same metric to minimize differences due to judges with unique ideas about what is important when rating task performance. The first step is to create rating benchmarks and standards for all variables to measure. For example, in a writing sample, it's possible to rate overall quality and also more specific constructs such as grammar, vocabulary, and style. To create these benchmarks, the collection of writing samples would be reviewed to find examples of various anchors on the scale, such as at 1, 3, and 5 on a 5-point scale. This process would be repeated for each of the variables the judges will rate. Once the benchmarks have been built, all the judges meet to work through a small sample of cases. For each of these cases, they would provide ratings for each variable and then discuss them until they came to a consensus (i.e., shared-mental model) on what a "5" looks like, what a "3" looks like, etc. Once the judges appear to be rating in a consistently similar way, the judges would proceed to review the entire sample and rate all cases. For the purposes of AI, several hundred rated samples would be needed, typically with at least three judges to ensure confidence that the "true" rating for each writing sample has been obtained. There may be a need to have periodic meetings and calibrations to account for things like rater drift (Harik et al., 2009). Once the judges have rated several hundred cases, an important check is to review inter-rater reliability, or the degree to which raters are consistently giving the same ratings on each variable of interest for each sample of writing. See Hall and Brannick (2008) for a good review of the various considerations when choosing a metric for inter-rater reliability, and Gibson and Mumford (2013) for an example of this rater-training process used in practice.

Of course, many judgment calls will need to be made about specifics in the process. We prefer to be conservative about ratings to ensure the highest quality data to train the model. For example, in some contexts, once judges have established sufficient agreement, only one judge may be needed to rate new data. This is more likely to be viable in cases with extremely high inter-rater agreement or when rating more concrete variables (e.g., a construct that has been very specifically operationalized). In other cases, raters may be allowed to have different ratings on a specific instance as long as, on average, they're in agreement (e.g., one rater gives a "2" and another gives a "5" on the same product). Given the relative newness of using these algorithms, we have opted to be more conservative, such as expecting ratings to have evaluations within one point of each other on a 1-to-5 rating scale. Thus, if the first rater evaluates a work sample as a "4," the third rater would need to have an evaluation of 3, 4, or 5, for that same work sample; otherwise, they would need to meet to discuss and draw a shared conclusion.

Now that labels have been created, the next step is identifying the algorithm to be used. While bag-of-words (Zhang, Jin, & Zhou, 2010) and long short-term memory recurrent neural networks (Hochreiter & Schmidhuber, 1997) are reasonable methods to use, they both have shortcomings that are beyond the scope of this chapter. Note that many of the approaches described in this chapter are new and are still actively being developed, so rather than dive into a technical guide of a given method, it is more useful to review broadly the considerations when choosing an analytic technique. More recently, Vaswani et al. (2017) introduced the transformer architecture, and Devlin et al. (2019) and Liu et al. (2019) expanded upon the transformer architecture to create the current state-of-the-art model: the bidirectional encoder representations architecture (BERT). This architecture has the capacity to take in embeddings as representations of the words and adjust those representations depending on the words coming both before and after it. These word embeddings hold information about the word or token's relationship to other words. This makes it possible for the language model to understand that in most cases, the words "customer" and "client" are very similar. While this architecture was originally very successful at predicting masked words (i.e., what a hidden word was most likely to be given surrounding words) and similarly creating state-of-the-art language translations (i.e., Google Translate), the researchers quickly realized it could also excel in downstream tasks, like question answering, predicting sentence sentiment, and more. These downstream tasks are tasks that the model wasn't explicitly trained for, but could be trained to do with new data and sufficient computational power. Because of its ability to produce accurate results on a number of complex natural language processing tasks, this architecture is ideal for the downstream task of replicating human evaluations on complex human behaviors within written work simulations.

The Architecture

BERT is freely and readily available via Hugging Face (huggingface.co). There are many variants to choose from, but we recommend RoBERTa's base model. There is also the need for hyperparameter tuning of the model. Hyperparameters are parameters outside of the algorithm itself that control how the algorithm performs. In neural networks, these can be things like learning rate, percentage of nodes/neurons that are dropped (dropout), optimization function, batch size, and more. We found success trying differing learning rates and dropout rates while using the largest batch size that could fit into memory.

The Data

An important point about machine learning and deep learning in general is that the algorithms are extremely powerful when it comes to learning from the data they were trained on. For this reason, users need some form of a holdout set to ensure predictions will generalize on responses outside of the specific responses the algorithm was trained on. When hyperparameter tuning, users will want to use a cross-validation strategy to prevent inadvertently overfitting to the holdout set. Hyperparameters, unlike model parameters, are parameters that need to be set outside of the model itself, but can have an impact on the quality of prediction. For this reason, it may make sense to test a variety of combinations to identify the best set for your given data. One common hyperparameter for the transformer architecture is learning rate, which is simply the size of the update step used for moving along the gradient. Another common hyperparameter is neuron dropout. This is the proportion of neurons within the hidden layers that are randomly set to zero. This is an extremely effective regularization technique for deep learning. First, test different hyperparameters on a *k*-fold cross-validation sample set. A *k*-fold cross-validation set involves slicing your data into *k*-folds; common *k*'s used would be 5 or 10. For example, given a dataset with a sample size of 1,000 and a 5-fold cross-validation, 800 responses and labels would be used to train the algorithm, 200 responses would be predicted, and then the process would be repeated with each set of 200 being used as the holdout set. When hyperparameters are found that provide satisfactory results, train the final model on a training set consisting of roughly 80% of the data and evaluate the model's performance on a holdout set of 20% of the data.

We also recommend data stratification, which consists of identifying differences in the data and ensuring those differences are consistent across the folds. The outcome/label is that it is important to ensure it is stratified across folds. Other things to consider may be the length of the responses, the population the sample was taken from, and any other parameters that may differ within the data. This stratification ensures that the model is consistently being trained and evaluated on very similar data.

An example we have used this on was virtual assessment center in-basket email responses used as part of a multi-method assessment for job selection. Candidates responded to a fictitious email from a colleague inquiring about a problem they were facing and asking for guidance on how to proceed. The candidate was asked to respond to the email with recommendations for handling the specific situation. This response was evaluated on a number of competencies, from effective communication to an ability to drive results. One thousand responses were labeled by two trained subject matter experts on each of the competencies operationalized using predetermined behaviorally anchored rating scales, and the two raters had to come to a consensus on the competency rating. The responses and labels were then stratified across label distributions into both 5-folds and a final unrelated 80/20 split for final algorithm training.

Several different dropout rates ranging from 0.05 to 0.15 and several different learning rates ranging from 1-e3 to 1-e6 were tested. The means and standard deviations of the correlations along with the means and standard deviations of the mean squared errors (MSEs) on the holdout folds were compared to one another and a final set of hyperparameters was chosen. MSE was chosen as the optimization function because our specific purpose involved a regression-based output. The best hyperparameters were then used to train a final model on the original 80/20 stratified split.

Output

The final result is a deterministic algorithm that can be used to make predictions on new work samples within milliseconds of when the candidates produce them. The algorithm is deterministic in the sense that, given identical inputs, it will always produce an identical output. This differs significantly from the stochastic nature of the deep learning training process, where given the way the model is trained each time you retrain it, you will end up with different parameter weights and thus different predictions. A current paper by some of the authors (Thompson et al., forthcoming) found correlations between SMEs and the algorithm predictions that averaged above 0.84 on seven separate competency/work simulations. We also found that, on average, predictions on the competencies were within one point of the consensus (on a 1–5 scale) SME evaluations 91% of the time and within 0.5 of the consensus SME evaluations 66% of the time, providing evidence that the algorithms are consistently and accurately replicating the SMEs' evaluation of the job candidate on these job-relevant competencies. In an effort to examine how good or bad these hit rates were, the algorithm was compared to the pre-consensus SMEs. This evaluation found that 75% of the time the SMEs were within one point of each other on a rating before consensus. This research suggests that not only is the algorithm evaluating the responses very similarly to the consensus rating provided by the two SMEs, but it is also producing more consistent ratings than any one SME.

Other Considerations

As was mentioned earlier, there are other options for replicating work simulation evaluations on natural language data. Our research found that the transformer architecture outperforms the bag-of-words and LSTM architectures using 25% of the sample size. With a training set of 250 responses, we found that the evaluations on the holdout of 250 responses were more accurate than when using 750 responses to train the bag-of-words model. This is almost certainly a result of the transfer learning these transformers provide. As briefly discussed earlier, the transformer architecture comes with data built into it via its language model. It naturally has a vocabulary and representations of that vocabulary from the original purpose of the architecture, which was to predict masked words and upcoming sentences. This encoded information allows the transformer architecture to make robust and generalizable predictions on several downstream tasks with fractions of the data of more naïve architectures/implementations.

While the above may not provide a step-by-step review of how to use the described tools, it should provide a conceptual overview of the process by which work simulations can be evaluated using SMEs and leveraging the transformer architecture to produce highly accurate predictions on never-before-seen responses to the same work simulation, effectively removing the human rater from the loop and automating what was once considered an extremely complex task that required human expert judgment and decision-making.

SCORING ACTIONS IN SIMULATED ENVIRONMENTS

Beyond unstructured text (i.e., text that does not have a predefined format), simulations capture detailed data about an individual's actions within the environment, including their motions, interactions with objects, and contextual information (e.g., time stamp, NPC). Event and motion data could have transformative potential and utility for scoring performance and providing feedback in virtual simulations. Actions within simulations are often logged by the software in a way that would be impossible for human raters to comprehend. Simulations produce a rich source of data on individuals, providing a considerable opportunity for training AI/ machine learning models to identify new metrics and features relevant to success. Nonetheless, this source of data also comes with additional complexity and considerations that makes it difficult to structure and analyze.

TRADITIONAL APPROACHES TO SCORING SIMULATIONS

Scoring and rating simulations still rely heavily on SMEs to generate scoring metrics and provide trainee feedback. The current gold standard for scoring such simulations is to leverage SMEs to develop a mapping from an individual's actions to the quality of their performance within the simulation, typically by providing ratings along the relevant dimensions of interest for the virtual simulation (Boyle et al., 2018; Oquendo et al., 2018). This is often performed during validation of a training simulation or when the simulation is used for evaluation purposes. Trained raters then use assessment tools (e.g., rating rubrics, checklists) to score and provide feedback to trainees.

Nonetheless, scoring approaches that use SMEs have a focused perspective that often uses only a small amount of the available simulation data. SMEs are especially valuable because they can identify and measure complex and abstract behaviors and constructs within simulated environments, but often at a high cost due to the required expertise and training. The reliance on SMEs for scoring and rating simulations is not cost-effective in providing scalable feedback and high-fidelity training to enable widespread adoption. AI/machine learning techniques are particularly good at identifying patterns and therefore could help supplement traditional scoring approaches with a scalable alternative. Expert ratings for unstructured, non-text data would be invaluable for training an AI to identify mistakes and potential errors in real time, which in turn would enhance feedback and skill acquisition.

An alternative approach for more scalable scoring is to produce simple metrics based on SME domain knowledge that can be easily calculated in an automated fashion. This manual feature engineering typically results in a small number of easily interpretable metrics. For example, the Fundamentals of Robotic Surgery (FRS) has offered standardization for scoring using a set of metrics, such as time-to-completion and deviation in cutting from a prespecified region. It is also commonly employed in game-based assessments using evidence-centered design (ECD), which specifies relationships between actions in a simulation with concepts the student is trying to learn (Mislevy, Steinberg, & Almond, 2003). This approach, while scalable and simple to deploy, suffers from multiple drawbacks. First, these methods are often poor indicators of actual performance on these tasks (Mills et al., 2017). Additionally, the extraction of SME knowledge requires the use of time-consuming methods, such as cognitive task analyses, to derive the simple scoring system. Finally, manually engineered metrics are often highly specified in a context, making generalization to new scenarios challenging. For example, we would likely observe large differences between ideal metrics for scoring heart surgery compared to bone surgery.

Ideally, simulations should provide a scalable method for the administration and scoring of a scenario relative to the size of the trainee body. The point of simulations is to provide easy and safe practice for skills training, where the quality of learning during simulations is dependent upon the quality of feedback provided to the learner. Research has demonstrated that machine learning-aided skills evaluation is a scalable and automated means for measuring and collecting data on multi-dimensional evaluation constructs (Vedula, Ishii, & Hager, 2017). However, as we note, the applicable approaches are dependent upon the collection and structuring of training data, which often includes a time-based component.

Data Representations for Modeling Simulations

Virtual simulations generate detailed logs recording the actions made by the user and events that occurred within the virtual environment. These logs provide a rich source of information about the user; however, they are typically stored in formats unsuited for direct application in machine learning models. The logs contain information unrelated to modeling goals, are sampled at excessively high frequencies (e.g., location at each millisecond), and are stored in a representation that is suboptimal for direct use in modeling. The first step of the machine learning pipeline is to preprocess this data into a format amenable to analysis. Simulators sample at a high frequency, as it is trivial to record the data and critical to capture every relevant action taken by the user, causing data files to quickly become extremely large (e.g., gigabytes per person). The high-frequency sampling rate causes many data points to be redundant, since the time between samples is so low that few actions have been performed. Down-sampling the data while ensuring critical actions are still recorded can dramatically speed up modeling time at little to no cost to predictive performance. It can also be necessary to adjust the sampling rate of data when merging across multiple sources, such as eye-trackers, videos, and tool motion (Vedula, Ishii, & Hager, 2017). The multi-modal data streams may be collected at different sampling rates through separate software, requiring adjusting the sampling rate of the disparate data streams to align time stamps for merging. Next, the log data is transformed to remove noise and make it easier for the model to learn the desired relationships. Transformations can occur across two axes: static and temporal. Static transformations translate representations on the observation level (i.e., each time point) from the goal of efficient storage for the simulator software to being more closely aligned with the objective of the model. For example, in game-based learning environments, it can be helpful to encode the user's goals, accomplishments, actions (e.g., talking to a non-playable character or NPC), and the entities with which they are performing the actions (e.g., the name of the NPC) (Geden et al., 2020; Min et al., 2017). It can also be useful for defining complex actions, such as the type of strokes made with a surgical tool in surgical simulations (Ahmidi et al., 2015). Additionally, temporal transformations remove trends and seasonal changes from the time series to remove undesirable artifacts from the data. A couple of common temporal transformations are taking the moving average, differencing, and detrending sessional components (Wei, 2006).

Manual feature engineering can also be used to improve the performance of a model by providing it with an SME's heuristic for interpreting the environment. While certain models are able to automatically generate features from raw data (e.g., deep neural networks), these approaches require large amounts of data to learn these complex relationships. SMEs can provide a curated set of features relevant to the task, allowing for the model to focus on the mapping from the heuristics to the outcome variable without having to learn the intermediate representation (Vedula, Ishii, & Hager, 2017; Uzuner, 2009; Kuhn & Johnson, 2019; Krajewski et al., 2009; Garla & Brandt, 2012). While potentially effective, manual feature engineering is a time-consuming and domain-specific process requiring SMEs to encode their knowledge, illustrating the continuing importance of SMEs in the development of new models even as AI tools are sought to replace them.

Machine Learning Methods for Scoring Simulations

A wide breadth of machine learning models have been successfully applied for scoring simulations across diverse domains such as education, medical, and transportation settings (Anh, Nataraja, & Chauhan, 2020; Henderson et al., 2020; Beninger et al., 2021). The diversity of machine learning models is partly fueled by the No Free Lunch theorem (Wolpert & Macready, 1997), which states that there is no single "best" model to use across all circumstances, requiring the researcher to explore multiple methods and tailor their solution to the structure that is unique to their problem. This makes it impossible to provide a simple prescription for the selection of a model, as it may depend upon a number of factors, including the volume of data available, the form that the data is represented in, the type and number of criterion variables (i.e., classification, regression), and the unique structure of the task. For example, the transformer method described in the previous section has been extremely successful when handling text data; however, it is not applicable to tasks without a sequential component (e.g., credit loans) or non-language tasks with a small sample size (i.e., cannot apply pre-trained models).

An important aspect of modeling simulator data is that criterion will rarely be available for each time point collected within the simulation. Instead, criterion data will be intermittently gathered during a window of time, such as the task performed for the last 10 minutes of the simulation. This creates a discrepancy in structure between the three-dimensional feature data (event × time-window × feature) and the two-dimensional criterion data (event × criterion). The relationship between multiple time points of the feature data and a single criterion impacts both how the researcher should structure their data and what models they can use. Broadly, there are two approaches available: traditional machine learning methods can be used if the feature data is compressed along the time axis to create a summarized representation aligned with the criterion, or time series methods can be used that natively handle the problem.

Static Methods Using Summarized Representations

The compression of the time series predictors from a three-dimensional structure (event × time series × predictors) to a two-dimensional summarized representation (event × time series) is typically accomplished using either simple summary statistics or manually crafted features. The predictive performance of the machine learning model is entirely dependent upon the quality of the summarized features, making it critical that the researcher thinks carefully about which statistics relate to the criterion of interest. For illustrative purposes, we will walk through this approach using three commonly employed supervised learning models: support vector machines (SVMs) (Cortes & Vapnik, 1995), random forests (Breiman, 2001), and deep learning models (Rumelhart, Hinton, & Williams, 1985). All methods can be easily found in many programming languages (e.g., R, Python, MATLAB).

SVMs are a robust, non-probabilistic, linear classifier that finds the decision boundary that maximizes the distance between classes. SVMs bear a strong resemblance to logistic regression, as both create predictions based on a linear combination of features; however, this is with one notable difference: the objective being optimized. Logistic regression minimizes the negative log-likelihood of the data, providing a probabilistic interpretation of the likelihood of each sample belonging to a particular class. SVMs use the hinge loss with a regularization term to maximize the distance between classes, encouraging the model to not only differentiate between classes but also to do so confidently. Due to the regularized hinge loss, SVMs provide a robust and scalable method that produces sparse solutions (e.g., coefficients encouraged to be 0). Zepf et al. (2019) used an SVM to predict drowsy driving in a simulated driving environment based on features automatically extracted from an EEG using principal frequency bands. Mirichi et al. (2019) used an SVM to create an interpretable model for predicting expertise in a VR simulation of a subpial tumor resection.

Random forests are an ensemble method constructed from a multitude of decision trees trained on random subsets of the data (Brieman, 2001). Decision trees are directed acyclic graphs that use simple binary rules (e.g., $X < 15$) to create predictions. Decision trees are able to map nonlinear structures; however, they have a tendency to overfit to the data and are very sensitive to outliers. Random forests address this limitation by combining many diverse decision trees to create a more stable, robust model. Beninger et al. (2021) used random forests, neural networks, and SVMs to predict inattention during driving in a simulated driving environment. They first preprocessed the data by lowering the sampling rate, sampling a 1-minute window of feature data before each event, normalizing the features within the window, and calculating summary statistics to flatten the data (e.g., minimum, maximum, median). In their evaluations, random forests outperformed the linear SVM and neural network. McDonald and colleagues (2014) used random forests to predict drowsy driving in a simulated driving environment based on features extracted from steering wheel motions.

Neural networks are directed acyclic graphs composed of layers with multiple nodes and are a universal function approximator (Rumelhart, Hinton, & Williams, 1985). Neural networks' extremely flexible structure has led to their widespread success and adoption, particularly on complex tasks with large amounts of data, such as text and image processing (Sun et al., 2017). The simplest neural network can be constructed from an input layer and an output layer with a single node and a linear activation function, which is the same structure as a linear regression. The most complex neural networks are composed of billions of parameters and thousands of layers (Wang et al., 2017; Devlin et al., 2019). Anh, Nataraja, and Chauhan (2020) demonstrated that deep neural networks were able to accurately assess surgical skill in suturing, knot tying, and needle passing. Richstone et al. (2010) used neural networks to predict expertise based on eye movements in simulated surgical environments.

Time Series Methods

Time series methods directly model the three-dimensional predictor data, sidestepping the need for creating a two-dimensional summarized representation. These methods typically require stronger assumptions about the structure of the time series data or the use of complex and flexible frameworks. Multivariate autoregression is a probabilistic model that predicts the criterion based on a linear combination of multiple previous time points of the predictors. Autoregressive methods require the researcher to specify the temporal dependence of the model (i.e., model order); this is usually found during exploratory data analysis by identifying trends and seasonal relationships in the data. Multivariate autoregression is an interpretable model which supports statistical inference; however, it does not natively support nonlinear relationships between features and criteria. Loukas and Georiou (2011) use multivariate autoregression to predict laparoscopic skills (i.e., knot tying and needle driving) during surgical training based on hand motions.

Another commonly employed method is recurrent neural networks (RNN). RNNs model temporal data by recursively calling a node based on feature data at the current time-step and intermediate data from the previous time-step of the RNN node (Rumelhart, Hinton, & Williams, 1985). RNNs make no assumptions about the data and can model nonlinear relationships; however, they are uninterpretable, black-box models that do not support inferential reasoning. RNNs can be difficult to train due to their recursive structure and struggle with long-term temporal dependencies, which led to the development of numerous variants. One of the most successful and well-known variants is long short-term memory (LSTM) networks inspired by the mechanics of human memory (Hochreiter & Schmidhuber, 1997). LSTMs modify RNNs by adding in the ability for the model to "forget" information, addressing the training stability issues of RNNs while allowing them to better model long-time dependencies. Hong and Wang (2020) used LSTMs to predict drowsy driving based on an individual's facial and steering features. Nguyen et al. (2019) use a modified form of LSTMs to predict surgical skill levels based on their hand motions within a surgical simulation.

Applications

With a growing body of literature concerning the collection and structuring of data, as well as the potential utility of various models, it is important to focus research attention on the pragmatic application of AI/machine learning techniques. To date, high-fidelity simulation (e.g., virtual reality) technology has seen wide adoption in military and medical applications, where environmental realism is worth the high price of technology. However, the last decade has seen a major expansion of virtual reality technology, including a boom in the gaming industry. Like all other technologies, improvements in hardware, a shared programming knowledgebase, and capital investments are making virtual reality technology less expensive and more accessible. Organizations are already using virtual reality for process planning and factory layout planning (Mujber, Szecsi, & Hashmi, 2004; Gong et al., 2019). Organizations can further use simulated environments to allow candidates to virtually tour a work facility with the ability to simulate daily activities and train new employees on the use of heavy equipment like forklifts and transport container cranes without risk of injury (Yuen, Choi, & Yang, 2010; Bruzzone & Longo, 2013; Choi, Ahn, & Seo, 2020).

Meaningful translation of technical advances in computer science to meet the specific contextual needs of human factors research could have a transformative influence on the field. There is increasing need for continued research concerning appropriate model and feature selection in addition to a nuanced understanding of when a model becomes confident enough to provide meaningful and stable feedback to a user or trainee. Additionally, traditional human factors topics concerning feedback communication should see revitalized attention in the new context of computer-simulated environments. The following sections highlight some important areas for consideration when adopting AI/machine learning techniques for applied settings.

Trainee Feedback

The major benefit of using simulations and simulated environments is the opportunity to provide trainee feedback in a structured, high-fidelity, and safe environment. Feedback is a long-standing topic of research in human factors, providing an abundance of relevant literature while leaving substantial room for new insights. Traditional topics, such as the design and evaluation of warning signals (Wogalter, Conzola, & Smith-Jackson, 2002), have seen revitalized attention given these new applications and contexts. The following sections highlight a few of the upcoming and important topics that are well-suited for examination through a human factors lens.

In general, AI/machine learning techniques have been criticized for lack of decision-making interpretability. Interpretability of model output has particular importance in providing feedback from training simulations (Mirichi et al., 2019). As we have discussed in this chapter, machine learning approaches are well-adapted for identifying meaningful patterns in data; however, deciphering the decision-making process of machine learning models remains difficult. For training feedback to be meaningful, trainees must understand why they received certain scores and how they can improve. Typical black-box approaches are not always well-suited for providing this level of feedback. Nonetheless, researchers have proposed means by which machine learning models can be used to provide meaningful feedback during and after simulations.

Early Prediction

Real-time detection has already demonstrated budding utility for static diagnosis and training simulations, but it also has applications beyond training, such as in computer-enhanced surgical assistance (Thai et al., 2020). In all instances, it is important to consider when and how a system should begin providing feedback to a user or trainee; accordingly, there are two distinct avenues of consideration when deploying this technology. The first is a technical perspective on when a model becomes confident enough to make a stable prediction of a future event or state (i.e., early prediction and model uncertainty). The second is the psychological consideration of how feedback should be delivered to a user or trainee in real time.

Early prediction specifically concerns confidence in prediction when working with temporal data. It concerns the question of when a model's prediction confidence is high enough (or uncertainty low enough) to make a stable prediction. For example, in medical research, early prediction would concern a model's ability to accurately detect an abnormality or disease at early stages of diagnosis or identify symptoms of early onset (e.g., Hsu & Holtz, 2019). Early prediction methods have shown tremendous utility for enhancing diagnostics across diverse circumstances, including predicting circulatory failure (Hyland et al., 2020), sepsis shock (Lin et al., 2018), and diabetes (Alam et al., 2019).

Real-Time Feedback

Advancements in early prediction pave the road for real-time feedback. Recent findings show tremendous potential for using motion data to provide meaningful feedback to trainees. For example, researchers have determined corollaries for motion inference in games-based settings and provided evidence that motion information could be used for early indications of events (Hart, Vaziri-Pashkam, & Mahadevan, 2020). A natural expansion of this research is the application of advanced machine learning/deep learning models that could infer dangerous motion and provide early warning indications. Researchers have used an adaptation of random forest and LSTM neural network models to create a real-time feedback tool for a temporal bone surgery simulator by identifying characteristics in drilling strokes that improved surgery performance (Ma et al., 2017a, 2017b). More abstractly, researchers have explored machine-learning approaches to predict early warning signs of critical transitions within dynamic systems (Lade & Gross, 2012; Füllsack, Kapeller, Plakolb, and Jäger, 2020). The concept of identifying predictors of critical transitions using machine learning models could have widespread applications for monitoring dynamic systems, such as predicting communication breakdowns in military squadrons or oncoming disequilibrium in a patient during operation.

Early prediction coupled with meaningful feedback would be useful across a variety of training settings, including team communication and education. However, for real-time feedback to be meaningful, it must be effectively received and processed by a user or trainee. Computerized systems should facilitate integration of established best practices, such as the use of personalized warnings (Wogalter, Racicot, Kalsher, & Simpson, 1994), meaningful use of alarms (Edworthy & Hellier, 2006), integration of multisensory warning signals (Ho, Reed, & Spence, 2007; Baldwin et al., 2012), and ensuring that warnings are maximally informative (Fagerlönn & Alm, 2010). Meaningful application of AI to training simulations will require an interdisciplinary perspective that can translate powerful analytics into products with pragmatic value.

Adaptive Simulations

AI's potential to produce real-time prediction and feedback will also enable advancements in adaptive simulations. The goal of AI-enhanced adaptive simulations would be to further increase the fidelity and learning opportunities within simulated training environments. Perhaps the most straightforward application is to adapt the difficulty of a simulation to create additional challenges when trainees are performing well. In addition to adaptive difficulty, simulations could include adaptive scenarios for dangerous events such as hydroplaning, losing control of an object when using machinery like cranes and forklifts, and emergency medical scenarios during surgery. Allowing simulations to adaptively introduce these events, especially in connection to real-time data about the environmental state, would increase simulation fidelity and better prepare trainees for low-frequency but high-risk events.

A recent review of adaptive simulations found that the most common simulation adaptation was adjustment to difficulty, such as adjustment to speed or resistance in rehabilitation exercises (Zahabi & Razak, 2020). The study also found that most simulations with adaptive content did not provide adaptive real-time feedback. In our discussion, we express the need to provide adaptation of controlled elements in the environment and the usefulness of adaptive feedback. Achieving these goals will require that AI gain sufficient knowledge about current states, desired states, and future states and be able to process this information quickly and efficiently. Codifying the knowledge of when an action should have been performed, and then adaptively providing feedback to redirect toward the desired state, is not a task that can be easily programmed from a flat representation of actions, especially when moving beyond simple motion data. Furthermore, as it pertains to the application of such technologies outside of training simulations, the implementation of AI image (video) information is much more difficult in applied settings than in simulations (Vedula et al., 2017).

CONCLUSION

The implications of AI for society are immense, and such techniques are already being used to change how we think about and score simulations. This chapter has discussed domains where AI can help with two of the greatest challenges in scoring simulations: NLP models for handling unstructured text data and various other techniques for dealing with unstructured data such as event and motion data. While we framed the techniques around real-world examples of their use, it should be noted that we have not tried to be exhaustive in terms of the possible analytic techniques or provided enough information so a reader could jump to using the above analytics directly. Each of these techniques likely deserves an entire chapter devoted to real-world usage, but such detail would be far beyond the scope of this chapter. Our goal was to outline both the implications of these technologies for how we score simulations and exposure to the various analytics techniques that could be, and often already are being, used in the field to score simulations. The truth is that the best techniques at a given time are rapidly changing, and any guide or cookbook for how to use a technique will become dated quickly. So rather than focus on the specifics, we wanted to outline consistent themes and limitations that should be considered whenever AI is being used and describe what may be possible.

A consistent theme across everything discussed in this chapter is that the importance of SMEs, and an understanding of the domain at hand, will continue to be of critical significance. While AI is capable of extraordinary things, getting the most from these tools requires collaboration between data scientists and domain-specific SMEs, as the model is only as effective as the quality of data and simulation being assessed. Broadly speaking, we see two primary areas where AI will forever change how we score simulations. One will be the automation of scoring unstructured text data that previously would require highly trained human raters, and the other would be new sources of data too complex for humans to process. We argue both areas have promising futures in the realm of simulations, whether for training or evaluative purposes. Lastly, we believe AI will create a golden age in the use of simulations as we are able to create models that help simulations become more accurate, scalable, and comprehensive.

REFERENCES

Ahmidi, N., Poddar, P., Jones, J. D., Vedula, S. S., Ishii, L., Hager, G. D., & Ishii, M. (2015). Automated objective surgical skill assessment in the operating room from unstructured tool motion in septoplasty. *International Journal of Computer Assisted Radiology and Surgery*, 10(6), 981–991.

Alam, T. M., Iqbal, M. A., Ali, Y., Wahab, A., Ijaz, S., Baig, T. I., Hussain, A., Malik, M., Raza, M., Ibrar, S., & Abbas, Z. (2019). A model for early prediction of diabetes. *Informatics in Medicine Unlocked*, 16, 1–6.

Anh, N. X., Nataraja, R. M., & Chauhan, S. (2020). Towards near real-time assessment of surgical skills: A comparison of feature extraction techniques. *Computer Methods and Programs in Biomedicine*, 187, 105234.

Baldwin, C. L., Spence, C., Bliss, J., Brill, C., Wogalter, M., Mayhorn, C., & Ferris, T. (2012). Multimodal cueing: The relative benefits of the auditory, visual, and tactile channels in complex environments. *Proceedings of the Human Factors and Ergonomics Society.*

Beninger, J., Hamilton-Wright, A., Walker, H. E., & Trick, L. M. (2021). Machine learning techniques to identify mind-wandering and predict hazard response time in fully immersive driving simulation. *Soft Computing*, 25(2), 1239–1247.

Bernardin, H. J., & Buckely, M. R. (1981). Strategies in rater training. *Academy of Management Review*, 6, 205–212.

Boyle, W. A., Murray, D. J., Beyatte, M. B., Knittel, J. G., Kerby, P. W., Woodhouse, J., & Boulet, J. R. (2018). Simulation-based assessment of critical care 'front-line' providers. *Critical Care Medicine*, 46(6), e516.

Breiman, L. (2001). Random forests. *Machine Learning*, 45(1), 5–32.

Brown, T. B., Mann, B., Ryder, N., Subbiah, M., Dhariwal, P., Neelakantan, A., Shyam, P., Sastry, G., Askell, A., et al. (2020). Language models as few shot learners. Arxiv: https://arxiv.org/pdf/2005.14165.pdf

Bruzzone, A., & Longo, F. (2013). 3D simulation as training tool in container terminals: The TRAINPORTS simulator. *Journal of Manufacturing Systems*, 32, 85–98.

Choi, M., Ahn, S., & Seo, J. (2020). VR-based investigation of forklift operator situation awareness for preventing collision accidents. *Accident Analysis and Prevention*, 136, 1–9.

Cortes, C., & Vapnik, V. (1995). Support-vector networks. *Machine Learning*, 20(3), 273–297.

Devlin, J., Chang, M. W., Lee, K., & Toutanova, K. (2019). BERT: Pre-training of bidirectional transformers for language understanding. *Proceedings of the 2019 Conference of the North American Chapter of the Association for Computational Linguistics. Human Language Technologies*, 4171–4186. Minneapolis, MN.

Edworthy, J., & Hellier, E. (2006). Alarms and human behaviour: Implications for medical alarms. *British Journal of Anesthesia*, 97(1), 12–17.

Fagerlönn, J., & Alm, H. (2010). Auditory signs to support traffic awareness. *IET Intelligent Transport Systems*, 4(4), 262–269.

Füllsack, M., Kapeller, M., Plakolb, S., & Jäger, G. (2020). Training LSTM-neural networks on early warning signals of declining cooperation in simulated repeated public good games. *MethodsX*, 7, 100920.

Franz, L., Shrestha, Y. R., & Paudel, B. (2020). A deep learning pipeline for patient diagnosis prediction using electronic health records. *BioKDD 2020: 19th International Workshop on Data Mining in Bioinformatics*. San Diego, CA.

Garla, V. N., & Brandt, C. (2012). Ontology-guided feature engineering for clinical text classification. *Journal of Biomedical Informatics*, 45(5), 992–998.

Geden, M., Emerson, A., Rowe, J., Azevedo, R., & Lester, J. (2020). Predictive student modeling in educational games with multi-task learning. *Proceedings of the AAAI Conference on Artificial Intelligence*, 34(1), 654–661.

Gibson, C., & Mumford, M. D. (2013). Evaluation, criticism, and creativity: Criticism content and effects on creative problem-solving. *Psychology of Aesthetics, Creativity, and the Arts*, 7, 314–331.

Gong, L., Berglund, J., Berglund, A., Johansson, B., & Borjesson, T. (2019). Development of virtual reality support to factory layout planning. *International Journal of Interactive Design and Manufacturing*, 13, 935–945.

Hall, S., & Brannick, M. T. (2008). Performance assessment in simulation. In D. A. Vincenzi, J. A. Wise, M. Mouloua, & P. A. Hancock (Eds.), *Human factors in simulation and training* (pp. 149–168). Boca Raton, FL: CRC Press.

Harik, P., Clauser, B. E., Grabovsky, I., Nungester, R. J., Swanson, D., & Nandakumar, R. (2009). An examination of rater drift within a generalizability theory framework. *Journal of Educational Measurement*, 46, 43–58.

Hart, Y., Vaziri-Pashkam, M., & Mahadevan, L. (2020). Early warning signals in motion inference. *PLoS Computational Biology*, 16(5), e1007821.

Haugeland, J. (1985). *Artificial intelligence: The very idea*. Cambridge, MA. The MIT Press.

Henderson, N., Kumaran, V., Min, W., Mott, B., Wu, Z., Boulden, D., … Lester, J. (2020). Enhancing student competency models for game-based learning with a hybrid stealth assessment framework. *International Educational Data Mining Society*, 13, 92–103.

Hochreiter, S., & Schmidhuber, J. (1997). Long short-term memory. *Neural Computation* 9(8), 1735–1780.

Ho, C., Reed, N., & Spence, C. (2007). Multisensory in-car warning signals for collision avoidance. *Human Factors: The Journal of the Human Factors and Ergonomics Society*, 49(6), 1107–1114.

Hong, L., & Wang, X. (2020). Towards drowsiness driving detection based on multi-feature fusion and LSTM networks. *International Conference on Control, Automation, Robotics and Vision*,732–736.

Hosny, A., Parmar, C., Quackenbush, J., Schwartz, L. H., & Aerts, H. (2018). Artificial intelligence in radiology. *Nature Reviews Cancer*, 18, 500–510.

Hsu, P., & Holtz, C. (2019). A comparison of machine learning tools for early prediction of sepsis from ICU data. *2019 Computing in Cardiology (CinC)*, 46, 1–4.

Huang, K., Altosaar, J., & Ranganath, R. (2020). ClinicalBERT: Modeling clinical notes and predicting hospital readmission. Arxiv: https://arxiv.org/pdf/1904.05342.pdf

Hyland, S. L., Faltys, M., Hüser, M., Lyu, X., Gumbsch, T., Esteban, C., Bock, C., Horn, M., Moor, M., Rieck, B., Zimmermann, M., Bodenham, D., Borgwardt, K., Rätsch, G., & Merz, T. M. (2020). Early prediction of circulatory failure in the intensive care unit using machine learning. *Nature Medicine*, 26(3), 364–373.

Krajewski, J., Sommer, D., Trutschel, U., Edwards, D., & Golz, M. (2009). Steering wheel behavior based estimation of fatigue. *Proceedings of the International Driving Symposium on Human Factors in Driver Assessment, Training and Vehicle Design*, 118–124.

Kuhn, M., & Johnson, K. (2019). *Feature engineering and selection: A practical approach for predictive models*. Abingdon, UK: Taylor & Francis Group.

Lade, S. J., & Gross, T. (2012). Early warning signals for critical transitions: A generalized modeling approach. *PLoS Computational Biology*, 8(2), e1002360.

Loukas, C., & Georgiou, E. (2011). Multivariate autoregressive modeling of hand kinematics for laparoscopic skills assessment of surgical trainees. *IEEE Transactions on Biomedical Engineering*, 58(11), 3289–3297.

Lin, C., Zhang, Y., Ivy, J., Capan, M., Arnold, R., Huddleston, J. M., & Chi, M. (2018). Early diagnosis and prediction of sepsis shock by combining static and dynamic information using convolutional-LSTM. *Proceedings of the International Conference on Healthcare Informatics*, 219–228.

Liu, Y., Ott, M., Goyal, N., Du, J., Joshi, M., Chen, D., Levy, O., Lewis, M., Zettlemoyer, L., & Stoyanov, V. (2019). RoBERTa: A robustly optimized BERT pretraining approach. Arxiv: https://arxiv.org/pdf/1907.11692.pdf

Ma, X., Wijewickrema, S., Zhou, Y., Zhou, S., O'Leary, S., & Bailey, J. (2017). Providing effective real-time feedback in simulation-based surgical training. *International Conference on Medical Image Computing and Computer-Assisted Intervention*, 566–574.

Ma, X., Wijewickrema, S., Zhou, Y., Zhou, S., Mhammedi, Z., O'Leary, S., & Bailey, J. (2017b). Adversarial generation of real-time feedback with neural networks for simulation-based training. *Proceedings of the International Joint Conference on Artificial Intelligence*, 3763–3769.

McDonald, A. D., Lee, J. D., Schwarz, C., & Brown, T. L. (2014). Steering in a random forest: Ensemble learning for detecting drowsiness-related lane departures. *Human Factors*, 56(5), 986–998.

Mills, J. T., Hougen, H. Y., Bitner, D., Krupski, T. L., & Schenkman, N. S. (2017). Does robotic surgical simulator performance correlate with surgical skill? *Journal of Surgical Education*, 74(6), 1052–1056.

Min, W., Mott, B., Rowe, J., Taylor, R., Wiebe, E., Boyer, K., & Lester, J. (2017). Multimodal goal recognition in open-world digital games. *Proceedings of the AAAI Conference on Artificial Intelligence and Interactive Digital Entertainment*, 13(1), 80–86.

Mirichi, N., Bissonnette, V., Yilmaz, R., Ledwos, N., Winkler-Schwartz, A., & Del Maestro, R. (2019). The virtual operative assistant: An explainable artificial intelligence tool for simulation-based training in surgery and medicine. *PLoS One*, 15(2), 1–15.

Mislevy, R. J., Steinberg, L. S., & Almond, R. G. (2003). Focus article: On the structure of educational assessments. *Measurement: Interdisciplinary Research and Perspectives*, 1(1), 3–62.

Moore, G. E. (1965). Cramming more components onto integrated circuits. *Electronics Magazine*, 38(8), 114–117.

Mosier, K. L., & Manzey, D. (2020). Humans and automated decision aids: A match made in heaven? In M. Mouloua & P. A. Hancock (Eds.), *Human performance in automated and autonomous systems: Current theory and methods* (pp. 19–42). Boca Raton: CRC Press.

Motowidlo, S. J., Dunnette, M. D., & Carter, G. W. (1990). An alternative selection procedure: The low-fidelity simulation. *Journal of Applied Psychology*, 75, 640–647.

Mouloua, M., & Hancock, P. (2020). *Human performance in automated and autonomous systems: Current theory and methods*. Boca Raton: CRC Press.

Mracek, D. L., Peterson, N., Barsa, A., & Koenig, N. (2021). DEEP*O*NET: A neural network approach to leveraging detailed text descriptions of the world of work. In K. Nei (Chair), *Demonstrating natural language processing for improving job analysis*. Symposium conducted at the meeting of the Society for Industrial/Organizational Psychology, New Orleans, LA.

Mujber, T., Szecsi, T., & Hashmi, M. (2004). Virtual reality applications in manufacturing process simulation. *Journal of Materials Processing Technology*, 155–156, 1834–1838.

Nguyen, X. A., Ljuhar, D., Pacilli, M., Nataraja, R. M., & Chauhan, S. (2019). Surgical skill levels: Classification and analysis using deep neural network model and motion signals. *Computer Methods and Programs in Biomedicine*, 177, 1–8.

Norman, G. R., Grierson, L. E. M., Sherbino, J., Hamstra, S. J., Schmidt, H. G., & Mamede, S. (2018). Expertise in medicine and surgery. In K. A. Ericsson, R. R. Hoffman, A. Kozbelt, & A. M. Williams (Eds.), *Cambridge handbooks in psychology: The Cambridge handbook of expertise and expert performance* (pp. 331–355). Cambridge: Cambridge University Press.

Oquendo, Y. A., Riddle, E. W., Hiller, D., Blinman, T. A., & Kuchenbecker, K. J. (2018). Automatically rating trainee skill at a pediatric laparoscopic suturing task. *Surgical Endoscopy*, 32(4), 1840–1857.

Parasuraman, R., & Mouloua, M. (1996). *Automation and human performance: Theory and applications*. New York: CRC Press.

Poola, I. (2017). How artificial intelligence is impacting real life every day. *International Journal of Advance Research, Ideas and Innovations in Technology*, 2., 96–100.

Richstone, L., Schwartz, M. J., Seideman, C., Cadeddu, J., Marshall, S., & Kavoussi, L. R. (2010). Eye metrics as an objective assessment of surgical skill. *Annals of Surgery*, 252(1), 177–182.

Roch, S. G., Woehr, D. J., Mishra, V., & Kieszczynska, U. (2012). Rater training revisited: An updated meta-analytic review of frame-of-reference training. *Journal of Occupational and Organizational Psychology*, 85, 370–395.

Rumelhart, D. E., Hinton, G. E., & Williams, R. J. (1985). *Learning internal representations by error propagation*. California Univ San Diego La Jolla Inst for Cognitive Science. Cambridge, MA: Bradford Books/MIT Press.

Rust, R. T., & Huang, M. (2014). The service revolution and transformation of marketing science. *Marketing Science*, 33, 206–221.

Ryman-Tubb, N. F., Krause, P., & Garn, W. (2018). How artificial intelligence and machine learning research impacts payment card fraud detection: A survey and industry benchmark. *Engineering Applications of Artificial Intelligence*, 76, 130–157.

Salkowski, L. R., & Russ, R. (2018). Cognitive processing differences in experts and novices when correlating anatomy and cross-sectional imaging. *Journal of Medical Imaging*, 5(3), 031411.

Schleicher, D. J., Day, D. V., Mayes, B. T., & Riggio, R. E. (1999). A new frame for frame-of-reference training: Enhancing the construct validity of assessment centers. Paper presented at the annual conference of the Society for Industrial and Organizational Psychology, Atlanta, GA.

Schwab, K. (2017). *The fourth industrial revolution*. New York: World Economic Forum.

Schwarting, W., Alonso-Mora, J., & Rus, D. (2018). Planning and decision-making for autonomous vehicles. *Annual Review of Control, Robotics, and Autonomous Systems*, 1, 187–210.

Sun, C., Shrivastava, A., Singh, S., & Gupta, A. (2017). Revisiting unreasonable effectiveness of data in deep learning era. *Proceedings of the IEEE International Conference on Computer Vision*, 843–852.

Sydell, E., Ferrell, J., Carpenter, J., Frost, C., & Brodbeck, C. C. (2013). Simulation scoring. In M. Fetzer & K. Tuzinski (Eds.), *Simulations for personnel selection* (pp. 83–107). New York, NY: Springer.

Thai, M., Phan, P., Hoang, T., Wong, S., Lovell, N., & Do, T. (2020). Advanced intelligent systems for surgical robotics. *Advanced Intelligent Systems*, 2, 1–33.

Thomson, D. R., Besner, D., & Smilek, D. (2015). A resource-control account of sustained attention: Evidence from mind-wandering and vigilance paradigms. *Perspectives on Psychological Science*, 10(1), 82–96.

Thompson, I., Koenig, N., Mracek, D. L., & Tonidandel, S. (forthcoming). Integrating deep learning and measurement science: Automating the subject matter expertise used to evaluate candidate work samples. *Journal of Applied Psychology.*

Tonidandel, S., Thompson, I. B., Mracek, D. L., & Koenig, N. (2020). Automating subject matter expertise used to evaluate candidate work samples. In E. Campion & M. Campion, (Chairs). *The Construct Validity of Computer-Assisted Text Analysis (CATA).* Symposium conducted at the annual conference of the Society for Industrial and Organizational Psychology, Austin, TX.

Uzuner, Ö. (2009). Recognizing obesity and comorbidities in sparse data. *Journal of the American Medical Informatics Association*, 16(4), 561–570.

Vaswani, A., Shazeer, N., Parmar, N., Uszkoreit, J., Jones, L., Gomex, A. N., Kaiser, L., & Polosukhin, I. (2017). Attention is all you need. *Proceedings of the 31st International Conference on Neural Information Processing*, 6000–6010.

Vedula, S. S., Ishii, M., & Hager, G. D. (2017). Objective assessment of surgical technical skill and competency in the operating room. *Annual Review of Biomedical Engineering*, 19, 301–325.

Wang, F., Jiang, M., Qian, C., Yang, S., Li, C., Zhang, H., … Tang, X. (2017). Residual attention network for image classification. *Proceedings of the IEEE Conference on Computer Vision and Pattern Recognition*, 3156–3164.

Wei, W. W. (2006). Time series analysis. In T. D. Little (Ed.), *The Oxford handbook of quantitative methods in psychology: Vol. 2.* Oxford, UK: Oxford University Press.

Wickens, C. D., Hollands, J. G., Banbury, S., & Parasuraman, R. (2016). *Engineering psychology and human performance* (4th ed). New York: Routledge.

Wogalter, M., Conzola, V., & Smith-Jackson, T. (2002). Research-based guidelines for warning design and evaluation. *Applied Ergonomics*, 33(3), 219–230.

Wogalter, M., Racicot, B., Kalsher, M., & Simpson, S., (1994). Personalization of warning signs: The role of perceived relevance on behavioral compliance. *International Journal of Industrial Ergonomics*, 14(3), 233–242.

Wolpert, D. H., & Macready, W. G. (1997). No free lunch theorems for optimization. *IEEE Transactions on Evolutionary Computation*, 1(1), 67–82.

Yuen, K., Choi, S., & Yang, X. (2010). A full-immersive CAVE-based VR simulation system of forklift truck operations for safety training. *Computer-Aided Design & Applications*, 7(2), 235–245.

Zahabi, M., & Razak, A. M. A. (2020). Adaptive virtual reality-based training: A systematic literature review and framework. *Virtual Reality*, 24, 725–752.

Zepf, S., Stracke, T., Schmitt, A., van de Camp, F., & Beyerer, J. (2019, December). Towards real-time detection and mitigation of driver frustration using SVM. *Proceedings of the International Conference on Machine Learning and Applications*, 202–209.

Zhang, Y., Jin, R., & Zhou, Z. (2010). Understanding bag-of-words model: A statistical framework. *International Journal of Machine Learning and Cybernetics*, 1, 43–52.

12 Dissecting the Neurodynamics of the Pauses and Uncertainties of Healthcare Teams

Ronald Stevens, Trysha Galloway, and Ann Willemsen-Dunlap

CONTENTS

INTRODUCTION

Simulation is widely accepted as an important educational tool for healthcare professionals in training and practice, and evidence for its positive impact on patient safety is growing (Schmidt et al., 2013). High-fidelity simulations provide opportunities for students to dynamically integrate their static knowledge, for professionals to acquire and maintain their professional skills, and for high-stakes testing (Lateef, 2010).

Clinical simulations of complex and evolving patient conditions often occur in high-fidelity environments using realistic mannequins and standardized participants, a degree of fidelity not always available to smaller healthcare facilities, therefore making it difficult for them to achieve the research-reported benefits of simulation training (Goodwin et al., 2021).

Today, virtual reality (VR) and augmented reality (AR) have arrived on the simulation scene and they are here to stay as they provide supplements/alternatives to

DOI: 10.1201/9781003401360-12

mannequin-based simulations (Creutzfeldt et al., 2016). For instance, VR is providing opportunities for surgeons to visualize complex operations and surgical treatment options both preoperatively and intraoperatively where they can make mistakes and learn from them but in an environment where there is no risk to the patient. Hospitals have also experimented with VR during the informed consent process as it allows patients and their families to virtually walk through their surgical plan, stepping into their own anatomy and diagnosis as they gain the deep understanding and trust needed to make complex medical decisions (Surgical Theater, 2021).

Simulations both support and are supported by a variety of stakeholders, each with a need to understand different facets of the simulation process. Professionals in training are generally thought of as the primary learners, but as shown above, patients can also benefit from VR simulations. Another stakeholder would be the training facilitators who support the learning of trainees by probing and identifying deficits in knowledge, attitudes, and skills that arise during debriefings as well as help trainees form new abstract concepts and generalizations (Maytin et al., 2015). In parallel, this group of simulation educators might also draw on their debriefing experiences and use them to train new facilitators or use their skills to design new simulations to focus on unaddressed knowledge gaps. Finally, program directors in charge of residency programs invest a significant amount of time and resources in the recruitment process, as well as maintaining efficiency and cost-effectiveness. Use of VR facility tours as an alternative to in-person tours of affiliate training facilities during a residency interview day is a viable and innovative option that can save time and money and favorably impact the applicant's impression of the program (Zertuche et al., 2020).

The diversity of stakeholders, and their need for understanding different levels of simulation processes, argue for automated systems that will provide rapid, objective, and quantitative representations of performance information that provide educationally meaningful evidence of short- and long-term learning tailored to the needs of the different stakeholders. Such systems built around cognitive frameworks would be applicable across content domains as well as simulation platforms created and delivered by established methods or by the emerging AR, VR, mixed reality (MR), or generative immersive scenario testbed (GIST) technologies.

A challenge for conceptualizing and delivering this next generation of capabilities is that the dynamics of learning during simulation training or even making estimates of when that learning might be occurring are poorly understood. Sometimes, the best that can be said is that learning likely occurred during the simulation, and perhaps more so during the debriefing. At the neuronal level, however, it is increasingly apparent that learning is driven by unexpected events, i.e., those that cause uncertainty (Gillon et al., 2021).

This chapter describes the progress our lab has made in developing increasingly understandable neurodynamic models that analyze the brainwaves of individuals and teams in real time, and report the frequency, magnitude, and duration of neural correlates of uncertainty. These neurodynamic measures and models quantitatively scale from neurons to teams providing the chance to report training gaps and performance levels to stakeholders throughout the healthcare simulation community whether they are the trainees, learning facilitators, simulation designers, or program directors.

The challenge in developing performance-based evaluations using neural measures is not with the EEG measures themselves. Since the discovery of brainwaves, many measures have been developed using EEG, i.e., the frequency, amplitude (power), phase, complexity, scalp topology, ERPs, etc. An equally large number of methods have been developed for collecting, preprocessing, and modeling the measures in real time (Delorme & Makeig, 2004; Mullen et al., 2015; Bigdely-Shamlo et al., 2015; Oostenveld et al., 2011).

The challenge of broad-scale neural modeling has been that bottom-up analytic approaches rapidly become complicated as most low-level neural processes are not in themselves directly causal to team performance but instead are the result of everyday cognitive activities that support seeing, listening, decision making, etc. It is when these activities are transiently amplified or modified by the context that they assume greater importance for understanding teamwork.

Higher-level representations of neural processes are needed where modifications to and amplifications of the micro-scale dynamics are allowed to change freely, while still providing the "best fit" (i.e., more stable) functional approximations for higher-level activities; in other words, abstract representations that have a basis in mechanism but where many of the micro-details don't need specifying (Flack, 2017a).

Information is one such abstraction. It has been proposed that biological systems, like teams, are hierarchies of information that are functionally organized across spatial and time scales (Flack, 2017). Uncertainty is the messenger on this hierarchy guiding information back and forth between the environment and the team (Flack et al., 2012), with ripples and islands in these information streams representing periods of changing organization (see Stevens et al., 2013 for team examples). This changing information helps the brain identify statistical regularities in the environment and use them to shape adaptations along the macroscopic and microscopic continuum of experience and learning (Daniels et al., 2017).

We have proposed that the informational structure of EEG rhythms might be a candidate representation for bridging the micro–macro scale gap as they have a basis in organization, not power or phase, and may be more likely to align with processes responsible for observable macro-scale organizations and behaviors.

Scalp brainwaves are defined by the amplitudes of EEG-determined frequencies and their phase with other brainwaves. Changes in EEG amplitude are seen for all frequencies in the 1–40 Hz EEG spectrum, and observable cognitive behaviors are increasingly being linked with these (short-lived) changes in different EEG frequencies and amplitudes. For instance, the meaning of EEG power is important. Alpha band oscillations (8–12 Hz) emphasize different functional priorities depending on whether their states are synchronized (aka activated or high power) or desynchronized (aka deactivated or low power). Low-power (i.e., often also called suppressed) states are seen during attentive reading (Lachaux et al., 2008) and tend to favor new memory encodings. Higher alpha power states may transiently suppress gamma rhythms and help protect the contents of working memory from being disturbed, thereby enhancing retention (Klimesch et al., 2006; Klimesch, 2012; Wianda &

Ross, 2019; Ossandon et al., 2011; Bonnefond & Jensen, 2015). Similar considerations might apply to (32–40 Hz) gamma waves (Sedley & Cunningham, 2013) or delta waves (1–4 Hz), which show an increase in power during the onset of fatigue, which decrease following challenge interruptions (Bodala et al., 2018).

THE SIGNIFICANCE OF STRUCTURE IN EEG AMPLITUDES

Detecting structure in data streams involves first deconstructing continuous data into discrete symbols. From the different cognitive activities described above, EEG amplitudes might exist in activated, deactivated, or neutral states, which could be assigned any three symbols that were easy to visualize. In our studies, activated states are assigned "3," deactivated states are assigned "–1," and neutral states are assigned "1." So during a second when the EEG amplitude was high, the symbol for that second would be 3; if it were low, the symbol would be –1 and 1 if in an average power state. The result is a data stream of 3's, 1's, and –1's. The number of states to divide the EEG amplitude into each second involves a trade-off between resolution and computation; dividing the amplitude each second into six or nine states might increase the resolution, but would slow the modeling.

Figure 12.1 shows a team of two persons where the EEG amplitudes were separated into three states each second (Figure 12.1A), or six states (Figure 12.1B), or nine states (Figure 12.1C). For Figure 12.1A, since there are two persons and three symbols in each person's data stream, the team data stream would have nine symbols. The temporal structure (not power) in this data stream can be estimated each second by measuring the mix of the nine symbols in a 60 s segment that slides over the data and is updated each second. If only one of the nine symbols was expressed in this 60 s segment, the entropy would be 0 bits; if there was an equal mix of the nine symbols, then the entropy would be 3.17 bits, which is the maximum. So, the fewer the symbols expressed in a window of 60 s, the more organized the team and the lower the entropy. Since it could be confusing having lower entropy mean higher organizations, the data plots are made more intuitive by calculating the neurodynamic information (NI), which is the bits of information when entropy values are subtracted from the maximum entropy for the number of unique symbols. So, an entropy value of 2.87 would have a NI level of 3.17 minus 2.87, or 0.4 bits. The resulting NI profile is shown to the right and the average NI of the team's performance was 0.16 bits. Similar calculations can be made when the amplitude was separated into six or nine states (Figure 12.1B and C). Although the NI values increased with additional symbols in each group, the NI profiles were similar, indicating that adding additional symbols had little effect on the dynamical structure of the data; for most studies we separate EEG data into three categories.

A general, and perhaps more important, point from this figure is that symbolically analyzing the structure of EEG amplitude creates a quantitative and bounded scale of EEG organizations. If there is no organization to the EEG data stream, the NI would be 0. If the EEG was maximally organized, i.e., all a single symbol, the NI would be the maximum for the number of symbols in the system, i.e., 4.75 bits

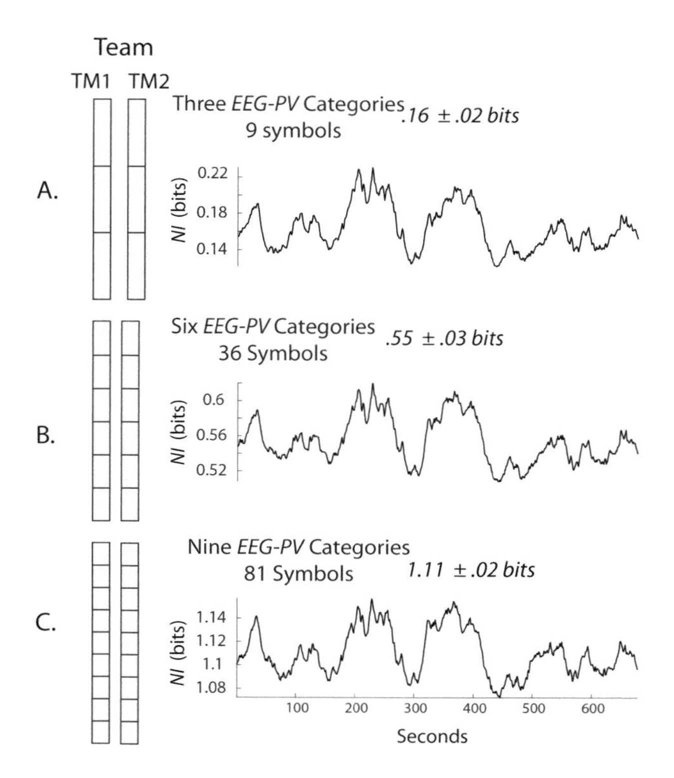

FIGURE 12.1 Symbolic modeling of neurodynamic data. The EEG was collected from two team members (TM1 and TM2) and each second the scalp averaged EEG amplitude was separated into three (A), six (B), or nine (C) divisions. The NI was calculated for the three models using a 60 s moving window that was updated each second.

for a 27-symbol three-person team, 3.17 bits for a 9-symbol dyad, or 1.59 bits for an individual.

These mathematical limits have implications for creating quantitative performance measures. It means that the neurodynamic information of any team of two persons who are performing any task where the EEG is separated into three levels will have NI levels between 0 and 3.17 bits. The average value of 0.16 bits for the team in Figure 12.1B is a value that can be quantitatively compared with other team performances, and these can be aggregated for a class of trainees, or used to compare one training protocol to another. If the average NI for a team member is calculated, this value can be quantitatively compared with that of other team members. It means that the neurodynamic organization of one brain region can be compared with that of another brain region and across the 1–40 Hz EEG spectrum. The same reasoning applies if the neurodynamic organization of a team is compared in the simulation scenario versus the debriefing, or across a critical healthcare event like intubation.

Such neurodynamic organizations contribute properties to the system not always possessed by the amplitude or phase of brainwaves alone. For instance, neurodynamic information has been shown to link with the organization of team activities (Stevens & Galloway, 2017) or speech (Gorman et al., 2016), or submarine (Stevens et al., 2017), or healthcare (Stevens & Galloway, 2021) team expertise. These options are explored in the next section of this chapter.

There is one other important question that needs exploring first: Why do neurodynamic organizations correlate with these team activities, while the EEG power values (EEG-PV) from which the NI was calculated do not? This is because it is not the power levels per se in an EEG data stream that link with behavioral activities, but the information or *degree of organization within* the *EEG-PV* that is important for making this connection. The differences between the two are shown in Figure 12.2. Figure 12.2A shows the NI of a neurosurgeon with five to seven discrete peaks during the 1700 s of a simulated operation. The information profile was developed from a frequency-averaged EEG data stream from the T4 sensor using a 30 s window where the EEG power was first divided into equal numbers of three discrete segments; these were –1 for below average levels, 1 for average levels, and 3 for above average levels as described earlier.

There were also parallel fluctuations in EEG-PV. If we use the symbols –1, 1, or 3 numerically, then equal numbers of these symbols over the performance would have an average of 1.0. As shown in Figure 12.1B, about half of the fluctuations were below the performance average power levels, some were above, and others were around the average. Whenever there were periods of persistent EEG-PV, either high

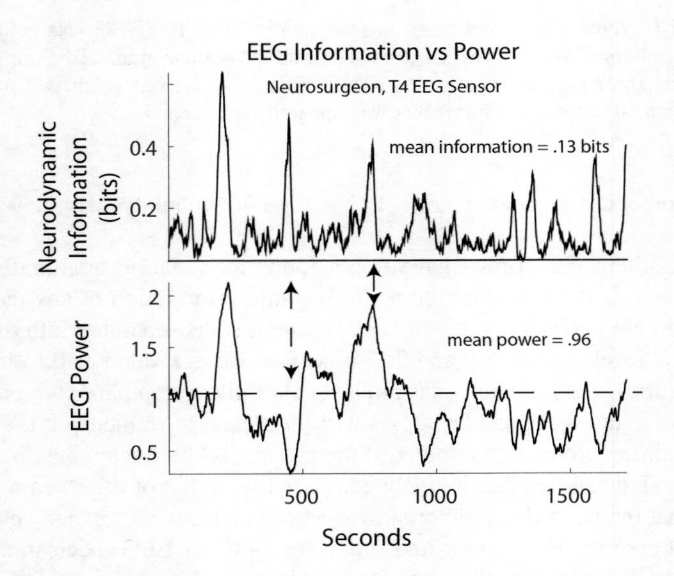

FIGURE 12.2 Neurodynamic information (NI) accumulates over time as teams experience periods of uncertainty. The EEG power values (EEG-PV) progress through periods of high and low power but don't accumulate like the NI.

or low, there was elevated NI. The result was that the overall EEG power levels were 0.96, i.e., around the expected average, but the levels of NI when averaged for the performance were significantly elevated as the information is agnostic to the details of what is being organized, just that it is organized.

A second interesting feature of NI versus EEG-PV was the strong link between periods of elevated NI and periods of uncertainty, which does not exist for EEG-PV (Stevens, Galloway, Halpin, & Willemsen-Dunlap, 2016; Stevens & Galloway, 2019).

NEURODYNAMIC CORRELATES OF UNCERTAINTY

Uncertainty is a fundamental property of neural computation that helps us estimate the (perceived) state of our world. The brain uses this uncertainty to continually access memories (the past) to imagine future possibilities and the actions that would give the best outcomes, outcomes that might be orders of magnitude away in the future.

Humans maintain low levels of uncertainty by operating in familiar environments and situations where well-rehearsed sequences of cognition can be exploited. As a result, we think and act in terms of chunks of several seconds up to a minute, which help streamline the moment-to-moment activities (Rumelhart, 1980; Schank & Abelson, 1977; Cooper & Shallice, 2000; Daw et al., 2006; Schneider & Logan, 2015). To the extent that the planning and execution of these routines meet the immediate task requirements, the future will be predictable, and so we avoid being surprised.

Occasionally, unfamiliar environments or unexpected outcomes increase our uncertainty about what to do next. When this happens, the brain switches from exploiting past experiences to exploring new possible approaches (O'Reilly, 2013; Soltani & Izquierdo, 2019; Domenech et al., 2020). In professional settings, this exploratory uncertainty and the pauses and hesitations that it generates are often early indicators of deteriorating performance (Ott et al., 2018). Currently, there is no good way to predict how long the uncertainty will last. The ability to rapidly and quantitatively measure uncertainty would have implications for educational and training efforts by supporting in-progress corrections, or generating forecasts about future disruptions, or using identified periods of uncertainty to target reflective discussions about past actions.

The links between elevated NI and uncertainty were shown to be common during short periods (~1 min) of verbalized uncertainty, or during submarine navigation, while the data needed to establish the submarine's position was being collected and shared among the navigation team (Stevens & Galloway, 2017). The links were extended to include medical students, hospital anesthesiologists, and operating room staff (i.e., circulating nurse, scrub nurse, and neurosurgeon, etc.) during simulation training when they experienced difficulties while ventilating a patient or deciding a course of patient management (Stevens et al., 2016).

While originally the elevated NI was described in the context of spoken uncertainty, these elevations more generally occur during stressful periods whether or not someone was speaking (Stevens et al., 2017). These associations were not a product of the simulation environment; they were also seen in two neurosurgeons and an anesthesiologist during a live-patient surgery (Stevens et al., 2019). Finally, these linkages have been made more explicit by training artificial neural networks to recognize pattern variations in the NI peaks associated with verbalized uncertainty (Stevens & Galloway, 2019, 2021).

ESTIMATING THE FREQUENCY MAGNITUDE AND DURATION OF UNCERTAINTY

The picture emerging is that as simulations (and real-world events) evolve, the neurodynamic information accumulates and the bits accumulated are a function of the frequency, magnitude, and duration of periods of uncertainty. More experienced teams accumulate less NI during a task by virtue of having fewer, smaller, and/or shorter periods of uncertainty than less experienced teams. These features are present in healthcare, military, and pre-college teams and appear to be a general property of human performance (Stevens & Galloway, 2017, 2019, 2021).

From a training and feedback perspective, important questions are: How frequently does uncertainty occur? What is the level of uncertainty? Where in the brain do elevated NI levels come from? How long will the uncertainty last?

More practically, how can these estimates be used to improve the efficiency and effectiveness of training? The team neurodynamic profiles in Figures 12.3 and 12.4 provide an analytic context for approaching these questions.

These figures show an experienced team that had previously worked together in the operating room. Brainwaves were collected from an anesthesiologist (AN), a circulating nurse (CN), a scrub nurse (SN), and a standardized participant (SP) acting in the role of the neurosurgeon.

There was minimal talk during the first 11 min of the simulation, while the patient and suite were prepared for neurosurgery. There were occasional low-level peaks of NI, while the AN intubated the patient, the surgeon was gloved and gowned, and the patient draped. Just before the surgery began, signs of malignant hyperthermia (MH) were recognized, indicating the start of a life-threatening event that can occur in persons with a rare genetic condition following the inhalation of anesthetic gasses.

Elevations in CO_2 levels combined with an increase in heart rate as seen in Figure 12.3 at 680 s were the first signs of this life-threatening hypermetabolic event, and this triggered the AN to issue requests to the team such as to order chemistries, contact the pharmacy for Dantrolene (a muscle relaxant), and obtain ice for cooling the patient. The AN spoke the most during the next 16 min as she coordinated these efforts (Figure 12.3A).

The AN also showed the greatest NI elevation at the onset of the MH event. The CN who was assisting the AN showed moderate NI levels until toward the end of the simulation when she prepared sodium bicarbonate and Dantrolene solutions for injection. The SN showed moderate NI while preparing the patient for surgery, and low NI levels for the remainder of the simulation. Across the four team members during the scenario, the NI levels were not significantly correlated (all p-values >0.16) (Table 12.1).

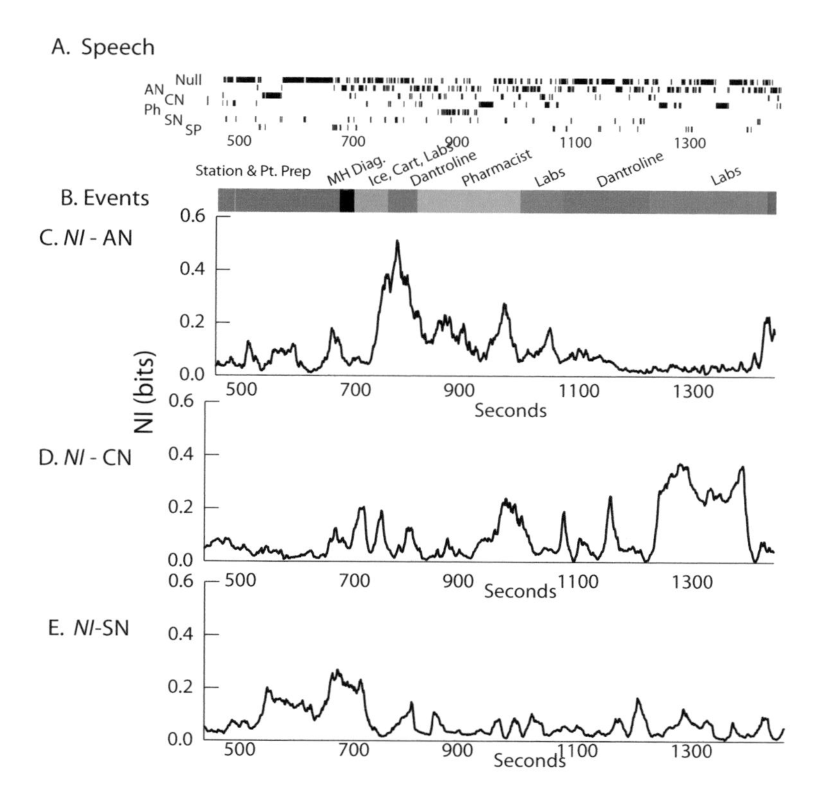

FIGURE 12.3 The scalp-averaged NI levels of an experienced medical team performing a simulated case of malignant hyperthermia. (A) Periods of speech are shown for the AN, CN, SN, the pharmacist, and standardized participant. (B) The segments highlight the major sections of patient management. (C, D, E) The individual NI profiles of the AN, CN, and SN are shown on the same scale.

To illustrate how the analysis can be moved to a lower and finer-grained level, scalp-wide NI levels were calculated for the AN using the 19 channels arranged from the front of the scalp toward the back (Figure 12.4A). The AN was selected for this analysis as she was responsible for directing the subsequent management of the patient (Figure 12.4). This composite map shows that elevated NI levels were not uniform over time or across the channels but varied with the evolving situation (highlighted in Figure 12.4B).

This composite begins at 680 s when the elevated HR and CO_2 levels were first mentioned. A small elevation of NI rapidly occurred followed by near baseline levels for ~30 s while the initial requests were made for chemistries and a cart supplied with sodium bicarbonate and the muscle relaxant Dantrolene. The NI then rapidly rose as the requests were addressed, with NI increases in the frontal (Fz), central (C3, Cz, C4), and temporal/parietal sensors (T4, Pz, P4). As described in more detail below, brain activities located below these channels are part of the default mode

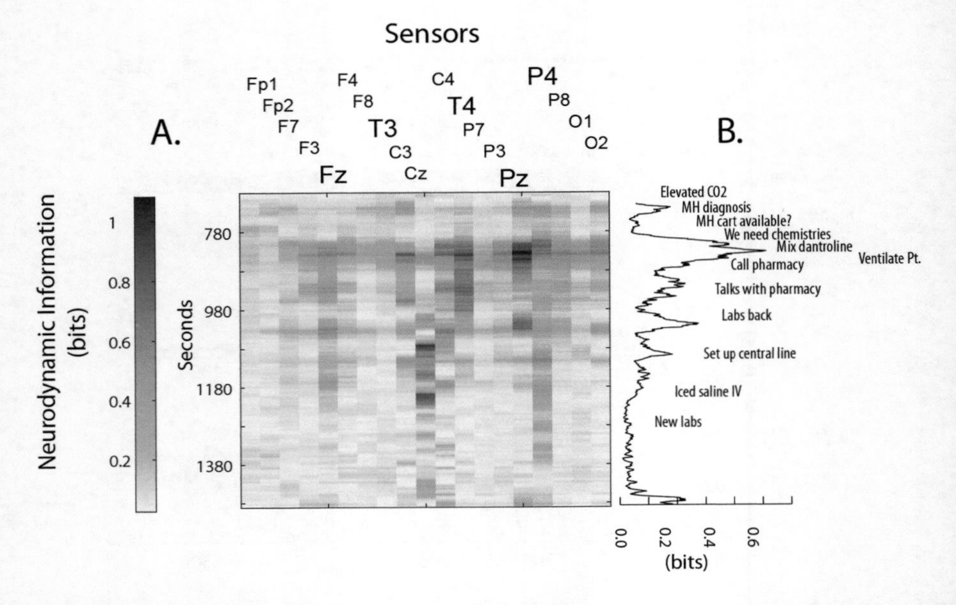

FIGURE 12.4 (A) The NI values for the AN in Figure 12.3 are plotted for the 19 EEG channels arranged from the front of the scalp toward the back according to the 10/20 labeling system. The channels in the larger bold font are the sites for default mode network activity. (B) The sensor averaged NI values are annotated with key events during the simulation.

TABLE 12.1
NI Correlations for the Team, *p*-Values –0.16

	AN	CN	ST	SP
AN	1.00			
CN	–0.24	1.00		
ST	–0.13	–0.12	1.00	
SP	–0.14	–0.12	–0.25	1.00

network (DMN), which is involved in planning, and the sensory-motor network involved in imagining and executing motor activities. The heterogeneity of brain regions involved suggests sub-goal exploration, which requires multitasking between cognitive processes as the solution path(s) for future sequential cognitive processes are established. This initial NI peak lasted ~100 s and included a smaller shoulder peak when the call was made to the pharmacy to request additional Dantrolene. These initial peaks would be considered prolonged as previous studies have shown that the NI duration of uncomplicated patient management averages around 40–45 s for healthcare simulations (Stevens & Galloway, 2021).

A major cognitive shift occurred around 1,000 s and the NI activity in the Fz, T4, and Pz channels decreased and NI activity in the Cz and P4 channels increased. These changes occurred when the lab results came back and when the team decided to set up an arterial line. This cognitive shift was accompanied by a slow NI decline toward baseline levels as the AN switched from planning the patient management to executing the plan.

This performance illustrates dynamical features of NI that have practical significance for training. The elevated NI at different channels suggests that different computations are being performed in these brain regions, each with its own degree of uncertainty. The first question is why would these computations be important for training? Previous studies have shown that difficulties executing motor activities (i.e., intubating a patient or controlling an instrument) are often accompanied by elevated NI levels in the central region (C3, Cz, C4), which provide an indication of the difficulties they are having (Stevens & Galloway, 2017, 2019, 2021; Stevens et al., 2016; Stevens, Galloway & Willemsen-Dunlap, 2018). The AN in this simulation rapidly intubated the patient with only a minor NI elevation around 550–600 s.

The current simulation also involved lengthy periods of mixing Dantrolene with saline prior to injection, and while the AN was not directly involved in the mixing, she was advising the CN and SN on the best procedure for doing so. Closely watching others perform a procedure can deactivate mu rhythms in the sensorimotor brain region, and as described previously, this would result in increased NI (Stevens & Galloway, 2021).

The parallel sensor activations in the Fz (anterior medial prefrontal cortex), Pz (posterior cingulate cortex), and T4 (angular gyrus) sensors, also known as the default mode network (DMN), during the early planning stages (Figure 12.4) illustrate uncertainty in areas that are associated with collecting and analyzing complex data (Čeko et al., 2015). An emerging view of the brain shows that much of its activity is spent in maintaining accurate models of past events and estimating the sensory results that would occur if particular action(s) were taken based on these models (Hawkins, 2021). The above-mentioned distributed set of brain regions collectively termed the default mode network, has been proposed to be a network where prior intrinsic information that is continuously accumulated over seconds to minutes is melded with arriving extrinsic sensory information (Yeshurun et al., 2021). This temporally extended accumulated information is tied to the context of the stimulus, much like when you read a story you continuously link the current chapter with events and characters in the previous chapters. In this simulation, the absence of increased NI levels in elements of the DMN might actually be a cause for concern.

AUGMENTING DEBRIEFINGS WITH NEURODYNAMICS

If simulations provide an opportunity for honing process skills, debriefings provide the opportunity for trainees to learn what they don't know. Debriefing is considered a key element for transferring experiences in simulation-based education into learning that can be applied to patient care. In order for this transformation to happen, experts agree that reflection, currently mediated through conversation, is necessary (Baker et al., 1997).

Fanning and Gaba (2007) defined debriefing as "a facilitated or guided reflection in the cycle of experiential learning." These facilitated conversations involve the group's identification of significant teamwork and clinical events that serve as exemplars to guide learning and adapt current behaviors to future challenges.

The literature on trainee simulation and debriefing has been evolving over the past 30 years (Lederman, 1992; Rudolph et al., 2006; Eppich & Cheng, 2015; Seelandt et al., 2021). Currently, useful frameworks, guidelines, and standards exist (International Association for Clinical Simulation & Learning Board of Directors, 2013); however, little is known about debriefing from a dynamical perspective that is linked to the most fundamental neural processes of learning. Debriefing frameworks have practical value, but provide few theoretical suggestions or quantitative data to accelerate reflective learning for simulation participants or the journey to competence for debriefers. There is a sense in the simulation and research communities that more dynamic models, objective metrics, and theories of debriefing are needed (Fanning & Gaba, 2007; Bowe et al., 2017), but empirical evidence about what these models might look like is sparse. Neurodynamic modeling as described in the first section of this chapter may help fill this gap, although perhaps not initially at the level of trainees. Successful adoption of new educational resources needs time and a full understanding by, in this case, the debriefers.

USING NEURODYNAMIC ANALYSES TO TRAIN THE TRAINERS

The time required to achieve competency as a debriefer is variable, requiring the expertise of an experienced mentor. Factors such as understanding key principles, practice opportunities, self-reflection, and expert feedback are all important. Just as simulation and debriefing have the potential to standardize clinician education, we believe that a combined analysis of team NI and video, in conjunction with discussion and expert feedback, has the potential to standardize the formation of debriefers and decrease time to competency.

As described earlier, NI may be thought of as streams of information capable of highlighting the frequency, magnitude, and duration of individual and team uncertainty. It is represented as bits of data. Figure 12.5A shows the NI profile for a single team member during the scenario of a simulation. The average NI for the performance was 0.085 bits. This NI level (dot in Figure 12.5B) is compared with 16 other prior performances and was slightly lower than the average of the other teams; i.e., this was a good performance. During the debriefing the instructor has just clicked on a peak with 0.25 bits of information and this triggers the playing of the team video starting 30 s before the peak and extending 30 s past the peak.

While the highlighted peak is shown as an illustration, during practice the debriefer would likely want to view the final 200 s and try to surface from the

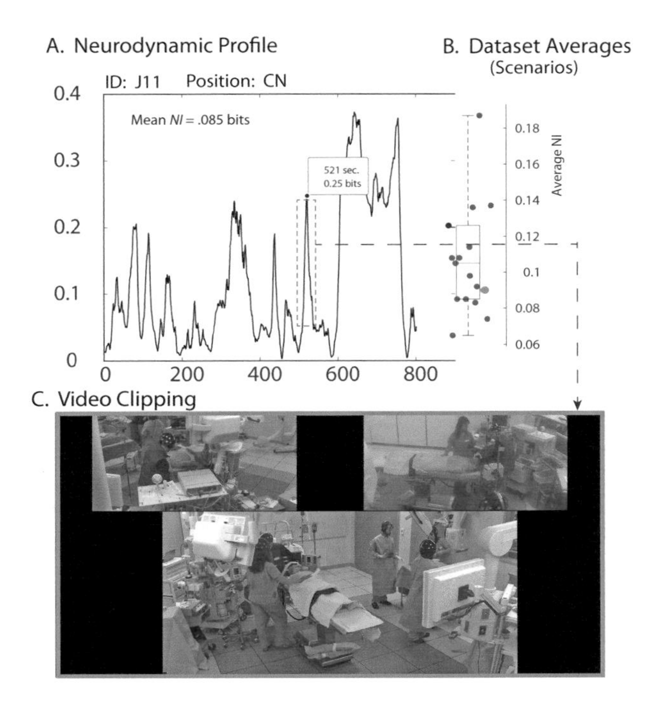

FIGURE 12.5 Sample dashboard for linking neurodynamics with team activities. (A) The NI profile of a team member participating in a simulation of malignant hyperthermia. (B) The scalp-average NI for 17 team members during simulation scenarios. (C) A video clip of the period highlighted by the dotted text box in Figure 12.4A.

participant what was happening during this period where there was elevated NI and little speech (Figure 12.3A).

While Figure 12.5 demonstrates NI for a single individual, dashboards are also available for each member of the team in the simulation, much like Figure 12.3, and these can be generated in near real time. In the next section, we offer several use cases for the use of NI data, combined with video review, to understand teamwork and individuals' contributions to the team in the context of debriefing.

Early Novices

Early novice debriefers would watch complete recordings of simulations. Debriefing mentors would ask them to note segments they would debrief, their reasons for selecting each segment, and how they tie those segments to pre-defined learning objectives. Mentors would then review the novice debriefer's responses with them and introduce NI data to explore the correlation between that data and actions observed during the performance video. The purpose of this is to identify portions of the simulation, based on both observational and quantitative data of team members' uncertainty, for which they could then practice formulating debriefing questions. Discussion and feedback between debriefing mentor and novice on the segments chosen for debriefing, what the novice debriefer observed, the debriefing questions, and their relationship to learning objectives are crucial to making this approach successful. A talk-out-loud review in which the mentor describes key behaviors they observe in the video, its linkage to NI data, and questions they might ask in debriefing may be beneficial.

Later Novices

Novice debriefers with a measure of experience could potentially benefit from a more advanced form of this exercise. This group would watch recordings in which simulation participants did not demonstrate neurodynamic uncertainty during scenarios that either showed potential or became critical due to inaction or incorrect action. Novice debriefers would be asked to formulate debriefing questions to determine whether participants' gaps were around knowledge, skills, or attitudes. An akin review and discussion, including a talk-out-loud approach, is expected to be similarly beneficial. Similar exercises, conducted by trained, peer debriefing coaches, would be used by intermediate and advanced debriefers to further hone and extend their debriefing skills.

Evolving the Technology

Currently, the NI data for training debriefers contain six simulations with videos and NI profiles, which makes them well-suited as a training tool mentors can use to extend and refine the skills of debriefers at all skill levels. It is also reasonable to expect these static snapshots of NI to evolve into real-time debriefing aids displaying dashboards with the NI of each team member, easy time-stamping, and annotation by debriefers. Future iterations of such technology will improve the ability to provide reliable assessments capable of comparing individuals and teams across time, training programs, and team member composition.

SUMMARY

In this chapter, we have illustrated a quantitative framework for parsing the neurodynamic correlates of a team's uncertainty into those of the team members, and then to the brain regions where the uncertainty arose and then to the brain rhythms active during periods of uncertainty at those sites. The quantitative scale of neurodynamic information is bounded (meaning it can never be less than 0 and never more than the maximum information of the symbol system), and generates the ability to aggregate the bits of information and compare team members of teams, or different teams, or the effects of different learning experiences on team members and teams. In other words, neurodynamic information is a causal intermediate between low-level neural processes and the organizations that we recognize as important for teams, and it is one that tracks closely with the observed hesitations and pauses that we associate with uncertainty.

REFERENCES

Baker, A. C., Jensen, P. J., & Kolb, D. A. (1997). In conversation: Transforming experience into learning. *Simulation & Gaming, 28*(1), 6–12. https://doi.org/10.1177/1046878197281002

Bigdely-Shamlo, N., Mullen, T., Kothe, C., Su, K. M., & Robbins, K. A. (2015). The PREP pipeline: Standardized preprocessing for large-scale EEG analysis. *Frontiers in Neuroinformatics, 9*, B153.

Bodala, I. P., Li, J., Thaor, M. V., & Al-Nashas, H. (2018). EEG and eye tracking demonstrate vigilance enhancement with challenge integration. *Frontiers in Human Neuroscience*, 10–273. https://doi.org/10.3389/fnhum.2016.00273

Bonnefond, M., & Jensen, O. (2015). Gamma activity coupled to alpha phase as a mechanism for top-down controlled gating. *PLoS One, 10*, e012866.

Bowe, S. N., Johnson, K., & Puscas, L. (2017). Facilitation and debriefing in simulation education. *Otolaryngologic Clinics of North America, 50*(5), 989–1001. https://doi.org/10.1016/j.otc.2017.05.009

Čeko, M., Gracely, J. L., Fitzcharles, M. A., Seminowicz, D. A., Schweinhardt, P., & Bushnell, M. C. (2015). Is a responsive default mode network required for successful working memory task performance? *The Journal of Neuroscience: The Official Journal of the Society for Neuroscience, 35*(33), 11595–11605.

Cooper, R., & Shallice, T. (2000). Contention scheduling and the control of routine activities. *Cognitive Neuropsychology, 17*(4), 297–338.

Creutzfeldt, J., Hedman, L., & Felländer-Tsai, L. (2016). Cardiopulmonary resuscitation training by avatars: A qualitative study of medical students' experiences using a multiplayer virtual world. *JMIR Serious Games, 4*, e22.

Daniels, B., Flack, J., & Krakauer, D. (2017). Dual coding theory explains biphasic collective computation in neural decision-making. *Frontiers in Neuroscience* 2017. https://doi.org/10.3389/fnins.2017.00313

Daw, N. D., O'doherty, J. P., Dayan, P., Seymour, B., & Dolan, R. J. (2006). Cortical substrates for exploratory decisions in humans. *Nature, 441*(7095), 876–879.

Delorme, A., & Makeig, S. (2004). EEGLAB: An open source toolbox for analysis of single-trial EEG dynamics including independent component analysis. *Journal of Neuroscience Methods, 134*(1), 9–21.

Domenech, P., Rheims, S., & Koechlin, E. (2020). Neural mechanisms resolving exploitation-exploration dilemmas in the medial prefrontal cortex. *Science, 369*, 1076.

Eppich, W., & Cheng, A. (2015). Promoting Excellence and Reflective Learning in Simulation (PEARLS): Development and rationale for a blended approach to health care simulation debriefing. *Simulation in Healthcare: Journal of the Society for Simulation in Healthcare, 10*(2), 106–115. https://doi.org/10.1097/SIH.0000000000000072

Fanning, R. M., & Gaba, D. M. (2007). The role of debriefing in simulation-based learning. *Simulation in Healthcare: Journal of the Society for Simulation in Healthcare, 2*(2), 115–125. https://doi.org/10.1097/SIH.0b013e3180315539

Flack, J. (2017). From matter to life: Information and causality. In S. I. Walker, P. C. W. Davies, & G. F. R. Ellis (Eds.), *Life's information hierarchy* (pp. 283–302). New York: Cambridge University Press.

Flack, J. (2017a). Coarse-graining as a downward causation mechanism. *Philosophical Transactions of the Royal Society A, 375*, 20160338.

Flack, J., Erwin, D., Elliot, T., & Krakauer, D. (2012). Timescales, symmetry, and uncertainty reduction in the origins of hierarchy in biological systems. In K. Sterelny, R. Joyce, B. Calcott, & B. Fraser (Eds.), *Cooperation and its evolution* (pp. 45–74). Cambridge, MA: MIT Press.

Gillon, C. J., Pina, J. E., Lecoq, J. A., Ahmed, R., Billeh, Y. N., Caldejon, S., ... & Zylberberg, J. (2021). Learning from unexpected events in the neocortical microcircuit. *BioRxiv*, 2021-01.

Goodwin, C., Velasquez, E., Ross, J., Kueffer, A. M., Molefe, A. C., Modali, L., Bell, G., Delisle, M., & Hannenberg, A. A. (2021). Development of a novel and scalable simulation-based teamwork training model using within-group debriefing of observed video simulation. *Joint Commission Journal on Quality and Patient Safety*. S1553-7250(21)00035-0. Advance online publication. https://doi.org/10.1016/j.jcjq.2021.02.006

Gorman, J. C., Martin, M. J., Dunbar, T. A., Stevens, R. H., Galloway, T. L., Amazeen, P. G., & Likens, A. D. (2016). Cross-level effects between neurophysiology and communication during team training. *Human Factors*, *58*(1), 181–199.

Hawkins, J. (2021). *A thousand brains: A new theory of intelligence*. New York: Basic Books. ISBN 1-5416-7581-9.

https://www.jumpsimulation.org/research-innovation/our-blog/2016/may/the-potential-of-virtual-reality-in-simulation

International Association for Clinical Simulation & Learning Board of Directors. (2013). Standards of best practice: Simulation. *Clinical Simulation in Nursing*, *9*(6, Supplement), 1–32. ISSN 1876-1399. https://doi.org/10.1016/j.ecns.2013.05.008

Klimesch, W. (2012). Alpha-band oscillations, attention and controlled access to stored information. *Trends in Cognitive Sciences*, *16*(12), 606–617.

Klimesch, W., Sauseng, P., & Hanslmayr, S. (2006). EEG alpha oscillations: The inhibition-timing hypothesis. *Brain Research Reviews*, *53*, 63–88. PMID 16887192 https://doi.org/10.1016/j.brainresrev.2006.06.003

Lachaux, J. P., Jung, J., Dreher, J. C., Bertrand, O., Minotti, L., Hoffman, D., & Kahane, P. (2008). Silence is golden: Transient neural deactivation in the prefrontal cortex during attentive reading. *Cerebral Cortex*, *18*, 443–450.

Lateef, F. (2010). Simulation-based learning: Just like the real thing. *Journal of Emergencies, Trauma, and Shock*, *3*(4), 348–352. https://doi.org/10.4103/0974-2700.70743

Lederman, L. C. (1992). Debriefing: Toward a systematic assessment of theory and practice. *Simulation & Gaming*, *23*(2), 145–160. https://doi.org/10.1177/1046878192232003

Maytin, M., Daily, T. P., & Carillo, R. G. (2015). Virtual reality lead extraction as a method for training new physicians: A pilot study. *Pacing and Clinical Electrophysiology*, *38*, 319–325.

Mullen, T. R., Kothe, C. A., Chi, Y. M., Ojeda, A., Kerth, T., Makeig, S., Jung, T. P., & Cauwenberghs, G. (2015). Real-time neuroimaging and cognitive monitoring using wearable dry EEG. *IEEE Transactions on Bio-Medical Engineering*, *62*(11), 2553–2567.

Oostenveld, R., Fries, P., Maris, E., & Schoffelen, M. (2011). FieldTrip: Open source software for advanced analysis of MEG, EEG, and invasive electrophysiological data. *Computational Intelligence & Neuroscience*, 2011, Article ID 156869, 9 pages. https://doi.org/10.1155/2011/156869.

O'Rielly, T. X. (2013). Making predictions in a changing world: Inference, uncertainty and learning. *Frontiers in Neuroscience*, *7*, 105.

Ossandon, T., Jerbi, K., Vidal, J. R., Bayle, D. J., Henaff, M. A., Jung, J., Minotti, L., Bertrand, O., Kahane, P., & Lachaux, J. P. (2011). Transient suppression of broadband gamma power in the default mode network is correlated with task complexity and subject performance. *Journal of Neuroscience*, *31*, 14521–14530.

Ott, M., Schwartz, A., Goldsmith, M., Bordage, G., & Lingard, L. (2018). Resident hesitation in the operating room: Does uncertainty equal incompetence? *Surgical Training, Medical Education*, *52*, 851–860.

Rudolph, J. W., Simon, R., Dufresne, R. L., & Raemer, D. B. (2006). There's no such thing as "nonjudgmental" debriefing: A theory and method for debriefing with good judgment. *Simulation in Healthcare: Journal of the Society for Simulation in Healthcare, 1*(1), 49–55. https://doi.org/10.1097/01266021-200600110-00006

Rumelhart, D. E. (1980). On evaluating story grammars. *Cognitive Science, 4*(3), 313–316. https://doi.org/10.1207/s15516709cog0403_5

Schank, R., & Abelson, R. (Eds.) (1977). *Scripts, plans, goals, and understanding: An inquiry into human knowledge structures*. Hillsdale, NJ: Lawrence Erlbaum Associates.

Schmidt, E., Goldhaber-Fiebert, S. N., Ho, L. A., & McDonald, K. M. (2013). Simulation exercises as a patient safety strategy: A systematic review. *Annals of Internal Medicine, 158*(5 Pt 2), 426–432. https://doi.org/10.7326/0003-4819-158-5-201303051-00010

Schneider, D. W., & Logan, G. D. (2015). Chunking away task-switch costs: A test of the chunk-point hypothesis. *Psychonomic Bulletin & Review, 22*, 884–889.

Sedley, W., & Cunningham, M. O. (2013). Do cortical gamma oscillations promote or suppress perception? An under-asked question with an over-assumed answer. *Frontiers in Human Neuroscience, 7*, article 595.

Seelandt, J. C., Walker, K., & Kolbe, M. (2021). "A debriefer must be neutral" and other debriefing myths: A systemic inquiry-based qualitative study of taken-for-granted beliefs about clinical post-event debriefing. *Advances in Simulation (London, England), 6*(1), 7. https://doi.org/10.1186/s41077-021-

Soltani, A., & Izquierdo, A. (2019). Adaptive learning under expected and unexpected uncertainty. *Nature Reviews Neuroscience, 20*, 435–544.

Stevens, R., & Galloway, T. (2017). Are neurodynamic organizations a fundamental property of teamwork? *Frontiers in Psychology,* May, 2017. https://doi.org/10.3389/fpsyg.2017.00644

Stevens, R., & Galloway, T. (2019). Teaching machines to recognize neurodynamic correlates of team and team member uncertainty. *Journal of Cognitive Engineering and Decision Making, 13*, 310–327. https://doi.org/10.1177%2F1555343419874569

Stevens, R., & Galloway, T. (2021). Parsing neurodynamic information streams to estimate the frequency, magnitude and duration of team uncertainty. *Frontiers in Systems Neuroscience*, 01 February 2021 Available from:// https://doi.org/10.3389/fnsys.2021.606823

Stevens, R., Galloway, T., Halpin, D., & Willemsen-Dunlap, A. (2016). Healthcare teams neurodynamically reorganize when resolving uncertainty. *Entropy, 18*, 427. https://doi.org/10.3390/e18120427

Stevens, R., Galloway, T., Lamb, J., Steed, R., & Lamb, C. (2017). Linking team neurodynamic organizations with observational ratings of team performance. In A. A. Von Davier, P. C. Kyllonen, & M. Zhu (Eds.), *Innovative assessment of collaboration* (pp. 315–330). Cham, Switzerland: Springer International Publishing.

Stevens, R., Galloway, T., & Willemsen-Dunlap, A. (2018). Quantitative modeling of individual, shared and team neurodynamic information. *Human Factors, 60*, 1022–1034.

Stevens, R., Galloway, T. L., & Willemsen-Dunlap, A. (2019). Advancing our understandings of healthcare team dynamics from the simulation room to the operating room: A neurodynamic perspective. *Frontiers in Psychology, 10*, 1660. https://doi.org/10.3389/fpsyg.2019.01660

Stevens, R., Gorman, J. C., Amazeen, P., Likens, A., & Galloway, T. (2013). The organizational neurodynamics of teams. *Nonlinear Dynamics, Psychology, and Life Sciences, 17*(1), 67–86.

Surgical Theater. (2021, April). *Virtual reality for surgery: Precison XR*. Retrieved April, 2021, from http://surgicaltheater.net/

Wianda, E., & Ross, B. (2019). The roles of alpha oscillations in working memory retention. *Brain and Behavior*, *9*, e01263. https://doi.org/10.1002.brb3.1263

Yishurun, Y., Nguyen, M., & Hasson, U. (2021). The default mode network: Where the idiosyncratic self meets the shared social world. *Nature Reviews Neuroscience*, *22*, 181–192.

Zertuche, J. P., Connors, J., Scheinman, A., Kothari, N., & Wong, K. (2020). Using virtual reality as a replacement for hospital tours during residency interviews. *Medical Education Online*, *25*(1), 1777066.

13 The Future of Simulation

P. A. Hancock

CONTENTS

PROEM

I argue that, in the coming decades, the conception of simulation will undergo a metamorphosis as the fundamental assumptions about what constitutes simulation evolve under the driving force of progressive technological innovation. The primary stimulus for development will come from the need to explore all processes through which humans interact with technology. Such future interaction will find operators working on representations of task spaces, presented via diverse forms of sensory display (see Mouloua et al., 2003). As the linkage between these display representations and actual system configurations will be contingent solely upon the software connections—and given that the metaphor for representation will be judged by its operational effectiveness, not the degree to which it replicates the appearance of the actual system—the difference between what is simulation and what is actual operation will disappear. The definition of simulation in such circumstances will depend solely on whether the operator actually effects change in the real-world system or is alternatively using, evaluating, or training at the time on exactly the same display connected to an electronic surrogate. In multiple-operator, multiple-system configurations, even this criterion will eventually fail to hold any permanent

DOI: 10.1201/9781003401360-13

distinction because momentary control of the action affecting the system will be passed between individuals at different times. At this juncture, simulation will have passed the Turing test for reality.

In systems where the only criterion difference between real and surrogate worlds is visual fidelity (conceived for enclosed control-room activities such as emergency response centers), it is feasible that visual projection capabilities will soon meet or exceed that which the eye encounters in the real world. It is to be anticipated that further progress in the visual realm will take us toward *supersimulation*, in which what can be seen in a computer-generated reality exceeds that which can be seen outside such a facility. Given our knowledge of the human visual system and the technical focus on improving visual graphical representations, we are now passing quickly into this evolutionary stage.

Supersimulation in vision will stimulate the desire for supersimulation in all other sensory modalities, and in the foreseeable future, we shall pass the comparable Turing tests for all of the major sensory systems. At that juncture, we shall be incapable of distinguishing between a computer-generated and a physically generated world. Human factors will have a critical voice in contributing to and evaluating these developments because many of the barriers along this avenue of progress are composed of questions about human–system interaction. To create such supersimulated worlds, we shall want to know much more about cognitive and emotional capacities and individual variations in human abilities and attributes.

Early improvements may well be seen in relation to working with handicapped individuals who will benefit most immediately from the blurring of reality and simulation. Some of these notions have been expressed in both science fiction and film media, and their aspirations tend to anticipate scientific progress. The fundamental problems that follow will then concern the very nature of experiential reality itself. This is a philosophical issue and is addressed at the end of this chapter, where I seek to show what constraints bind us to our present reality, and what forces may emerge to divorce us from what we have previously been pleased to call the "real world."

The intentions of a tool are what it does. A hammer intends to strike, a vise intends to hold fast, a lever intends to lift. They are what they are made for. But sometimes a tool may have other uses that you don't know. Sometimes in doing what you intend, you also do what the (tool) intends without knowing (Pullman, 2000).

THE FUNDAMENTAL AND PRACTICAL REASONS FOR SIMULATION

Similar to robotics and artificial intelligence (AI), the bedrock impetus behind simulation is its capacity to support the efforts of human beings to artificially recreate themselves, and to control the world around them. Unlike robotics and AI, simulation is neither necessarily anthropocentric nor fundamentally anthropomorphic. Nevertheless, simulation has, throughout its history, been focused largely on the creation and recreation of "real" environments, often under the mandate of either entertainment or more serious practical necessities. The most obvious stimulus, and therefore the source of support for simulation, will continue to be the interest in gaming and in the training of individuals and teams for subsequent performance in real and often dangerous situations. Consequently, in the immediate future, simulation is likely to continue on its present course and follow these established trends.

One growing concern, however, will emanate from the need for researchers to explore complex dynamic processes in disciplines ranging from chemistry and physics to geology, mathematics, and medicine. In fact, virtually all areas where humans pursue understanding can and will benefit from the dynamic and malleable representations that simulation renders. This will be a burgeoning aspect of the world of applied simulation. Even in light of this growing range of applications, the fundamental motivation behind future simulation will remain essentially involved with our never-ending quest to attain mastery over that which we can presently perceive, but over which we cannot at present exert control (Hancock, 1997a). Given such traditional and emerging drives, such as the desire for control and a pragmatic need for exploration in surrogate or transferred environments, it is an exercise in both logic and imagination to distill the future of simulation.

SIMULATIONS IN THE PAST

Those who want to see far into the future must first look well into the past. So, before we address what might be coming, a glimpse into the past can help us set our quest in motion. The first thing we need to recognize is that simulation is not a modern invention. Indeed, models and representations for practical employment have been around for many centuries. For example, the architectural model for San Petronio in Bologna, Italy, built in 1390, was 59 ft long and allowed people to walk inside to visualize what the finished building would be like. There is even evidence that Greek architects employed models in a similar manner some five centuries before the birth of Christ (see King, 2001). Clearly then, modeling and simulation are nothing new. Dependent upon the degree to which we let our definitional boundaries of the word "simulation" dissolve, we can even include artifacts such as religious icons as representations or simulations. One particularly interesting example can be seen in some of the great European Gothic cathedrals where, on the floor of the nave, was inlaid a maze, the one remaining at Chartres in northern France being an outstanding example (Figure 13.1). The primary purpose of these constructions was to act as surrogates for pilgrimages to Jerusalem. For those too old, too sick, too poor, or too pressed for time to accomplish the actual journey, the maze allowed completion of the journey symbolically. Eventually, of course, we shall recover this symbolic aspect for our more technically replete simulations.

The true origin of modern technical simulation may be attributed to Brunelleschi's demonstration of perspective. A Florentine architect, Filippo Brunelleschi (1377–1446) is best known for his construction of the dome of Santa Maria del Fiore, which remains the visual icon representative of Florence to the present day. Prior to Brunelleschis interest and demonstration in perspective, medieval painting was predominantly nonrepresentational and, to us, a strange mixture of two-dimensional and fractal depictions (and see the discussion of the use of the camera obscura technique by Hockney, 2001). Brunelleschi's concern was not expressed in terms of the

FIGURE 13.1 The maze at Chartres Cathedral, one of the few remaining ones. It was used as a surrogate representation for symbolic pilgrimages to the Holy Land. An early "simulation," its validity depended upon the credence ascribed by the user. Our future simulations will also eventually embrace this functional symbolism.

word "simulation," but this was exactly what he achieved. With respect to the object of simulation, he chose one of the city's most famous sites, one which would be immediately recognizable to his fellow Florentines—the Baptistry of San Giovanni. Standing just inside the door of the Cathedral of Santa Maria del Fiore (with which his name would forever be linked), Brunelleschi painted a small panel of the baptistry outside in the correct perspective, with the cathedral doorway as a frame. Using the vanishing point as his reference location, he created a small hole in the panel. Thus, Brunelleschi could replicate exactly what individuals would see when they looked out of the door, by replacing the real scene with a static painted representation. This element of the simulation worked excellently. However, Brunelleschi faced the problem of simulation dynamics. It was acceptable that people did not appear in his painted scene; after all, the locale was not always populated with pedestrians. But what of the sky? Although there might be no moving terrestrial objects, were not the clouds always moving? Brunelleschi came up with an ingenious solution, which was characteristic of his highly inventive capacities. He solved the problem by making the display a hybrid one. The top part of the display panel was covered with a mirror-like surface which reflected the actual sky above—surely one of the first examples of mixed reality (see Figure 13.2). The illusion was so convincing that it changed how people (and particularly how artists) represented their world (see also Hockney,

FIGURE 13.2 Illustration of Brunelleschi's mixed reality simulation. This illustration of the middle 1400s shows the situation as it would appear to the observer. At the center top of the door, one can see the eye-point and the hole through which the observer views the scene. In the mirror that the right hand is holding, one can see the painted panel secured by the fingers of the left hand. The mirror, in this case, shown as a circular one, can be temporarily removed and the observer then sees the real scene, shown in the figure at the edges of the mirror, for example, where the chin appears. The painted panel can be repositioned over the real scene to provide the artist's surrogate. The panel surface above the painted building is a reflective one so as to allow the sky to change dynamically and further complete the illusion. Arguably the origin of formal simulation, this was a magical wonder when first created. (Illustration by Lauriann Jones.)

2001). Today, we should applaud this inventor as perhaps the spiritual godfather of simulation.

With the impact of Brunelleschi's demonstration, it was little wonder that subsequent competitions awarding commissions to construct specific parts of the Cathedral of Santa Maria del Fiore required the proposer to produce a model so that the city fathers could judge between different conceptions. One has the feeling that Filippo Brunelleschi would have been very happy at today's computer-aided design (CAD) station, visualizing the innovations he is now recognized for. But his unique simulation was for just this one fixed scene, and if any of the circumstances changed, such as the observer turning around, the illusion was immediately lost. The freedom to explore completely simulated worlds had to wait hundreds of years, until the present era. Indeed, only recently have we arrived at the point where we can now prognosticate the future of simulation.

ON PREDICTING THE FUTURE

Predicting the future has many advantages. First, many of the developments that we are bound to see are direct, linear extrapolations of currently existing trends. For example, display resolution will improve, and networked computational systems will increase in capability and will be used to enhance multimodel experience. The design and facility of head-mounted units will undergo significant improvements. Indeed, for those who have ever seen the "Dayton Grasshopper," this latter line of progress will already be evident. With respect to computational capacity, although we have not reached the clear physical barriers to computer speed, it is important to understand that we shall soon have to be concerned with the impending approach of these inherent limits. Despite the continuing progress in recent decades (see Moravec, 1988), it is evident that we cannot continue to double computational capacity at the rate we have been doing. However, the point in time when fundamental physical computational constraints will start to curtail simulation enacted on a network of parallel machines is perhaps a little too far in the future to affect the expected lifespan of the present chapter. Other linear forms of progression will include much greater research efforts on the integration of visual input with input from other sensory systems to enhance the overall experience of "presence." Audition, tactile stimulation, and olfaction will each play greater roles in improving the fidelity of all simulation. As a consequence, we shall have to make significant methodological progress in understanding and measuring the sense of presence in order to gauge the evolving state of the art.

Thus, the trick in prognostication is to attach some spurious numeracy to these predictions of linear or geometric progression and then feign surprise and publicize to the greatest possible extent the moments when such states of development are reached. Recognition then as a futurist depends upon a fallacy in readers' memories in which the author emphasizes all the successful predictions and relies on the frailty of human memory in failing to recall all the predictions made by the futurist that failed to reach fruition. I shall not engage in such legerdemain; but I am warning the interested reader that touts of all sorts, be they astrologers, financial advisors, or tabloid seers, all rely upon this fallibility of human memory (see Gardner, 2000; Schacter, 2001). However, if I am right in my predictions, be prepared to hear about it on multiple occasions; if I am wrong, I shall hope the publisher prices this book at the usual exorbitant rate!

Linear extrapolations are not the problem in predicting the future; it is the nonlinear leaps of progress, often fueled by a concatenation of unusual and often unforeseeable developments, which represents the great challenge to the seer. In the realm of simulation, this challenge is significant indeed as simulation lies at the confluence of so many volatile sciences. Obviously, any major stride in computer science directly affects simulation and its progress. This is equally true for both hardware and software innovations. However, computation alone is not the be-all and end-all of simulation. Psychology and the cognitive and neurosciences have an enormous influence on what will develop in simulation because the laws of simulated worlds are largely psychological rather than physical. Indeed, it could be argued that the central point of simulation is to eventually free individuals from these physical and psychological constraints, a brief discussion of which I present later. But this is not the end of the confluence of forces. Many of the future developments in simulation will depend upon artists and those in the entertainment field whose conceptions and ideas are far removed from the groove of scientific thought. The minds of such individuals entering into the mix provide many quirks and bijouteries, and this is perhaps the major reason for their inclusion. However, all contributions need to be tempered to a scientific understanding. We can add much more to the factors influencing simulation's development. Among these, we have not yet included customers and their diverse desires, issues, and challenges, which will fuel the financial segment and so help shape the agenda of progress. The forces involved in the cauldron of simulation development are so varied and so tangential to each other that it is almost inevitable that they will produce hiccups and left-field notions that will have mainstream influence on all simulation technologies.

One of the possible nonlinearities of progress in simulation already being pursued is the way in which we get information into the brain itself. It may well be that, in the near future, we find that the eye is a relatively limited channel through which to get information to the visual cortex. The complexities, inversions, and selectivity of the retina might well prove to be the bottleneck in moving information across the cortical barrier. Where, then, is the sense of spending millions of dollars improving visual graphics systems for marginal overall gains when direct neural stimulation provides some orders of magnitude better return? To illustrate the possibilities involved in all forms of short-circuiting of the simulation process, it is critical to have an underlying theoretical framework of goal-directed behavior. One such framework is illustrated in Figure 13.3, which deviates from the normal linear stimulus–response information-processing perspective to emphasize the recursive nature of the perception–action/action–perception process (see Flach, Hancock, Caird, & Vicente, 1995; Hancock, Flach, Caird, & Vicente, 1995). In dealing with a recursive (in this case, actually a spiraling) process, one can essentially start and finish anywhere in the circular representation, but for the sake of convenience and clarity let us use the environment as the initial point of departure. In contemporary, electronic simulation, we replace the proximal (immediately perceived) world with its computational surrogate. Previous forms of simulation (such as those used in movie production) used physically erected sets that looked solid, but had, in actuality, little substance. The purpose of each of these forms, similar to the purpose of stage magic, is to "fool" the sensory systems. Often this is done by taking advantage of foibles and illusions well-known to perceptual psychologists. This constitutes the past and present of simulation with which we have already dealt.

Obviously, the next developmental stage of simulation is a hybrid one. Similar to the part-real, part-electronic combinations discussed later (see Figures 13.4 and 13.5), the synthesis of part-electronic and part-direct cortical stimulation will be the next breakthrough stage in evolution. The logical step to full simulation via cortical stimulation will be an obvious, feasible, and embraced challenge. However, notice that the goal of all these efforts is to communicate the perceived world to the user.

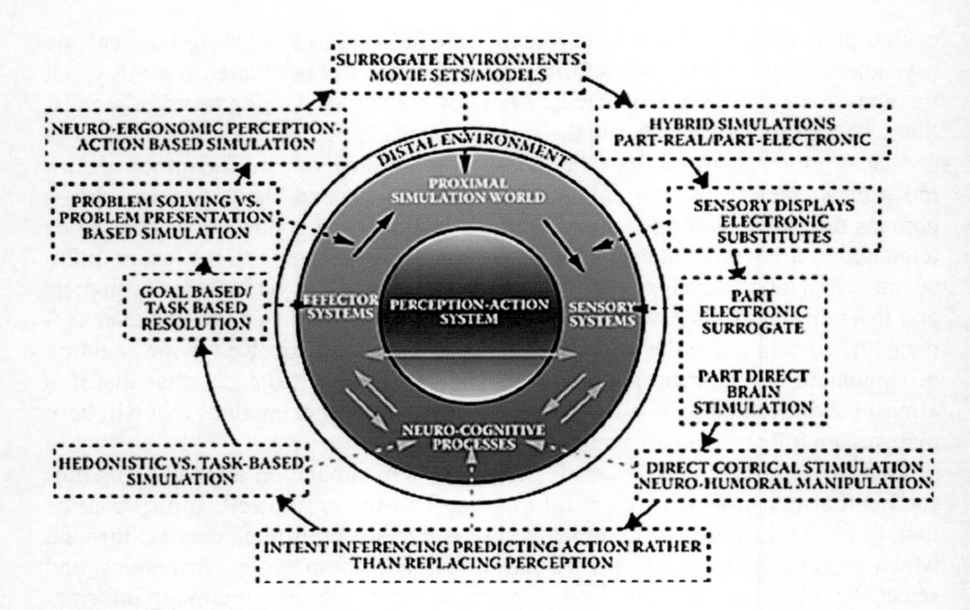

FIGURE 13.3 The future of simulation expressed as a function of the site of simulation and its representation. This sequence shows the fundamental purpose of simulation as—breaking the barrier between the perceiver and the perceived, and between the same actor and the act.

FIGURE 13.4 Illustration of a virtual forest, showing individuals experiencing a hybrid reality composed of a physical substrate with virtual overlays. (Copyright 2003, Media Convergence Laboratory, University of Central Florida, reproduced with permission.)

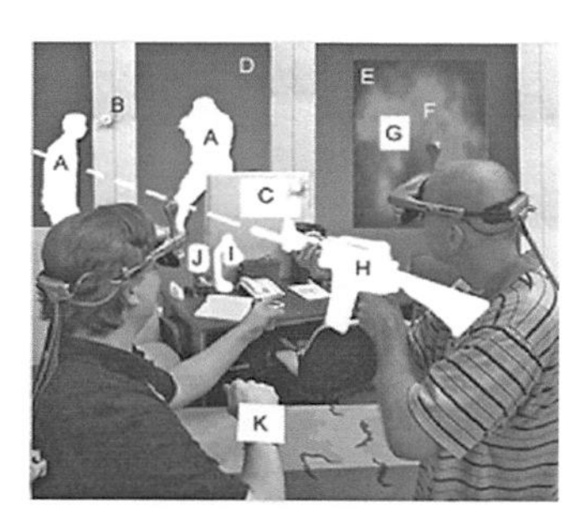

FIGURE 13.5 At left is the physical appearance of the environment, and at right the complete environment with the virtual elements that provide the totality of the experience. (Copyright 2003, Media Convergence Laboratory, University of Central Florida, reproduced with permission.)

The next important conceptual step is to understand that simulation needs to make a transition from perception to action. For this progression, we will need to know much more about the tasks, the goals, and, more generally, the intent of the user. At such a point in time, our concern becomes much more directed as to why simulation is necessary. If we are controlling, or practicing to control, complex systems in dangerous situations, the need to perform overrides the imperative to represent. If we can project simulations directly to the cortex, can we not similarly extract response directly from it? This being so, the surrogate and the experiential reality can be simply set together in a closed-loop relation (see Bush, 1945). There are movements toward creating such technologies as companions to input simulation, in which the neurological state initiates action in the real world via machine activation. This area, given the name *neuroergonomics*, is, at present, in its earliest infancy (see Hancock & Szalma, 2003; Parasuraman, 2003). In future, it promises to complement the advances in the perceptual augmentations of simulation by creating a way for the brain to directly affect the world. As we more fully understand human response in the context of expanding technological capacities, we shall see how simulation is a crucial component in dissolving the barrier between virtually expressed intention and subsequentially realized reality. At such a juncture one has, collectively, to be extremely wary of what one wishes for, as intent will almost immediately be translated into action. Simulation will be critical in showing us the outcome of our intentions prior to their physical expression. However, this is for the far-off future. Let us return to a little more near-term scenario.

THE PRACTICALITIES OF SIMULATION

Up to the present, I have largely engaged in speculation about the possible futures of simulation and the fundamental driving forces behind said progress. However, to those involved in the practicalities of simulation, such discourse may well seem almost completely superfluous. After all, those who are beset by pragmatic needs are concerned with the advantages simulation can render tomorrow and not in the far-off horizon of the decades to come. Therefore, it is incumbent upon me to consider these near-term issues because simulation science has, for most of its existence, been a very practical concern.

Those looking to employ simulation in the near term have to ask very pointed questions about whether it can save them time, money, and/or even lives in return for their investment of resources when compared to alternative approaches to their respective problems. Simulation is almost always appreciably cheaper than operations on the real system. Almost any technology we can think of, such as process control, aviation, commercial vehicle operations, and others, is much more expensive to train on the system itself than on a valid simulation. Of course, the problem of effective skill transfer always remains, but for technologies such as the space shuttle, one simply cannot train on the system itself as there are just too few of them in existence. For technologies such as nuclear reactors that are in constant operation, training on the actual system is neither feasible nor advisable. It is here that the cost-effectivity of simulation steps to the fore and makes its case.

Being able to achieve one's goal of operator training in the cheapest possible manner actually drives most practical decisions in our cost-conscious world today. Indeed, it was this focus on training that served historically to separate simulation from mainstream in human factors. As Chapanis Garner, Morgan, and Sanford (1947) noted, training and thus simulation were largely the domain of the personnel psychologist who evolved into the industrial/organizational specialist of mainstream psychology. In contrast, human factors and human engineering scientists come from the roots of experimental psychology. This irrational division between human factors and simulation science is still evident to some degree in, for example, the comparable evolution of the Human Factors and Ergonomics Society and the American Psychological Association's Division 21 on Applied Experimental Society, although the latter organization has by itself a complex history. Fortunately, this rift has been healed by a much closer association that should now serve to bring human factors and simulation applications much closer together, hopefully to the strong benefit of each (R. S. Kennedy, personal communication, 2004).

Although the simple fact of cost often dictates decisions, such deliberations are also often constrained by time. Building actual systems, or even full-scale mock-ups and models, can take an extensive amount of time, and in business worlds where time and money are considered interchangeable, the argument again comes down to the fundamental cost to achieve the goal. Simulations can be run so as to allow the consideration of an almost endless number of "what-if" scenarios. For actual systems or even models in which catastrophic consequences put an end to the physical entities, the cost of running a failure scenario can be prohibitive (as those who used model boards for early flight simulation know only too well). In the software world, the cost is essentially zero.

Allied to this argument is the question of risk exposure. Whereas actual models and even systems themselves can be rebuilt, human beings cannot. It is often the case that the practicalities of simulation are specifically for training in high-risk circumstances where failure is not an option. In simulation, we can engage in multiple attempts. In comparable real-world systems, these become one-shot trials in which one or more lives are clearly on the line. Again, this is a protection of resources and, in business terms, falls back to cost. However, when, for example, an instructor pilot and a student are killed in a crash, more than money is lost. Thus, for simulation to continue to function healthily in the future and even to burgeon, it needs to make very explicit its advantages in terms of time, money, and safety to the customers. Quantifying these advantages through mutually inspective and reliable assessment methods remains an important goal of the simulation scientist.

SIMULATION AND TRAINING

Simulation has always supported training. Giving individuals surrogate experiences in advance of events has always been considered an important advantage of simulation. However, for future simulations, much harder questions will be asked about the cost and effectiveness of such training. This implies that the behavioral scientist and the simulation technologist will have to work in even closer association to perpetuate simulation's advantage. In the near future, it will not be sufficient to merely present scenarios that approximate expected conditions and then expose individuals to these general circumstances. There will have to be a much more targeted approach as to what specific skills need to be trained and what component elements of such skill development can best be served by exposure to simulations. Thus, we could construct situations to familiarize the individual with general contextual information as to the global environment of operations. Alternatively, we could provide support for the assimilation of specific strategic skills such as decision making. For example, we could provide simulation support specifically for purpose-directed psychomotor skills of extremely high value, such as surgical expertise. It is evident that the nature of the cues that compose such simulations vary according to the skill set that is required for transfer. As has been previously pointed out (Kantowitz, 1992), this means that future simulations need to focus on the psychological variables necessary for skills support and cease to be driven solely by appearance and technological innovations. It may well be that the best simulation for supporting specific continuous psychomotor skill transfer (e.g., driver training) is simply a dynamic line display, and all other elements of the sensory display (such as texture) actually distract from such transfer. Thus, future researchers will have to ask very hard questions about the goal of simulation use and tailor the simulation created to those specific goals.

In this line of evolution, we may well get some surprising paradoxes such as the one described earlier, where reducing the apparent fidelity of simulation improves its specific utility. Such developments await a much more systematic understanding of what facilitates simulation-based skill assimilation for subsequent real-world transfer (see also Morris et al., 2004).

Practical simulation is not only used for training (see Aldrich, 2004). Indeed, one can argue that the role of simulation in the process of design and systems acquisition might, in the future, be even greater than training (cf., AGARD, 1980; Lane et al., 1994). It has been noted that the degree to which training is needed in human–machine systems reflects the degree of flaw in the original design. As an absolute statement, this is certainly subject to debate. However, it is true that some design improvements obviate the need for more extensive training, and so, in the search for ever-more cost-efficient systems, the wise procurer will think carefully about up-front investment in design. Here, of course, simulation again steps to the fore. The physical and cognitive operational characteristics of any system can be presented first in simulation and quickly changed to search for optimal, or at least improved, design options. However, this is a somewhat static view of the design process. I have argued previously (Hancock, 1997b) that, in the future, design will be a much more fluid and interactive process. Because systems themselves will be much more dynamic, generative, and evolutionary, the linkage between simulation and design will also be a much more flexible and interactive one. Indeed, as noted earlier, operators of the future need not necessarily know whether they are working on the actual system or practicing on a perfect surrogate; this same blurring will happen with the process of design, also. In other words, fluidity will be of much higher value, and the idea of the fixity of final products or finished simulations will begin to fade. The days of nomothetic, or generalized, simulations are numbered. The future will require much greater flexibility and a much greater focus on the individual as designer, operator, user, trainee, or customer. The future watchword of simulation will be customization or, perhaps more realistically, individuation (see Hancock, 2003a).

DISCOURSE BETWEEN TWO WORLDS

Whatever the trajectory of future simulation, I think it is safe to assert that there will be two fundamental considerations. Both concern the nature of the simulated worlds. I predict, without much temerity, that we will see a continued focus on improving real-world replication. In contrast, there will be a steadily growing interest in representing artificial environments. Since the very earliest forays into simulation, one of the main concerns has been with the practical issue of presenting simulacra of the real world. Obviously, the roots of simulation lie in training, in which it has been assumed that the most effective course is to expose individuals to a complete representation of an environment before they have to face the real situation. The less forgiving the real-world situation, the greater the utility of simulation. Little wonder then that the military has always been at the forefront of simulation technology as it is the quintessential institution engaged in exposing individuals to real-world threats. One of the mantras of this line of development has been "the better the simulation, the more effective the transfer of developed skills to the real world." The issue here, as noted previously, lies in the nature of the necessary skills. Simulation developers have been computer scientists and engineers, not psychologists. Engineers use a face validity, "existence-proof" measure of simulation quality. In seeking to control issues such as frame rate, polygon frequency, texture mapping, stair stepping, and the like (each of which is largely issues in computational speed), they focus on the appearance of the world, rather than on its psychological characteristics. Their traditional yardstick for improvement is to enter the simulation, and if it looks better and more like the real world, that is progress. The problem focus is on the technical barriers and glitches that prevent this metric of appearance from reaching better levels. Unfortunately, simulations are built for people and built for a purpose, and often such purpose and subject interaction are subtly at odds with the appearance metric. Let me give one illustration, although the reader will find many other examples in this text. In constructing wraparound ground-vehicle simulations, it is often the case that attempts are made to provide a high-fidelity visual appearance of the whole field, which can extend to a 360° field of view. Problematically, the human visual system does not register such levels of detail throughout its range, and indeed devotes detailed processing only to a very narrow 2°, the foveal visual field. Providing highly realistic, detailed information in the person's visual periphery is therefore not only computationally wasteful, it can also induce simulation sickness and therefore negate the very reason for the simulation in the first place (see Stanney, 2002; and see Stanney, Kennedy, Hale, and Champney in chapter 6, the original book). Few people can acquire important skills to be transferred to real-world operations while they are nauseous. Fortunately, as simulation research has begun to foster much greater interdisciplinary contributions and is now often encountered as a team effort, such issues as nominal realism versus psychological composition are receiving significant and deserved attention. It is a simple prediction that such interactive teams will make significant strides not merely in the improvement of simulation effectiveness but also in distilling composite measures of progress that include both computational and psychological parameters.

HYBRID SIMULATION WORLDS

Before passing on to the issues of assessment, it is important here to comment directly on the most interesting initiative that concerns simulations of mixed or hybrid forms. It is easy to see that simulation has made great strides in replicating certain forms of sensory stimulation, with vision being the outstanding one. It is equally easy to see that simulation has made relatively little progress in some comparable areas such as tactile stimulation. Here, I do not go into the reasons as to why this is so, but it does reflect, in part, the state of knowledge and research progress, as well as the result of practical resource allocation, in the immediate past. Be that as it may, we are left with a conundrum that simulation is close to fooling some sensory systems, whereas others are hardly touched at all. The interim solution to this problem is the use of mixed or hybrid worlds in which the visual and auditory cues are computer-generated, whereas the tactile cues remain firmly rooted in our present reality. Such a compromise results in the use of many overlay technologies such as blue screens, which have been pioneered and long-used in the entertainment industry. There is great near-term promise for these technologies, which are now beginning to have numerous practical uses. The illustrations shown in Figures 13.4 and 13.5 are from the University of Central Florida's Media Convergence Laboratory and are taken from its mixed reality innovation test bed. As noted earlier, not only are these important steps in simulation evolution, they also mark the way to other hybrid forms once further technical developments have been accomplished. Consequently, the way they integrate information sources is a most instructive development.

ASSESSING THE PROGRESS OF SIMULATION TECHNOLOGIES

To comprehend the problem of measuring the progress of simulation technologies, let us begin in a very roundabout manner; namely, by considering why (I think) ventriloquists are so awful. As a child and later as an adult, I have had to squirm in embarrassment when watching a "nominal" (amateur or, unbelievably, even professional) ventriloquist go through the shtick of "throwing" their voice to an unconvincing dummy. First, the voice is rarely thrown very far. In fact, in most cases, there are only 4–5 inches between the performer and the clacking mouth of a wooden puppet, or even more horrendously, a decorated sock! Fortunately, this pursuit seems to have disappeared from our major media, but still persists in what I prefer to call distributed pockets of psychological illness. The failure of the ventriloquists' illusion is evident in the decreasing age of their audience. Nowadays, unless restrained, even 5-year-olds hoot such performances off the stage. Perhaps it can now quietly and reverently return from whence it came: the adult-to-baby game of peek-a-boo.

The problem is quite evident. The so-called artists seek to suspend reality by relocating causation from themselves to alternate entities. But the illusion simply does not work. Ventriloquists can never throw their voice sufficiently far so that a facing audience can "see" the voice as if it were coming from a different location (i.e., the dummy). Hence, the dummy is always designed to be visually attractive, with a large moving mouth. The putative entertainers augment this nominal suspension of belief by adopting the stereotypical funny voice and by the minimal movement of their own mouths. Similar to that of a bad magician whose rabbits and pigeons are leaping and flying from the dress-coat, we see through and deride the failing illusion of the sad performer. In a general sense, ventriloquists are trying, albeit however slightly, to pervert the laws of physics and produce the appearance of sentience and causation in an obviously inanimate object. Note that the conversational and comedic aspects of ventriloquism are exactly parallel to the two-man stand-up comedy act, but in the latter, we do not experience any dissonance as the two individuals are both clearly seen as sources of causation. In essence, the sad ventriloquist is the equivalent of the failed simulation software and hardware engineer. They have tried to produce an illusion and manifestly failed. However, how and why the illusion fails are crucial questions for the future of simulation.

THE TURING TEST OF SIMULATION

To a degree, all simulations fail. Perhaps this failure is what we actually mean by simulation, that is, a degraded version of reality. However, soon this degree of failure or degradation will become so disappearingly small that those exposed to simulation will be hard put to tell the difference. All human sensory systems have a limit to their resolution capacities, and simulation capabilities will sequentially approach these individual limits for each specific sensory system. Paradoxically, because vision is a distal sense, in that it decodes information that is largely remote from the actual retinal site of activation, the visual capacities of simulation are rapidly approaching the level of reality. I have labeled this as the "Turing test" of simulation because, analogous to the way that Alan Turing constructed a test for machine intelligence, the inability to tell the real from the simulated world constitutes a watershed threshold (Turing, 1950). For audition, this threshold is a little further removed. Not only are auditory displays serial representations, they are omnidirectional and very sensitive to change in locality. Also, quite bluntly, we know much more about vision, having dispensed many Nobel Prizes for vision research. Compared with the funding for vision, audition research is the poor cousin. Unfortunately, the most proximal sense, that of touch, receives virtually no funding in comparison. The few brave souls who labor in the field of tactile kinesthesis are made aware of this lack on a daily basis but—and again paradoxically—it is the barriers to the simulation of touch that are likely to be the last to fall in the search for the final passage of simulation's Turing test. Subjectively, we attempt to assess this overall experience through a "subjective" sense of "presence." This latter construct represents the degree to which we are willing to suspend our disbelief of the simulation shortfall. As presence improves, so we approach closer to passing the Turing test.

One method of achieving this passage early is by finessing the problem of touch. Instead of artificial actuators attempting some form of direct stimulation, one can merely substitute the actual real world for this part as in the hybrid simulations discussed in the previous section (see Figures 13.4 and 13.5). Transportation simulation is already well along the way to crossing the reality threshold. Using the strong advances of visual and auditory projection, the problem of touch is circumvented by having the individual sit in a cab of some sort, often taken from an actual vehicle. Providing stationary tactile cues is now taken care of, and the issue devolves to a question of providing appropriate dynamic motion cues to match events represented in the visual and auditory displays. In current high-end, single-seat aircraft simulators, this experience gets close enough to reality to make pilots sweat in hazardous situations. Of course, one can finesse the problem of vision and audition also by simulating naturally degraded worlds such as those occupied by fog in which both visual and auditory displays are, naturally, reduced. However, depending on what it is we are intending to simulate, the feasibility of passing the Turing test increases daily.

SUPERSIMULATION

Human sensory systems have finite resolution capabilities. They address only a restricted portion of the electromagnetic range and have intrinsic limitations in both spatial and temporal acuity. Thus, what we experience as reality is a highly restricted "window" into a much richer world. Simulation is not bound in the same way. Indeed, it has been noted that whereas the laws of physics bound real worlds, it is the laws of psychology that bound virtual ones. So, if we pass the simulation Turing test, what then? Due to the fact that many physical constraints can be fractured in simulation, it is both feasible and practicable to generate representations beyond real-world fidelity; these capacities, which I have labeled "'supersimulation," are the next evolutionary step. Of course, we already have several forms of supersimulation. We can present the individual with displays of information taken from infrared and ultraviolet ranges, and scale them to be displayed via intrinsic visual capacities. We could also render the same isomorphisms in both auditory and tactile-kinesthetic worlds in which stimulation beyond the normal range can be mapped into the spectrum of normal capacities. This would, for example, allow us to hear what dogs experience, and via teleoperation we could "touch" worlds where no human has yet been. Variations in temporal presentation are also very familiar to us through entertainment media in which slow-motion or time-lapse photography either slows down or speeds up real-time events, respectively. The variation in visuospatial resolution is rather akin to the use of dynamic binoculars that provide a radically increased magnification resolution in some specified part of the visual field (where, of course, auditory and tactile-kinesthetic analogs are also feasible). In some sense, the latter manipulation then replicates visual structure and function of the fovea and periphery in the retina where an area of especially high sensitivity is bound by a surrounding region of lower resolution. In supersimulation, the boundaries of space, time, and electromagnetic range of innervations are no longer immalleable, and I fully expect to see the exploitation of these dimensions continue in the future to an even greater degree. Finally, as with earlier observations, the actual content of such worlds is bound only by imagination as we have already seen in many entertainment "worlds."

THE MORAL DIMENSION OF SIMULATION

As well as the technical future of simulation, there is also a growing concern for its moral dimension, which it is important to consider here. In recent decades, we have seen the growth of the World Wide Web and the massive impact it has exerted in a number of fields. However, even the most cursory survey of the Web shows that one of the dominant themes is pornography, or more properly, explicit sexual content, since pornography is very much in the eye of the beholder (Lawrence, 1929). In public, many individuals deplore this tendency, and yet such images and activities must find a widespread and ready market; otherwise, they would not be so plentiful. Our attitude toward this issue often reflects our own individual moral grounding, and questions of usage and censorship vary according to the different regions of the world and the cultural diversity in which such use is embedded. Many individuals agree that limitations on some aspects—for example, on child pornography—are needed, and this consensus allows groups to designate certain activities as illegal. However, is all public condemnation to extend to purely private circumstances, and what role does enforcement play in such situations? Such questions represent the horns of moral dilemmas with which future simulation scientists must inevitably wrestle (see Hancock, 1998, 2003b). What is clear is that sex is a fundamental human drive and, like other forms of physical activity, sexual activity, too, can be represented in simulation situations. Given the fundamental nature of this drive, it can be anticipated that much effort will go into producing surrogates. However, what does this say about reality? If simulation were to reach a level of sufficient viability in reproducing tactile cues as well as visual and auditory stimulation, would this represent a breakdown in real-world sexual activity, which would pale by comparison with its unlimited and unconstrained electronic substitute? By extension, would this mean the curtailment of other social relationships? Teenagers already spend enormous amounts of time playing in their own rooms in electronic game worlds. Simply imagine the alternatives that could be rendered by advanced simulation possibilities. Indeed, how many of us would return to a mundane real-world existence given the choice of ultimate fantasy worlds?

It is not solely this "virtual isolation" which is of concern. What do we do about those individuals who wish to conduct illegal, illicit, or threatening activities and seek to use the facility of simulation to advance these goals? Indeed, the terrorists of September 11, 2001, took extensive advantage of simulation to achieve their ends (Hancock & Hart, 2002). Do we take the scientists' traditional dissociation excuse and indicate that the fruits of science are morally neutral until they are put to specific ends through social and political implementation (see Hancock, 2003b; Parasuraman, Hancock, Radwin, & Marras, 2003)? It was this very dilemma that had to be faced by Oppenheimer in the creation of the atomic bomb, which was certainly never a neutral technology. Thus, the future of simulation is not merely a technical challenge but promises to be one of radical social import. The creators of technology can no longer legitimately claim moral neutrality, and perhaps simulation science is an area that can address this thorny but vital issue. Can we, indeed should we, find and impose limits?

A PHILOSOPHICAL VALEDICTION

I cannot leave a discourse on the future of simulation without explicit speculation as to these wider personal and social implications. Such projections are, of necessity, eventually to be founded in philosophical discussions. I know that such deliberations can be anathema to a segment of readers who are pragmatic and practice-oriented, who are hereby excused, without prejudice, to proceed directly to the final, summary section. However, the pillars of philosophy are the foundations of society, and as the future of simulation promises to shake these very pillars, this brief excursion is more than justified.

Perhaps the most relevant place to begin is with considerations of the nature of reality. The British empiricists Locke (1690), Berkeley (1710), and Hume (1739) were among the vanguard of modern philosophers to question the fundamental nature of experience itself. It is evident that our moment-to-moment experience is derived overwhelmingly through our immediate sensory stimulation, and the basic question is whether all experience is contingent upon this ongoing stream, as other potential contributions are themselves contingent upon remembered experiences extracted from the same source. Locke's original comments on newborn children, as tabulae rasae upon which nature writes as if on an empty page, is one radical position in the ongoing nature–nurture contention. Contemporary biogenetics tells us that the situation is much less absolutist than this, and that newborns are equipped with many innate capacities to help them deal with the challenges of the environment. In essence, Locke considered individual memory as an important player in the totality of experience, but did not, at that time, comprehend the notion of inherited genetic capacities (essentially a genetic memory) that would help frame the very earliest forms of experience. Our contemporary knowledge helps us understand that, indeed, reality is more than the stimulation of the moment.

Following Descartes, the question of reality as illusion became an important philosophical issue. Could it be, independent of the nature of all previous experience, that what was perceived as reality was actually illusion? Descartes could imagine an all-powerful entity (which he expressed as a devil, but in simulation terms might merely be an exceptionally capable computer) that could present to his senses a sufficiency of stimulation so that he was fooled. His only comfort was the fact that such a computer could not deal with the pure intuition that made it possible for him to doubt that illusory reality. His famous aphorism, "*cogito ergo sum*," assumes that the very essence of self is thought, not perception—although in this he was also mistaken to some degree. Berkeley identified that all-powerful being as God and asked the most sensible of questions as to whether matter actually existed. His argument is most instructive, especially for future simulation scientists. Berkeley, himself a Bishop of the Church, was a believer in the omnipotence and infallibility of God—omnipotence, meaning God could do anything; infallible meaning God was perfect. Since God was perfect, God would not make mistakes. Thus, God would not engage in any action that itself was not the most economic and efficient method of achieving a specific aim. From this, Berkeley argued that God could put directly into the mind of each individual the experience of reality, omnipotence allowing this difficult but conceivable action to take place. This being so, creating matter, that is, creating an intermediary mechanism for the perception of reality, is not really needed. We do not need to go through the step of "perceiving" external objects because God can project such an image directly to the brain. Therefore, God being both infallible and omnipotent, matter is unnecessary—*quod erat demonstrandum* (QED). This wonderful solipsist conundrum has never been resolved and, indeed, there are good grounds to believe that empirical resolution may be impossible. Today, we do not believe in this solipsist assertion, not because of any proof to the contrary but rather because of much greater doubts about the presence, actions, and capacities of any particular deity. Berkeley's argument however remains unassailed.

The question for the future of simulation in this context is—can we play the role of God? In actuality, this is logically equivalent to the passage of the Turing test for simulation reality. At present, when we are in a simulation, we remain very much aware that we are in that simulation, not least in part because we remember entering that environment and agreeing, at least to some extent, to suspend our disbelief. The sense of reality would be much enhanced if, for example, we woke up to such a simulation, where the power of the continuous stream of memory played a diminished role. Similar to other concerns about the basic nature of individual existence, developing simulations that slowly and surreptitiously introduce artificially mediated parts of the environment into naturally occurring situations may help fully suspend our disbelief.

Contemporary psychologists and neuroscientists are very aware that the content of consciousness is a dynamic interplay between the stream of incoming information interacting with the centralized, largely memory-mediated processes. In this manner, to fabricate a true reality, simulation will have to embrace much more than simple surrogate sensory displays. It will have to dig deeply into the nature of memory and the facets of individual differences that connote personal identity and idiographic experience. Thus, the creation of "constrained" realities, that is, those in which the individual personally, voluntarily, and knowingly "buys" into the premises and constraints of the surrogate world, is not far off, and for many gaming situations, it is already here. A convincing replacement world for a non-cooperative individual is further away, and designs will continue to rely on support from actual real-world surroundings for some time to come. Although the barriers are daunting, such problems are not insurmountable, and as we understand more about problematic issues such as tactile-kinesthetic stimulus replacement and the integrated experience of consciousness, we will eventually achieve other realities; but what then?

Simply creating persuasive other worlds is only the first step. Since our present concern is with the philosophical issues, let us cast aside, albeit temporarily, the pragmatic drivers that will power the future of simulation and ask the greater social questions. Let us suppose we can now create infinite alternative universes (and by this, I mean that the users will be able to dynamically construct and control any object or agent in their "worlds" and that some method has been found to port the material essentials, for example, food, oxygen, water, etc., into these worlds). In such worlds, the user will be able to instantly satisfy any physical or cognitive desire. Who would wish to occupy such a world? Having occupied it, who would wish to return to this one? Coming back to "reality" would mean encountering individuals who are odd, unpleasant, incompliant, uncaring, polemic, sadistic, and even murderous. Who would take these characters over a collective of purpose-built electronic substitutes having infinite empathy, pity, love, and caring? In essence, what would happen to the fabric of human society when the necessity of social cohesion is fractured? The philosopher Rousseau asked much the same questions some centuries ago, arguing that any social bond is one in which one trades a degree of security for a restriction in autonomy (Rousseau, 1762). Further, he had a recommendation for those who wished to exercise untrammeled autonomy because they could not accept the restrictions of civilized society—he recommended they go to America! There, he argued, they would find a "new" world where the pressure of population on land was sufficiently small that people could "do their own thing." This transmigration of the discontent was not, of course, the preferred immigration policy of the Native Americans of the time. In today's world, Rousseau's contract is no longer voluntary and, therefore, no longer a viable contract. An individual born today cannot effectively decide to opt out, find unclaimed but productive land, work on that land, and remain socially isolated. There are individuals and small groups who try to sustain such isolation, and some (such as the Amish) have a degree of success. Other individuals and groups are, for differing reasons, not successful, especially when they directly encounter the unsympathetic power of society at large. In this sense, the cases of the Uni-Bomber and the Branch Davidian group are particularly instructive and recent examples (Reavis, 1998). Given the prior claim on all lands of the world, today's social contract is imposed on individuals by force variously disguised in order to maintain a relatively stable collective. In this inherent tension, is there an opportunity for simulation science? I suggest there is.

Given the current population, it is evident that we will not reach any planetary system sufficient to support human sustenance and expansion in the time we have available (although growth is actually slowing in some regions). Neither, despite all the optimistic prognostications, are we liable to find a voluntary balance in the global population. Given that there is little actual (effective) real estate for pioneers to explore and that the pressures on land continually increase, can the simulation sciences represent the "new" world? In the film *The Matrix*, we are shown warehoused individuals stockpiled for their power generation capabilities (a very doubtful premise). However, it may be that such warehousing is much more aligned with excessive population. Indeed, we already warehouse those elements of society who are termed criminal. Two hundred years ago, we ported such individuals to "new" worlds, such as the forced migration of "criminals" to Australia (see Hughes, 1987; Rees, 2002) by the English. Today, we essentially have no such remote lands to be exploited. Could we port such individuals to simulated worlds? Immediately, the questions of cruelty and purposelessness come to the fore, but like the colonies of old, simulation worlds need not be inherently unproductive, and much of value may be brought back from these electronic potentialities to the one that will remain "mother reality."

Sufficiently advanced, future simulations can therefore act to change society once more in order to present choice to individuals. The problem with this is that as human beings are inherently structured to accept options involving the least effort, who would be left to tend the machinery, advance the technology, and frame mother reality? My answer would be that it will, as it has always been, be those who embrace challenge, see opportunity, and turn adversity into progress. As America once was, as the West once represented, as the Apollo program once exemplified, future simulation can represent the new frontier.

SUMMARY AND CONCLUSION

The future of simulation is bright. There continue to be many circumstances in which we wish to train individuals, and yet not expose them immediately to dangerous situations. The traditional custom of military forces will continue, and the ever-burgeoning demand for entertainment will also drive simulation technology to improve. However, as well as progressing along these traditional lines, simulation will begin to expand and, in some ways, dissolve. I expect to see dissolution in the actual process itself. Since advanced system operators already act on representations of systems, and not directly on systems themselves, they already (to a degree) act on simulations. That this simulation changes between the actual system and a computer surrogate (and back again) could be easily achieved, and the substituted remain opaque to the operator. Such technological progress is feasible, and I expect to see it engaged in various forms in the very near future. However, the flexibility and change in simulation will not stop there.

Simulation has often been used in the design process. The future will see a much more interactive role for simulation here, and in the same way we will find it difficult to parse simulation for training from simulation for operations; we will find it similarly difficult to parse simulation for design from simulation in operation. If the future continues to emphasize speed and flexibility, these dissolutions of definition will also be accelerated. Further, we will see much greater customization. In general, human factors have gone from their earliest forms in which individuals built and customized their own tools, through eras of mass production to adaptive systems, and are now finally returning to individualization (see Hancock, 2003a). In the near future, everyone will expect their respective simulation(s) to adapt to themselves and their own personal settings. Simulation will have to follow this trend and also show ever-increasing cost-effectiveness through a much greater focus on what the goal of the simulation is and how to achieve that goal at the least cost. The future looks bright—but don't worry! it is coming whatever you do!

ACKNOWLEDGMENTS

I would very much like to thank Robert Tyler, Brian Goldiez, John Wise, and Robert Kennedy for their comments on an earlier version of this chapter. Preparation of this chapter was facilitated by grants from the U.S. Army. The first was the Multiple University Research Initiative-Operator Performance under Stress (MURI-OPUS) Grant (#DAAD19-01-0621). The second was from the Advanced Decision Architecture Consortium (# DAAD 19-01-0009). The views expressed in this chapter are those of the author and do not necessarily reflect the official policy or position of the Department of the Army, the Department of Defense, or the U.S. government. The author would like to thank Elmar Schmeisser, Sherry Tove, and Mike Drillings for providing administration and technical direction for the first grant, and to Mike Strub for the second grant.

REFERENCES

AGARD. 1980. *Fidelity of simulation for flight training.* AGARD Advisory Report No. 159, Harford House, London.

Aldrich, C. 2004. *Simulations and the future of learning.* Wiley: San Francisco.

Berkeley, G. 1710. *Treatise concerning the principles of human knowledge.* Tonson: London.

Bush, V. 1945. As we may think. *The Atlantic Monthly,* 176(1), 101–108 .

Chapanis, A., Garner, W. R., Morgan, C. T., & Sanford, F. H. 1947. *Lectures on men and machines: An introduction to human engineering*. Systems Research Laboratory: Baltimore, MD.

Flach, J., Hancock, P. A., Caird, J. K., & Vicente, K. (Eds.). 1995. *Global perspectives on the ecology of human-machine systems*. Lawrence Erlbaum: Mahwah, NJ.

Gardner, M. 2000. *Did Adam and Eve have navels?* W. W. Norton: New York.

Hancock, P. A. 1997a. *Essays on the future of human-machine systems*. Banta: Eden Prairie, MN.

Hancock, P. A. 1997b. On the future of work. *Ergonomics in Design*, 5(4), 25–29.

Hancock, P. A. 1998. Should human factors prevent or impede access. *Ergonomics in Design*, 6(1), 4.

Hancock, P. A. 2003a. *Individuation: Not merely human-centered but person-specific design*. Paper presented at the 47th Annual Meeting of the Human Factors and Ergonomics Society. Denver, CO.

Hancock, P. A. 2003b. The ergonomics of torture: The moral dimension of evolving human-machine technology. *Proceedings of the Human Factors and Ergonomics Society*, 47, 1009–1011.

Hancock, P. A., Flach, J., Caird, J. K., & Vicente, K. (Eds.). 1995. *Local applications in the ecology of human-machine systems*. Lawrence Erlbaum: Mahwah, NJ.

Hancock, P. A., & Hart, S. G. 2002. Defeating terrorism: What can human factors/ergonomics offer? *Ergonomics in Design*, 10(1), 6–16.

Hancock, P. A., & Szalma, J. L. 2003. The future of neuroergonomics. *Theoretical Issues in Ergonomic Science*, 4(1–2), 238–249.

Hockney, D. 2001. *Secret knowledge*. Viking Studio, Penguin: New York.

Hughes, R. 1987. *The fatal shore*. Knopf: New York.

Hume, D. 1739. *A treatise of human nature*. Noon: Cheapside, London.

Kantowitz, B. H. 1992. Selecting measures for human factors research. *Human Factors*, *34*(4), 387–398.

King, R. 2001. *Brunelleschi's dome*. Penguin Books: New York.

Lane, N. E., Kennedy, R. S., & Jones, M. B. 1994. Determination of design criteria for flight simulators and other virtual reality systems. *Proceedings of the IMAGE Conference*, Tucson, AZ 12–17 June.

Lawrence, D. H. 1929. *Pornography and obscenity*. Faber and Faber: London.

Locke, J. 1690. *An essay concerning human understanding*. Basset: London.

Moravec, H. 1988. *Mind's children: The future of robot and human intelligence*. Harvard University Press: Boston, MA.

Morris, C. S., Hancock, P. A., & Shirkey, E. C. 2004. Motivational effects of adding context relevant stress in PC-based games training. *Military Psychology*, 16(2), 135–147.

Mouloua, M., Gilson, R., & Hancock, P. A. 2003. Designing controls for future unmanned aerial vehicles. *Ergonomics in Design*, *11*(4), 6–11.

Parasuraman, R. 2003. Neuroergonomics: Research and practice. *Theoretical Issues in Ergonomic Science*, 4(1–2), 5–20.

Parasuraman, R., Hancock, P. A., Radwin, R. A., & Marras, W. 2003. Defending the independence of human factors/ergonomics science. *Human Factors and Ergonomics Society Bulletin*, 46(11), 1, 5.

Pullman, P. 2000. *The amber spyglass*. Random House: New York.

Reavis, R. J. 1998. *The ashes of Waco: An investigation*. Syracuse University Press: New York.

Rees, S. 2002. *The floating brothel*. Hyperion: New York.

Rousseau, J. J. 1762. *The social contract: On principles of political right*. Translation G. D. H. Cole. www. consitution.org/jjr/socon.htm
Schacter, D. L. 2001. *The seven sins of memory*. Houghton-Mifflin: Boston.
Staney, K. M. 2002. (Ed.). *Handbook of virtual environments: Design, implementation and applications*. Lawrence Erlbaum: Mahwah, NJ.
Turing, A. M. 1950. Computing machinery and intelligence. *Mind*, 59, 433–460.

Appendix A: Glossary of Modeling Terms

Compiled by Michael G. Lilienthal, Ph.D., CPE, CTEP, senior analyst, Electronic Warfare Associates, Government Systems, Inc., and William F. Moroney, Ph.D., CPE, Professor Emeritus, Human Factors Program, University of Dayton (Ohio).

Term	Definition
Activity models	A process model that describes the functional activity under examination in terms of inputs, transforms, outputs, and controls.
Analytical model	A model consisting of a set of mathematical equations, e.g., a system of solvable equations that represents the laws of thermodynamics or fluid mechanics.
Black box model	A model whose inputs, outputs, and functional performance are known, but whose internal implementation is unknown or irrelevant. For example, a model of a computerized change-return mechanism in a vending machine that is in the form of a table indicating the amount of change to be returned for each amount deposited.
Computational model	A model consisting of well-defined procedures that can be executed on a computer. For example, a model of the stock market in the form of a set of equations and logic rules.
Conceptual data model	A model that documents the business information requirements and structural business process rules of the architecture and describes the information that is associated with the information of the architecture. Included are information items, their attributes or characteristics, and their inter-relationships.
Conceptual data model	The description of what the model or simulation will represent, the assumptions limiting those representations, and other capabilities needed to satisfy the user's requirements. A collection of assumptions, algorithms, relationships, and data that describe a developer's concept about the simulation.
Continuous model	A mathematical or computational model whose output variables change in a continuous manner.
Data model	In a database, the user's logical view of the data in contrast to the physically stored data or storage structures. A description of the organization of data in a manner that reflects the information structure of an enterprise.

(*Continued*)

Term	Definition
Descriptive model	A model used to depict the behavior or properties of an existing system or type of system. For example, a scale model or written specification used to convey to potential buyers the physical and performance characteristics of a computer.
Deterministic model	A model in which the results are determined through known relationships among the states and events, and in which a given input will always produce the same output. For example, a model depicting a known chemical reaction.
Digital elevation model	A numerical model of the elevations of points on the earth's surface. Digital records of terrain elevations for ground positions at regularly spaced horizontal intervals.
Discrete model	A mathematical or computational model whose output variables take on only discrete values; that is, in changing from one value to another, they do not take on the intermediate values. For example, a model that predicts an organization's inventory levels based on varying shipments and receipts.
Dynamic model	A model of a system in which there is change, such as the occurrence of events over time or the movement of objects through space. For example, a model of a bridge that is subjected to a moving load to determine characteristics of the bridge under changing stress.
Emulation	A model that accepts the same inputs and produces the same outputs as a given system.
Enterprise model	Information model(s) that presents an integrated top-level representation of processes, information flows, and data.
Environment effect model	A model representing the impact or effect that an environmental feature has on a simulation entity, component or process.
Error model	A model used to estimate or predict the extent of deviation of the behavior of an actual system from the desired behavior of the system. For example, a model of a communications channel used to estimate the number of transmission errors that can be expected in the channel.
Executable model	A model that instantiates the conceptual model of a system as its design specification.
Federated Object Model (FOM)	A specification defining the information exchanged at runtime to achieve a given set of federation objectives. This information includes object classes, object class attributes, interaction classes, interaction parameters, and other relevant information. The FOM is specified to the runtime infrastructure (RTI) using one or more FOM modules. The RTI assembles a FOM using these FOM modules and one Management Object Model (MOM) and Initialization Module (MIM), which is provided automatically by the RTI or, optionally, provided to the RTI when the federation execution is created.

(*Continued*)

Term	Definition
Graphical model	A symbolic model whose properties are expressed in diagrams. For example, a decision tree used to express a complex procedure. Cf. mathematical model, narrative model, soft- ware model, and tabular model.
Hierarchical model	A model in which superior/subordinate relationships are represented, often as trees of records connected by pointers.
Human behavioral model	Model of a human activity in which individual or group behaviors are derived from the psychological or social aspects of humans. Behavioral models include a diversity of approaches; however, computational approaches to human behavior modeling that are most prevalent are social network models and multi-agent systems.
Iconic model	A physical model or graphical display that resembles the system being modeled. For example, a nonfunctional replica of a computer tape drive used for display purposes.
Graphical model	A symbolic model whose properties are expressed in diagrams (e.g., a decision tree used to express a complex procedure).
Hierarchical model	A model in which superior/subordinate relationships are represented, often as trees of records connected by pointers.
Human behavioral model	Model of a human activity in which individual or group behaviors are derived from the psychological or social aspects of humans. Behavioral models include a diversity of approaches; however, computational approaches to human behavior modeling that are most prevalent are social network models and multi-agent systems.
Human, social, cultural, and behavioral representation	A model of the structure, interconnections, dependencies, behavior, and trends associated with any collection of individuals ranging from the small unit level (e.g., tribes, militias, small military units, terrorist cells) to the macro level (e.g., of nations, religions, cultures, ethnic groups, and international organizations), and the integrated relationships between and among them.
Information model	A model that represents the processes, entities, information flows, and elements of an organization, and all relationships between these factors.
Logical data model	A model that provides a common dictionary of data definitions to consistently express models wherever logical-level data elements are included in the descriptions.
Markov chain model	A discrete, stochastic model in which the probability that the model is in each state at a certain time depends only on the value of the immediately preceding state.
Mathematical model	A symbolic model whose properties are expressed in mathematical symbols and relationships. For example, a model of a nation's economy expressed as a set of equations. Cf. graphical model, narrative model, software model, and tabular model.

(*Continued*)

Term	Definition
Metamodel	A model of a model or simulation. Metamodels are abstractions of the M&S being developed that use functional decomposition to show relationships, paths of data and algorithms, ordering, and interactions between model components and subcomponents. Metamodels allow the developer to abstract details to a level that subject matter experts can validate.
Mock-up	A full-sized model, but not necessarily functional, built accurately to scale, used chiefly for study, testing, or display.
Model	A physical, mathematical, or otherwise logical representation of a system, entity, phenomenon, or process.
Modeling and simulation (M&S)	The use of models, including emulators, prototypes, simulators, and stimulators, either statically or over time, to develop data as a basis for making managerial or technical decisions.
Natural model	A model that represents a system by using another system that already exists in the real world. For example, a model that uses one body of water to represent another.
Numerical model	(a) A mathematical model in which a set of mathematical operations is reduced to a form that is suitable for solution by simpler methods such as numerical analysis or automation. For example, a model in which a single equation representing a nation's economy is replaced by a large set of simple averages based on empirical observations of inflation rate, unemployment rate, gross national product, and other indicators. (b) A model whose properties are expressed by numbers.
Parametric model	A model using parametric equations that may be based on numerical model outputs or fits to semiempirical data.
Petri net model	An abstract, formal model of information flow, showing static and dynamic properties of a system defined by places, transitions, input function, and output function. It graphically depicts the structure of a distributed system as a directed bipartite graph with annotations.
Physical data model	A model that defines the structure of the various kinds of system or service data that are utilized by the systems or services in the architecture.
Physical model	A model whose physical characteristics resemble those of the system being modeled. For example, a plastic or wooden replica of an airplane; a mock-up.
Physical based model	Mathematical models in which the equations that constitute the model are those used in physics to describe or define physical phenomenon being modeled.

(*Continued*)

Term	Definition
Predictive model	A model in which the values of future states can be predicted or are hypothesized. For example, a model that predicts weather patterns based on the current value of temperature, humidity, wind speed, and so on, at various locations.
Prescriptive model	A model used to convey information regarding behavior or properties of a proposed system. For example, a scale model or written specification used to convey to a computer supplier the physical and performance characteristics of a required computer.
Probabilistic model	See: stochastic model.
Process model	A model that defines the functional decomposition and the flow of inputs and outputs for a system.
Qualitative model	A model that provides results expressed as a non-numeric description of a person, place, thing, event, activity, or concept.
Queuing model	A model consisting of service facilities, entities to be served, and entity queues (e.g., a model depicting teller windows and customers at a bank) ng to be served. For example, a model depicting a network of shipping routes and docking facilities at which ships must form queues to unload their cargo.
Queuing model	A model consisting of service facilities, entities to be served, and entity queues (e.g., a model depicting teller windows and customers).
Reliability model	A model used to estimate, measure, or predict the reliability of a system. For example, a model of a computer system that is used to estimate the total down time that will be experienced.
Representation	Models of the entity or phenomenon associated and its effects. Representations using algorithms and data that have been developed or approved by a source having accurate technical knowledge are often considered authoritative.
Scale model	A physical model that resembles a given system, with only a change in scale. For example, a replica of an airplane one-tenth the size of the actual airplane.
Simulation object model (SOM)	A specification of the types of information that an individual federate could provide to High Level Architecture (HLA) federations as well as the information that an individual federate can receive from other federates in HLA federations. The SOM is specified using one or more SOM modules. The standard format in which SOMs are expressed facilitates determination of the suitability of federates for participation in a federation.

(*Continued*)

Term	Definition
Static model	A model of an entity or system in which there is no change. For example, a scale model of a bridge that is provided for its appearance rather than for its performance under varying loads.
Stochastic model	A model in which the results are determined by using one or more random variables to represent uncertainty about a process, or in which a given input will produce an output according to some statistical distribution. For example, a model that estimates the total dollars spent at each of the checkout stations in a supermarket, based on probable number of customers and probable purchase amount of each customer. Syn. probabilistic model. See also: Markov chain model. Cf. deterministic model.

Source: Department of Defense Modeling and Simulation Glossary. M&S Glossary (msco.mil) https://www.msco.mil/MSReferences/Glossary/MSGlossary.aspx

Appendix B: Glossary of Simulation Terms

Compiled by T. Chris Foster, Ph.D., MSC, USN. Military Director, Human Systems Engineering, Naval Air Warfare Center Aircraft Division, Patuxent River, MD and Michael G. Lilienthal, Ph.D., CPE, CTEP, Senior Cybersecurity Analyst, Electronic Warfare Associates.

Term	Definition
Activity-based simulation	A discrete simulation that represents the components of a system as they proceed from activity to activity. For example, a simulation in which a manufactured product moves from station to station in an assembly line.
Agent-based simulation	A simulation that focuses on the implementation of agents and the sequence of actions and interactions of the agents over periods of time.
Aggregate Level Simulation Protocol	A family of simulation interface protocols and supporting infrastructure software that permit the integration of distinct simulations and war games. Combined, the interface protocols and software enable large-scale distributed simulations and war games of different domains to interact at the combat object and event level.
Augmented reality (AR)	A class of technology that enables the user to interact with the real environment while overlaying or otherwise adding information from a virtual environment to enhance the user's experience with the real environment. Drascic and Milgram (1996) state that 'AR describes a class of displays that consists primarily of a real environment, with graphic enhancements or augmentations.'
Augmented virtuality (AV)	A class of technology that enables the user to interact with the virtual environment while overlaying or otherwise adding information from the real environment to enhance the user's experience with the virtual environment.
Computer simulation	A simulation that is executed on a computer, with some combination of executing code, control/display interface hardware, and, in some cases, interfaces to real-world equipment.
Constrained simulation	A simulation where time advances are paced to have a specific relationship to wall clock time. These are commonly referred to as real-time or scaled real-time simulations. Human-in-the-loop (e.g., training exercises) and hardware-in-the loop (e.g., test and evaluation simulations) are examples of constrained simulations.
Constructive simulation	Simulations involving simulated people operating simulated systems. Real people can be allowed to stimulate (make inputs) to such simulations. See: live, virtual, and constructive simulation.

(*Continued*)

Term	Definition
Discrete event simulation	A simulation where the dependent variables (i.e., state indicators) change at discrete points in time referred to as events.
Distributed interactive simulation (DIS)	A time and space coherent synthetic representation of world environments designed for linking the interactive, free-play activities of people in operational exercises. The synthetic environment is created through real-time exchange of data units between distributed, computationally autonomous simulation applications in the form of simulations, simulators, and instrumented equipment interconnected through standard computer communicative services. The computational simulation entities may be present in one location or may be distributed geographically.
Event-driven simulation	A simulation in which attention is focused on the occurrence of events and the times at which those events occur. For example, a simulation of a digital circuit that focuses on the time of state transition.
Extended reality (XR)	A blanket term encompassing virtual reality (VR), augmented reality (AR), and mixed reality (MR).
Hardware in the loop simulation	Simulation and simulators that employ one or more pieces of operational equipment (to include computer hardware) within the simulation/simulator system.
Human in the loop simulation	Simulation and simulators that employ one or more human operators in direct control of the simulation/simulator or in some key support function.
Instructional simulation	A simulation that provides stimuli in the synthetic environment, for the purpose of training.
Interval-oriented simulation	A continuous simulation in which simulated time is advanced in increments of a size suitable to make implementation possible on a digital system.
Live simulation	A simulation involving real people operating real systems. See: live, virtual, and constructive simulation.
Live, virtual, and constructive simulation	A broadly used taxonomy describing a mixture of live simulation, virtual simulation, and constructive simulation.
Mixed reality (MR)	MR can be conceptualized as a balanced mix of the real environment and a virtual environment in which neither predominates and purposeful interaction between the two is enabled. Note: While this definition intentionally excludes AR and AV from the definition of MR, there are researchers that include AR and AV on the MR continuum.
Modeling and simulation (M&S)	The use of models, including emulators, prototypes, simulators, and stimulators, either statically or over time, to develop data as a basis for making managerial or technical decisions.
Monte Carlo simulation	A simulation in which random statistical sampling techniques are employed to determine estimates for unknown values (i.e., making a random draw).
Real environment	The real world with which individuals interact using their own body and senses without augmentation.

(Continued)

Term	Definition
Real-time simulation	Simulated time advances at the same rate as actual time. Faster than real time is when simulated time advances at a rate greater than actual time. Slower than real time is when simulated time advances at a rate less than actual time.
Simuland	The system being simulated by a simulation.
Simulation environment	The operational hardware, software including databases, communications, and infrastructure in which a simulation operates.
Simulation game	A simulation in which the participants seek to achieve some agreed upon objective within an established set of rules. For example, a management game, a war game. Note: The objective may not be to compete, but to evaluate the participants, increase their knowledge concerning the simulated scenario, or achieve other goals.
Simulation object model (SOM)	A specification of the types of information that an individual federate could provide to High Level Architecture (HLA) federations as well as the information that an individual federate can receive from other federates in HLA federations. The SOM is specified using one or more SOM modules. The standard format in which SOMs are expressed facilitates determination of the suitability of federates for participation in a federation.
Simulation time	(a) A simulation's internal representation of time. Simulation time may accumulate faster, slower, or at the same pace as sidereal time. (b) The reference time (e.g., Universal Coordinated Time) within a simulation exercise. This time is established by the simulation management function before the start of the simulation and is common to all participants in a particular exercise.
Simulator	A hardware or software device that provides input into an operational system or subsystem.
Stimulate	To provide input to a real system or subsystem to observe or evaluate the response.
Stimulation	The use of simulations to provide an external stimulus to a real system or subsystem. An example is the use of a simulation representing the radar return from a target to drive (stimulate) the radar of a missile system within a hardware/software-in- the-loop simulation.
Stimulator	A hardware or software device that provides input injects into system or subsystem platforms and environment that are not physically present.
Time-step simulation or time- interval simulation	Simulations in which simulation time is advanced by a fixed or independently determined amount to a new point in time, and the states or status of some or all resources are updated as of that new point in time. Typically, these time steps are of constant size, but they need not be. For example, a model depicting the year- by-year forces affecting a volcanic eruption over a period of 100,000 years.
Virtual environment (VE)	A virtual environment requires (1) a computer model, (2) a representation of that model which stimulators the user's senses (e.g., visual, auditory, haptic), (3) a user or users, and (4) a way in which the user(s) can interact with the computer model.

(*Continued*)

Term	Definition
Virtual reality (VR)	A class of technology that seeks to immerse the user(s) in a virtual environment and detach the user(s) from the real environment. Heim (1998) defines VR as a 'technology that convinces the participant that he or she is actually in another place by substituting the primary sensory input with data produced by a computer.'
Virtual simulation	A simulation involving real people operating simulated systems.

Source: https://www.msco.mil/MSReferences/Glossary/TermsDefinitionsN-R.aspx

Appendix C: Glossary of Verification, Validation, and Accreditation Terms

Compiled by Michael G. Lilienthal, Ph.D., CPE, CTEP, Senior Cybersecurity Analyst, Electronic Warfare Associates, and William F. Moroney, Ph.D., CPE, Professor Emeritus, Human Factors Program, University of Dayton (Ohio).

Term	Definition
Accreditation	The official certification that a model or simulation and its associated data are acceptable for use for a specific purpose.
Accreditation agent	The organization or individual designated to conduct an accreditation assessment for a model, simulation, and their associated data for a particular application.
Accreditation authority	The organization or individual responsible to approve the use of models, simulations, and their associated data for a particular application.
Accreditation criteria	A set of standards that a particular model, simulation, or federation must meet to be accredited for a specific purpose.
Accreditation plan	The plan of action for certifying a model, simulation, or federation of models and simulations and its associated data as acceptable for specific purposes. The accreditation plan specifies the reviews, testing, and other accreditation assessment processes.
Data certification	The determination that data have been verified and validated. Data user certification is the determination by the application sponsor or designated agent that data have been verified and validated as appropriate for the specific M&S usage. Data producer certification is the determination by the data producer that data have been verified and validated against documented standards or criteria.
Data verification and validation (V&V)	The process of verifying the internal consistency and correctness of data and validating that it represents real-world entities appropriate for its intended purpose or an expected range of purposes.
Face validation	The process of determining whether a model or simulation seems reasonable to people who are knowledgeable about the system under study, based on the model's performance. This process does not revie the software code or logic, but rather reviews the inputs and outputs to ensure they appear realistic or representative.

(*Continued*)

Term	Definition
Independent verification and validation	The conduct of verification and validation of a model, simulation, and associated data by an individual, group, or organization that did not participate in the development and is not in the same chain of command or organization as the developer.
Validation	The process of determining the degree to which a model or simulation and its associated data are an accurate representation of the real world, from the perspective of the intended uses of the model.
Verification	The process of determining that a model or simulation implementation and its associated data accurately represent the developer's conceptual description and specifications.
Verification and validation agent	The individual, group, or organization designated to verify and validate a model, simulation, and associated data.

Source: DoD Modeling and Simulation (M&S) Glossary, https://www.msco.mil/MSReferences/Glossary/TermsDefinitionsN-R.aspx

Index

P

R

S

T

U

V